# Lecture Notes in Artificial Intelligence 12877

Subseries of Lecture Notes in Computer Science

More information about this subseries at http://www.springer.com/series/1244

Antonio A. Sánchez-Ruiz ·
Michael W. Floyd (Eds.)

# Case-Based Reasoning Research and Development

29th International Conference, ICCBR 2021
Salamanca, Spain, September 13–16, 2021
Proceedings

 Springer

*Editors*
Antonio A. Sánchez-Ruiz
Universidad Complutense de Madrid
Madrid, Spain

Michael W. Floyd
Knexus Research Corp.
National Harbor, MD, USA

ISSN 0302-9743 ISSN 1611-3349 (electronic)
Lecture Notes in Artificial Intelligence
ISBN 978-3-030-86956-4 ISBN 978-3-030-86957-1 (eBook)
https://doi.org/10.1007/978-3-030-86957-1

LNCS Sublibrary: SL7 – Artificial Intelligence

This Springer imprint is published by the registered company Springer Nature Switzerland AG
The registered company address is: Gewerbestrasse 11, 6330 Cham, Switzerland

# Preface

This volume contains the papers presented at the 29th International Conference on Case-Based Reasoning (ICCBR 2021), which was held virtually during September 13–16, 2021. ICCBR is the premiere annual meeting of the Case-Based Reasoning (CBR) research community and serves an important role in disseminating the newest findings and developments in the field. The previous ICCBR, ICCBR 2020, had originally been planned to be held in Salamanca, Spain, but was instead held virtually due to the COVID-19 pandemic. We had hoped that Salamanca could instead serve as the venue for ICCBR 2021 but, unfortunately, the difficult decision was made to again hold it virtually due to ongoing health, safety, and travel concerns.

Previous ICCBR editions, including the merged European Workshops and Conferences on CBR, were held in the following locations: Otzenhausen, Germany (1993); Chantilly, France (1994); Sesimbra, Portugal (1995); Lausanne, Switzerland (1996); Providence, USA (1997); Dublin, Ireland (1998); Seeon Monastery, Germany (1999); Trento, Italy (2000); Vancouver, Canada (2001); Aberdeen, UK (2002); Trondheim, Norway (2003); Madrid, Spain (2004); Chicago, USA (2005); Fethiye, Turkey (2006); Belfast, UK (2007); Trier, Germany (2008); Seattle, USA (2009); Alessandria, Italy (2010); Greenwich, UK (2011); Lyon, France (2012); Saratoga Springs, USA (2013); Cork, Ireland (2014); Frankfurt, Germany (2015); Atlanta, USA (2016); Trondheim, Norway (2017); Stockholm, Sweden (2018); Otzenhausen, Germany (2019); and Salamanca, Spain, with virtual attendance (2020).

ICCBR 2021 received 85 submissions from 18 countries, spanning Europe, North America, and Asia. Each paper was initially reviewed by the conference co-chairs to determine whether it met the submission criteria and conference topics. Submissions that passed the initial review were then reviewed by at least three Program Committee (PC) members. If a consensus was not reached among the Program Committee reviewers, additional meta-reviews were provided by members of the Advisory Board. Of the 85 submissions, 21 were accepted for oral presentation. Although poster presentations have often been included in previous ICCBR editions, none were included in the ICCBR 2021 program as the presentation is more difficult in a virtual setting.

ICCBR 2021 took place over four days, from September 13 to 16, 2021. On the first day of the conference, the workshop program was held with meetings dedicated to specific sub-areas of CBR. Following the workshops, the first technical session of the main conference was held, with a virtual social event afterwards in the evening. The second day of the conference started with a keynote invited talk by Professor Santiago Ontañón Villar of Google Research and Drexel University, USA. The majority of the day was comprised of oral presentations, concluding with another virtual social event in the evening. The third day started with oral presentations and concluded with a virtual cooking class, serving as a gala dinner, where attendees were able to create Spanish dishes in their homes. This event brought a taste of Spain to the attendees, even though we were unable to physically meet in Salamanca. The final day of the conference began with an invited keynote talk by Professor Kerstin Bach of the Norwegian University of Science and Technology. Following the keynote talk, a panel was held to remember the life and career of Professor Dr. Michael M. Richter. Michael had a significant and lasting impact on the CBR community, including his research, his involvement with ICCBR, and the numerous CBR researchers who studied under him. A final technical session consisting of oral presentations followed. The day, and the conference, concluded with a private Program Committee meeting followed by a general community meeting to discuss future plans for ICCBR.

The task of organizing ICCBR 2021 was supported by the tireless efforts of many people. Juan Manuel Corchado and Fernando de la Prieta, along with their team, handled all aspects of the local organization – both the in-person planning for a physical conference in Salamanca as well as the online organization when the conference transitioned to a virtual event. Hayley Borck and Viktor Eisenstadt organized an excellent workshop program for the conference and coordinated with the various workshop chairs to plan those meetings. Stewart Massie and Stelios Kapetanakis chaired the Doctoral Consortium, an invaluable annual event that helps nurture the next generation of CBR researchers. We are extremely grateful for the guidance, advice, and support of the ICCBR Advisory Board members: Belen Díaz-Agudo, David W. Aha, David Leake, Barry Smyth, Rosina O. Weber, and Nirmalie Wiratunga. Additionally, the thoughtful, thorough, and constructive reviews provided by the PC and additional reviewers assisted us greatly in making decisions for the ICCBR 2021 program. We thank all those mentioned for their effort and assistance in making ICCBR 2021 a success.

July 2021

Antonio A. Sánchez-Ruiz
Michael W. Floyd

# Organization

## Program Chairs

Antonio A. Sánchez-Ruiz     Universidad Complutense de Madrid, Spain
Michael W. Floyd     Knexus Research, USA

## Local Chairs

Juan Manuel Corchado     University of Salamanca, Spain
Fernando de la Prieta     University of Salamanca, Spain

## Workshop Chairs

Hayley Borck     Smart Information Flow Technologies, USA
Viktor Eisenstadt     DFKI, Germany

## Doctoral Consortium Chairs

Stewart Massie     Robert Gordon University, UK
Stelios Kapetanakis     University of Brighton, UK

## Advisory Board

Belen Díaz-Agudo     Universidad Complutense de Madrid, Spain
David W. Aha     Naval Research Laboratory, USA
David Leake     Indiana University, USA
Barry Smyth     University College Dublin, Ireland
Rosina O. Weber     Drexel University, USA
Nirmalie Wiratunga     Robert Gordon University, UK

## Program Committee

Klaus-Dieter Althoff     DFKI and University of Hildesheim, Germany
Kerstin Bach     Norwegian University of Science and Technology, Norway
Ralph Bergmann     DFKI and University of Trier, Germany
Isabelle Bichindaritz     State University of New York at Oswego, USA
Hayley Borck     Smart Information Flow Technologies, USA
Derek Bridge     University College Cork, Ireland
Sutanu Chakraborti     Indian Institute of Technology Madras, India
Alexandra Coman     Capital One, USA
Dustin Dannenhauer     Parallax Advanced Research, USA

| | |
|---|---|
| Sarah Jane Delany | Technological University Dublin, Ireland |
| Viktor Eisenstadt | DFKI, Germany |
| Peter Funk | Mälardalen University, Sweden |
| Ashok Goel | Georgia Institute of Technology, USA |
| Mehmet H. Göker | ServiceNow, USA |
| Odd Erik Gundersen | Norwegian University of Science and Technology, Norway |
| Vahid Jalali | Indiana University Bloomington, USA |
| Stelios Kapetanakis | University of Brighton, UK |
| Mark Keane | University College Dublin, Ireland |
| Joseph Kendall-Morwick | Missouri Western State University, USA |
| Luc Lamontagne | Laval University, Canada |
| Jean Lieber | Université de Lorraine, France |
| Stewart Massie | Robert Gordon University, UK |
| Mirjam Minor | Goethe University Frankfurt, Germany |
| Stefania Montani | Università del Piemonte Orientale, Italy |
| Emmanuel Nauer | Université de Lorraine, France |
| Santiago Ontañón | Drexel University, USA |
| Enric Plaza | IIIA-CSIC, Spain |
| Luigi Portinale | Università del Piemonte Orientale, Italy |
| Juan A. Recio-García | Universidad Complutense de Madrid, Spain |
| Pascal Reuss | University of Hildesheim, Germany |
| Jonathan Rubin | Philips Research North America, USA |
| Frode Sørmo | Amazon Alexa AI, UK |
| Ian Watson | University of Auckland, New Zealand |
| David Wilson | University of North Carolina at Charlotte, USA |

## Additional Reviewers

| | |
|---|---|
| Ömer Ibrahim Erduran | Kyle Martin |
| Glenn Forbes | Ditty Mathew |
| Devi Ganesan | Adwait Parsodkar |
| Miriam Herold | S. Renganathan |
| Maximilian Hoffmann | Jakob Michael Schoenborn |
| Srashti Kaurav | Felix Theusch |
| Mirko Lenz | Ashish Upadhyay |
| Lukas Malburg | |

# Sponsors

University of Salamanca, Spain
Knexus Research, USA
Universidad Complutense de Madrid, Spain
BISITE Research Group, Spain
IoT Digital Innovation Hub, Spain
AIR Institute, Spain

# Contents

# The Bites Eclectic: Critique-Based Conversational Recommendation for Diversity-Focused Meal Planning

Fakhri Abbas[(⊠)], Nadia Najjar, and David Wilson

University of North Carolina at Charlotte, Charlotte, NC 28223, USA
{fabbas1,nanajjar,davils}@uncc.edu

**Abstract.** Diet diversification has been shown both to improve nutritional health outcomes and to promote greater enjoyment in food consumption. CBR has a rich history in direct recommendation of recipes and meal planning, as well as conversational exploration of the possibilities for new food items. But more limited attention has been given to incorporating diversity outcomes as a primary factor in conversational critique for exploration. Critiquing as a method of feedback has proven effective for conversational interactions, and diversifying recommended items during exploration can help users broaden their food options, which critiquing alone may not achieve. And all of these aspects together are important elements for recommender applications in the food domain. In this paper, we introduce DiversityBite, a novel CBR approach that brings together critique and diversity to support conversational recommendation in the recipe domain. Our initial user study evaluation shows that DiversityBite is effective in promoting meal plan diversity.

**Keywords:** Diversity · Recipe recommendation · Critique-based · Case-based Reasoning

## 1 Introduction

Diet diversification has been linked to positive health outcomes such as reducing incidence of cancer or mortality [7]. Moreover, by promoting a variety of healthful food choices, diet diversification can make food consumption more enjoyable [15]. Our research is investigating how recommender systems can help to promote dietary diversity through the use of exploration. We have designed DiversityBite, a Case-based Reasoning (CBR) approach that incorporates diverse recipe recommendation during the exploration process. DiversityBite employs a conversational CBR recommendation approach with diversity-focused dynamic critiquing, in order to support users in exploration of recipes and creation of diverse meal plans.

From a recommender system perspective, incorporating diversity provides a number of advantages. First, it reduces the problem of "filter bubble" effects [17].

© Springer Nature Switzerland AG 2021
A. A. Sánchez-Ruiz and M. W. Floyd (Eds.): ICCBR 2021, LNAI 12877, pp. 1–16, 2021.
https://doi.org/10.1007/978-3-030-86957-1_1

Second, diversity enables the user to explore alternative options that could be healthier which can increase the dietary diversity for individuals [7]. And finally, it increases user awareness and knowledge of existing recipes by providing more recipes that could be explored in different cultures, cuisines, or communities [10]. The problem of incorporating diversity mechanisms in recipe recommendation in order to generate diverse sets that also meet user requirements remains an important open research challenge [10].

Primary aspects of the challenge stem from the differences and similarities between recipes (e.g., cuisine, ingredients, nutrition, meal type, preparation), as well as users' perceptions of those differences and similarities. For example, a person may like fried chicken but not grilled chicken, or may prefer chicken for lunch but not for dinner. Due to this complexity in food selection, recommender systems should enable the user to shape the direction of the recommendations. This in turn, helps to reduce the effect of contextual factors that are hard to capture such as time, cultural background, food knowledge, and current user's needs. Therefore, we focused on a conversational recommender approach, in which the user can provide iterative critique feedback on recommendations.

This research extends our previous work [1,2] and investigates incorporating diversity in a critique based conversational recommender system for recipes. In this paper we introduce DiversityBite, a novel way of dynamically generating critiques, in which the generated critique leads the user to a more diverse set of recipe recommendations - and outcomes. In essence, this can be thought of as a "like this, but more diverse" approach. This paper reports on our initial investigation for the approach, and the main research question we address is:

- **RQ 1** In critique-based conversational recommendation, how does diversity-focused critique impact diversity in terms of user outcomes?

To address this overall research question, we have developed an implementation of the DiversityBite approach and conducted a lab-based user study, in which users were asked to prepare a weekly meal plan by exploring recipes. To help understand our general research question, the study investigates the following specific research questions:

- **RQ 1.1** Can a critique-based recommender result in finding more diverse recipes compared to a non critique-based recommender?
- **RQ 1.2** Can users compile a more diverse meal plan in a critique-based recommender compared to non critique-based recommender?
- **RQ 1.3** Can users perceive the modeled diversity of the recommended recipes?

This paper first discusses related work in Sect. 2. Section 3 presents the DiversityBite approach, and Sect. 4 describes our user study evaluation. The paper concludes in Sect. 5 with discussion and future directions.

## 2 Background

This research draws upon previous work in the areas of diversity, conversational recommendation - more specifically critiquing, and the domain of recipe recommendation.

### 2.1 Diversity

Bradley and Smyth [4] described diversity as the complement of similarity. Smyth and McClave [26] argue that diversity is as important as similarity in case-based recommender systems. They have suggested measuring the diversity of the list as the average pairwise distance [26] as shown in Eq. 1:

$$Diversity(R) = \frac{\sum\limits_{i \in R} \sum\limits_{j \in R/\{i\}} dist(i,j)}{|R|(|R| - 1)} \tag{1}$$

where $R$ is the recommended list of items, and $dist(i,j)$ is the distance between item $i$ and item $j$. Using this definition, Fleder and Hosanagar conducted a follow up study, showing that recommender systems reduce diversity by focusing primarily on accuracy [9]. In [26], Smyth and McClave applied a linear combination of relevance and diversity as an objective function to retrieve items. In contrast, Mcsherry [22] argues that increasing diversity with small sacrifice on similarity may not be applicable in every scenario. Example of such scenarios are items that are available for a short period of time such as jobs, and apartments. In such cases, similarity should be given priority such that any increase in diversity should not affect similarity. To address this problem, they presented a retrieval approach that increases diversity while preserving similarity.

Kelly and Bridge [16] applied a greedy reranking strategy in a conversational recommender system, diversifying results in each iteration cycle after feedback from the user. Item relevance was generated from a collaborative filtering recommender, and distance computed as the hamming distance between item rating vectors. McGinty and Smyth [21] incorporated diversity in the conversational recommender system (CRS) while balancing the tradeoff between diversity and relevance. The authors described a system where at each cycle, the user selects a critique which is used for the next iteration cycle. The selected item carried over the next recommendation cycle and displayed along other recommended items. If the user selects the carried over item again the system assumes that no progress has been made and a more diversified list is recommended on the next cycle. However, if the user selects a different item, then the system assumes positive progress has been made and generates results with less diversity and more relevance for the next cycle.

### 2.2 Critiquing

Recommender systems are most often considered as a type of one shot interaction, in which the system recommends a set of items and the user navigates through that set to find an item of interest. Typically, the system monitors discrete user interactions over time and tailors recommendations to consolidated user interests [5]. Conversational recommenders take a different approach, providing a richer interaction with the user through iterative feedback and refinement of results. During such iterations the system can elicit current user's preference, and context. This in return has a positive impact on enabling users to better understand the search space, and reduce the effect of the cold start problem [14]. Conversational recommenders use two different strategies to help the user in navigation, *navigation by asking*, and *navigation by proposing* each relies on different form of feedback. In navigation by asking, the user is asked to provide feedback on a feature. Conversely, in navigation by proposing the system proposes a set of items and asks the user to provide feedback on the recommended items. For example, the user may *critique* a feature, or may provide a *preference*, or may provide a *rating* for a set of the recommended items [27].

Smyth and McGinty noted four primary forms of feedback used in CRS, mainly, Value Elicitation, Critiquing, Ratings-Based, and Preference-Based feedback [27]. In this paper, we focus on the critiquing form of feedback. In critiquing feedback, the user provides a directional preference over a feature of recommendation [19]. For example, in a conversational car recommender, the user might ask for a smaller engine than the currently recommended car. Burke et al. [6] pioneered conversational recommenders with the FindMe approach of system-suggested critique. The critiquing in FindMe posed two challenges. First, there was a pre-designed set of critiques within the user interaction session, so called *static critique*. Second, each critique addressed constraints on one feature, so called *unit critique*. To address the first challenge, Reilly et al. [24] showed that standard critique can be extended to cover multiple features for *compound critique*. To address the second challenge, McCarthy et al. [18] developed a dynamic critique approach, in which the system combines the feature depending on the available items in the search space. McCarthy et al. [20] did address diversity in critiquing, but the focus was on creating diversity in the repoitoire of critiques rather than diversity in conversational outcomes. In this paper, we propose a novel approach to generate dynamic unit critiques in which each recommended item has an individually tailored set of possible critiques toward diversity in outcomes.

### 2.3   Recipe Recommendation and Diversity

The importance of diversity in recipe recommenders has several advantages such as: providing meals with varied sources of nutrition for a balanced meal diet [8], increasing user awareness of existing recipes [23], and covering a wide variety of options that could reduce the cold start problem [3]. Grace et al. [11] proposed the Q-Chef system that encourages dietary diversity by generating and recommending recipes based on models of surprise and novelty of the ingredients. While Q-Chef focused on helping the user find new, surprising recipes, the set of recommended recipes itself is not necessarily diverse. In a series of studies, Zeyen et al. [29,30] proposed a new approach to explore a collection of cooking recipes as cooking workflows. They have built a conversational retrieval method CookingCAKE in which users are guided through the search space by answering posed questions. While diversity has not been introduced in the adaptation process the authors indicated the importance to provide a more diverse and customized workflows. In [8], Elsweiler et al. acknowledged the importance of diversity in meal plans as a way to provide health, though their proposed meal planner did not specifically engineer diversity into the recommendations.

## 3   DiversityBite: A Conversational Dynamic Critique-Based Recommender

The DiversityBite approach adopts an initial zooming stage [6] based on cuisine type followed by a series of conversational interactions in the form of critique-based *recommend-review-revise* cycles [24]. This is enabled by two primary components: retrieval and critique which complement each other to address the similarity vs. diversity balance [26] in meal planning. The DiversityBite framework is modular and can support a variety of representations for user profiles or cases, as well as different metrics for similarity, retrieval, and critique. In this paper, we focus on analysis of an initial approach across these aspects, described in the following sections.

### 3.1   Recipe Case Representation and Similarity

Recipes can be represented using a variety of different features such as ingredients, preparation steps, nutrition details, or user ratings. In this research, we focus on a content-based case representation. Recipe cases are represented with three distinct sets of features: ingredients, nutritional information, and flavor characterization. More specifically, recipes are represented as a vector of 3807 potential ingredients (binary - presence or absence). Nutritional features (10) include: saturated fat, trans fat, fat, carbohydrate, sugar, calories, fiber, cholesterol, sodium, and protein (% recommended daily per serving). Flavour features (6) include: saltiness, sourness, sweetness, bitterness, spiciness, and savoriness (numerical 0–1 for intensity). Section 4.1 provides more details on the case data.

In this study, the similarity measure considers cuisine type, ingredient match, and meal course. Initial user preference for cuisine type is a hard constraint, and degree of similarity is based on ingredient match and meal course match. For a current reference case $(c_r)$ and candidate case for retrieval $(c_c)$:

$$sim(c_r, c_c) = \begin{cases} 0 & \text{if } c_c \text{ not selected cuisine} \\ sim_{ingr}(c_r, c_c) + sim_{crse}(c_r, c_c) & \text{otherwise} \end{cases}$$

The $sim_{ingr}(c_r, c_c)$ metric is straightforward cosine similarity across ingredients. The $sim_{crse}$ metric is the proportion of user-specified course types matched by the candidate case.

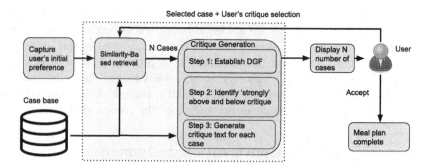

**Fig. 1.** DiversityBite model, starts with user initial preference and ends with user acceptance. The shaded area represents retrieval and diversity adaptation cycle through critique.

## 3.2    Diversity-Focused Conversational Critique

The DiversityBite approach, illustrated in Fig. 1, consists of three main components: Zooming Stage, Diversity-Goal Footprint, and Critique Generation.

**Zooming Stage.** The conversation starts by capturing the user's initial interest. In this study, this is represented by selecting one or more cuisine type(s) and meal course(s) of interest. Cuisine type is a hard constraint on case retrieval, constraining the search space throughout the conversation. Initial retrieval starts by recommending the closest $N$ cases to the centroid of the cuisine-constrained case-base, where the centroid represents the average across ingredient features.

**Diversity-Goal Footprint.** Once the user's initial preference has been entered, DiversityBite establishes a diversity-goal footprint (DGF). This is a set of $S$ cases that (1) meet the user's baseline preference and (2) are selected for high diversity within the set. The DGF approximates the maximal potential diversity among cases within the current case-base for the user's query. So, when a user selects a critique of a current case (essentially, "like this, but more diverse"), the DGF provides a reference for selecting directions to move toward in the case-base that are expected to increase diversity in recommendations. Our initial DGF approach follows Vergas et al. [28], who note that maximum diversity can be approximated through random case selection. To establish the DGF, we randomly select $S$ number of cases and measure average pairwise distances. This is repeated $R$ number of times, selecting the list with the highest diversity score.

**Critique Generation.** Given a list of recommended cases (either from initial zooming or previous conversation step), a set of potential critiques is dynamically generated for each case. These are selected as possible pathways for the user toward greater case diversity in the next step of the conversation. Only critiques toward greater diversity are presented as actionable options for the user.

In this study, potential critique dimensions correspond to flavor and nutrition features (e.g., more spicy or fewer calories). The DGF set is used as a reference point for critique direction (more/less) across the critique dimensions. First, the average value of each critique dimension across all cases in the search space is taken as a baseline threshold. Second, for each critique dimension the percentage of cases in the DGF that are above or below the baseline threshold are recorded. The **M** highest-percentage critique dimensions above and below the threshold are considered to be 'strongly' above and below the threshold for purposes of critique activation. This represents a directional vector used to guide critique toward greater diversity in results.

Third, the current case's value on each critique dimension is checked in relation to the overall threshold and the DGF profile for that dimension. If the case value is below the threshold and the DGF percentage for that dimension is strongly above the threshold, then a 'more' critique for that dimension is activated. Conversely, if the value is above the threshold and the DGF percentage for that dimension is strongly below the threshold, then a 'less' critique for that dimension is activated.

Finally, DiversityBite displays a list of cases along with its critique set. The user has the options to: (1) make selections from the list to add to the meal plan, and (2) select a case + critique that will be used as feedback for the next round of conversation. An applied critique serves as a filter that is applied to similarity-based retrieval - filtering cases from the top **N** that do not satisfy the critique with respect to the reference case.

## 4   Evaluation for DiversityBite

In this section we first describe the recipe dataset used in this study followed by a description of the evaluation study, and then a discussion of the main findings.

### 4.1   Recipe Dataset

In this work, we chose a recipe dataset that has a potential of diversity. In [25], Sajadmanesh et al. prepared a dataset with 120K recipes crawled from yummly.com, a personalized recipe recommender platform. The dataset consists of recipes from 204 countries. Each recipe has average review rating, ingredients, preparation time, course type, nutritional values, and flavor features. The raw data contains 11,113 ingredients. The course type feature has values related to the recipe type such as afternoon tea, bread, breakfast etc. The nutritional values features are saturated fat, trans fat, fat, carbohydrate, sugar, calories, fiber, cholesterol, sodium, and protein of a recipe per serving. Recipes are identified by six flavors, namely, saltiness, sourness, sweetness, bitterness, spiciness, and savoriness. The flavour features are represented on a scale from 0 to 1.

Given the close coupling between ingredients, flavor, and nutritional values, for this study we use the ingredients to represent the recipes directly to calculate the diversity scores, while the flavor and nutritional features are used as critique

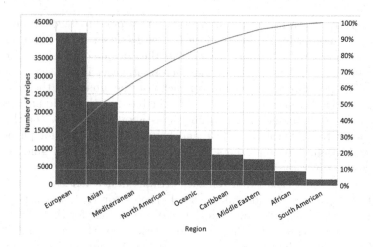

**Fig. 2.** The distribution of recipes over the region in the dataset used for the experiment

features between recipes. To reduce overall sparsity in ingredients we have used FOODON [12] ontology to map each ingredient to a food concept. The mapping reduced the number of unique ingredients from 11,113 ingredients to 3,807 ingredients. To facilitate user interaction and capturing the user initial interest we have grouped recipes by region into 9 regions: Caribbean, North America, South America, Europe, Africa, Middle East, Mediterranean, Asia, and Oceanic. Figure 2 shows the distribution of recipes over these regions.

## 4.2  Evaluation Study

To evaluate DiversityBite we implemented a web-based recommender application for users to interact with. We conducted a user study to evaluate the effectiveness of using dynamic critique to recommend more diverse recipes. Figure 3, shows a screenshot from DiversityBite displaying a list of recipes recommended to the user, as well as an example of the expanded view where the user can display more information for a particular recipe. The user can explore more recipes by selecting one of the displayed critiques, they can also dislike a recipe so it will not appear in any upcoming iterations while exploring. For the recommendation we chose $N = 10$, while the parameters for the proposed algorithm was set to: $S = 10$, $R = 50$, $M = 8$. The parameters were set through empirical lab experiment to ensure reasonable computation time during user interaction with the website.

We have implemented two variations of DiversityBite: dynamic critique recommender (*Dynamic-Rec*), and static critique recommender (*Static-Rec*). A similarity-based recommender (*Sim-Rec*) was used as a baseline. *Dynamic-Rec* generates dynamic critique as discussed in Sect. 3. *Static-Rec* displays same set of critique for each recipe, and finally *Sim-Rec* is a similarity-based recommender.

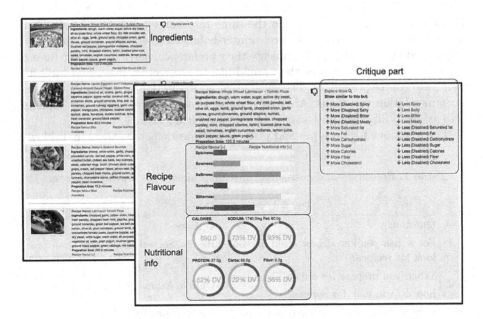

**Fig. 3.** A screenshot of *Dynamic-Rec* variation of DiversityBite. It shows four recipes, the enlarged image shows the recipes details i.e. flavour features, nutritional features, and critiques. The critique part shows only critique that can lead to more diverse cases.

In *Dynamic-Rec* and *Static-Rec* the user explores more recipes by using the critique, while in *Sim-Rec* the user explores more recipes by navigating through several pages of recommended recipes.

The study utilized a within-subject design where each participant experienced recommendations from each variation. The study was conducted in a lab setting, and underwent IRB approval. The participants interacted with the application that collected information about their preferences and displayed recommendations. The total duration of the study was on average 40 min.

Participants were asked to fill out a pre-survey about their demographic information (age, gender, and education), and their online behaviour while looking for recipes. After the initial survey, participants were asked to use the recommender system to prepare a week long meal plan. The task prompt was: "For the next three system variations, prepare a meal plan for a week". During the task, the application interface displays two progress bars. The first progress bar indicates exploration progress and its maximum value achieved after 7 exploration, while the second progress bar indicates the meal plan completion and its maximum value achieved after adding 7 recipes to the meal plan. However, participants can explore and add more recipes to the meal plan but the progress bars ensures a minimum of 7 cycles and 7 meals are added to the plan before ending the session and moving to the next variation. Since the study is a within-subject design, each participant had to do the same task three times. The view order of the recommender variations were counter-balanced to eliminate participants' fatigue and learning effects. To indicate their preference, participants were asked to select one or more regions they would like to see recipes from, and indicate their meal course preference. The same preference selection was used in all three recommendation variations. After each recommendation variation, participants were asked to fill out a post-survey based on their experience.

The questions for both pre and post survey questions are shown in Table 1.

**Table 1.** Pre-survey and post-survey questions along with available answers

| # | Question | Options |
|---|----------|---------|
| **Pre-survey questions** | | |
| 1 | For a dish you know, how often do you look for recipes? | Rarely/Sometimes/Often |
| 2 | When you prepare for a dish you know, how do you look for recipes? | Online/Asking Relatives/Others |
| 3 | How often do you look for new recipes? | Rarely/Sometimes/Often |
| 4 | How do you look for new recipes? | Online/Asking Relatives/Others |
| 5 | list some websites do you use? | Free text |
| 6 | What are the most important criteria do you look for when deciding on a recipes? | Free text |
| **Post-survey questions** | | |
| 1 | Did you find the recipe you were looking for? | Yes/No |
| 2 | Did you find new recipes ? | Yes/No |
| 3 | Among the recipes you liked, are you willing to try one of them? | Yes/Some of them/None |
| 4 | Do you think recipes were similar to each other in each displayed list? | Yes, recipes were similar with small variations in ingredients |
| | | No, recipes were different from each other |
| 5 | What was your main decision when you selected to see similar recipes? | Flavor |
| | | Nutritional facts |
| | | Preparation time & number of ingredients |

## 4.3 Evaluation Results

Twenty-six participants were recruited from students, staff and faculty at a U.S. public university. Gender distribution were 19 female, and 7 males. Most of participants age range was between 18 and 29 years and the majority of participants had at least a bachelor degree. All participants use online resources to look for new recipes or to refresh their memory regarding a recipe they know. Additionally, all participants indicated that they frequently look for new recipes. Regarding the online resources they use, the most frequent resources are: Google search, YouTube videos, and social network. Recipe ingredients, preparation time, and balanced dish were the main criteria participants look for when deciding on a recipe. This suggests that our participants had a good exposure to online resources when looking for recipes. Among the chosen regions, Asia, Mediterranean, Middle Eastern, and North America were the most frequently chosen regions while the least chosen regions were Caribbean, and Caucasus. The most frequently chosen meal course were main dish, appetizers, and lunch while the least frequently chosen ones are beverages such as tea, and cocktail. On average participant spent around 8 min using each variation, and viewed on average 8 different recipe lists in each variation; 80 different recipes in each variation.

**Meal Plan Size and Number of Disliked Recipes.** To understand the usefulness of our approach we looked at the meal plan size and number of disliked recipes. Our intuition is that, given the same number of recommended recipes, the recommmender is more useful if the user is able to compile a larger meal plan while disliking less recipes. The rationale is that the more recipes the user disliked the more likely the recommender system was not able to satisfy the user's need. Figure 4, shows the average percentage of meal plan size to the number of recommended recipes (left), and the average percentage of disliked recipes to the total number of recommended recipes (right). The results in Fig. 4 indicates that, participants added 20% of recommended recipes to their meal plan in *Sim-Rec* while only around 13% being added in the case of *Dynamic-Rec* and *Static-Rec*. For the number of disliked recipes, participants disliked 3% of recipes recommended by *Sim-Rec* and 2% in the case of *Dynamic-Rec* and *Static-Rec*. We note here that users completed roughly the same number of iterations ($\sim$7.5) and spent the same amount of time ($\sim$7 min) in each recommender variation. A one-way repeated measure ANOVA test shows no statistical significance for the number of disliked recipes ($F(2,50) = 0.68$, p-value $= 0.51$), suggesting that all variations seemed to be equally similar in meeting participants expectation. However, for the meal plan size there's a significant difference among the different variations ($F(2,50) = 3.79$, p-value $< 0.05$). Tukey's post hoc test shows that *Sim-Rec* had significantly higher meal plan size than other variations. We attribute this to fact that the *Sim-Rec* recommender is driven by user preference and is more likely to generate recommendations that are compelling to the user and added to the meal plan. The similarity in the number of disliked items suggest that the ability to generate useful recommendations was not affected by the recommender variation.

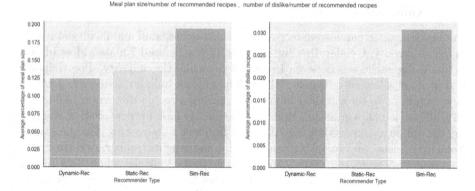

**Fig. 4.** The average percentage of meal plan size to the number of recommended recipes (left), and the average percentage of disliked recipes to the total number of recommended recipes (right)

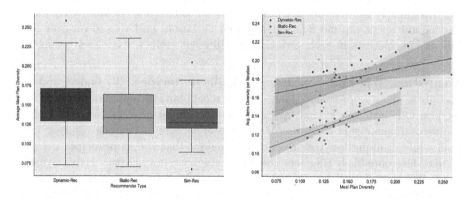

**Fig. 5.** Meal plan diversity score for each variation (left), the relation between meal plan diversity and diversity of recommended recipes (right)

**Diversity in Meal Plan vs Diversity in Recommended Recipes.** To address RQ 1.1 and R1.2 we first examined the relationship between the diversity of the meal plan and the diversity of the recommended recipes. The Meal plan diversity for each variation is summarized in Fig. 5 (left). These result indicate that *Dynamic-Rec* has a higher diversity score compared to *Static-Rec* and *Sim-Rec*. A one-way repeated measure ANOVA test shows there's a significant difference in the meal plan diversity ($F(2,50) = 3.8$, $p < 0.05$). Tukey's post hoc test shows that diversity in *Dynamic-Rec* is significantly higher than *Sim-Rec*. Suggesting that, the dynamic critique approach was able to allow participants create a more diverse list of meal plan. While the previous finding shows that participants created a larger meal plan in *Sim-Rec* compared to *Dynamic-Rec*, this findings shows that participants were able to create a more diverse meal plan in *Dynamic-Rec*.

Figure 5 (right) shows a scatter plot between the diversity of the meal plan on the horizontal axis and the average diversity score of recommended recipes. In all three variations, there's a direct relation between the recommended recipes and the meal plan. There's a strong correlation between meal plan diversity and diversity of the recommended recipes. The Pearson correlation for each variation is: *Sim-Rec* r = 0.7, p < 0.05, *Static-Rec* r = 0.6, p < 0.05, *Dynamic-Rec* r = 0.4, p = 0.06. The results suggest that for *Sim-Rec* and *Static-Rec* the diversity of recommended recipes effect the diversity of the meal plan diversity. While in the *Dynamic-Rec* the exploration process led the participant to a more diverse meal plan rather than the diversity in the recommended recipes.

**Reflection Survey.** To address RQ 1.3 we analyzed the results from the reflection survey. Each participant had to answer a set of questions after using each recommender variation as shown in Table 1 in post-survey part. For Q1 and Q2, all participants indicated that they found new recipes or found recipes they were looking for. In all variations, Q1 received on average less positive responses compared to Q2, Q1 received 19 positive responses compared to 25 for Q2. For Q3, none of participants indicated in any variation that they will not try any of the recommended recipes. A statistical test for Q1, Q2, and Q3 shows no significant difference between all variations in recommending useful, novel, and valuable recipes. While the aim of this study is not to focus on novelty, usefulness, and valuable finding but the results of Q1, Q2, and Q3 provide an indication about the quality of the recommendation.

To capture participants perception of diversity, we asked participants in Q4 if they thought recipes were similar to each other. According to [13], there are two types of diversity *categorical diversity* and *item-to-item diversity*. Q4 addresses the *item-to-item diversity* which aligns with the diversity style we are introducing. A chi-square test of independence showed that there was no significant association between recommender type and user perception of diversity $\chi^2(2, N = 26) = 5.2, p = 0.07$. This result aligns with the finding of [13] in which participants were not able to perceive item-to-item diversity. Therefore, we conclude that introducing diversity in the recommendation while exploring will not make a noticeable difference to participants but can result in a diverse selection.

The last question (Q5) in the post-survey asks participants about the main criteria they have used to on selecting to see similar recipes. We have received mixed answers of Flavor, Nutritional facts, preparation time, and number of ingredients. This aligns with participants' response at the post survey question Q6. A chi-square test of independence showed that there was no significant association between recommender type and the criteria applied to see similar recipes $\chi^2(4, N = 26) = 2.75, p = 0.60$. We have confirmed these results by looking at the user critique selection in *Dynamic-Rec* and *Static-Rec* by analysing user logs. The logs show no clear distinction in the frequency of using flavor and nutrition critique. This finding suggests that participants were trying to utilize the available exploration option with no bias toward one type of critique.

## 5   Discussion and Future Work

We have presented DiversityBite, a CBR framework that generates critique dynamically to enable diet diversification through exploration. The generated critique guides the user through the search space to explore more diverse cases. We have implemented DiversityBite in the domain of recipes and conducted a user study to evaluate the effect of DiversityBite on diversity.

Our user study compared between two variations of DiversityBite (*Dynamic-Rec*, and *Static-Rec*) and a baseline (*Sim-Rec*). Participants were able to create a larger meal plan size in *Sim-Rec* compared to the other two variations. However, the diversity of meal plan were statistically significant higher in *Dynamic-Rec* compared to the other two variations. Our interpretation to this is that participants were looking for a more diverse meal plan in each variation and they were able to satisfy their needs using *Dynamic-Rec* by compiling a shorter meal plan. We have also noted there is no strong correlation between the diversity in the meal plan and the displayed recipes in *Dynamic-Rec* which is not the case in *Static-Rec*, and *Sim-Rec*. This indicates that in *Dynamic-Rec* the exploration process led participants to a more diverse meal plan rather than the diversity in the recommended recipes only. We have also explored the relation between participants perception of item-to-item diversity and the exploration type. In all three variations there was no correlation between both variables, this findings aligns with the finding of [13] in which participants were not able to perceive item-to-item diversity.

Despite that DiversityBite enables users to explore diverse recipes, an actual system should allow users to drill into more similar recipes for example having a critique of 'show me recipes similar to this recipe'. However, the focus of this study is to evaluate the diversity aspect of DiversityBite. Another limitation is the evaluation of DGF, DGF evaluated using a random function for computational reasons. A more efficient evaluation can be applied by using another optimization algorithm such as greedy algorithm.

## References

1. Abbas, F.G., Najjar, N., Wilson, D.: Critique generation to increase diversity in conversational recipe recommender system. In: The International FLAIRS Conference Proceedings, vol. 34 (2021)
2. Abbas, F.G., Najjar, N., Wilson, D.: Increasing diversity through dynamic critique in conversational recipe recommendations. In: Proceedings of the 13th Workshop on Multimedia for Cooking and Eating Activities (2021)
3. Anderson, C.: A survey of food recommenders. arXiv preprint arXiv:1809.02862 (2018)
4. Bradley, K., Smyth, B.: Improving recommendation diversity. In: Proceedings of the Twelfth Irish Conference on Artificial Intelligence and Cognitive Science, Maynooth, Ireland. Citeseer (2001)
5. Bridge, D., Göker, M.H., McGinty, L., Smyth, B.: Case-based recommender systems. Knowl. Eng. Rev. **20**(3), 315–320 (2005)

6. Burke, R.D., Hammond, K.J., Young, B.C.: Knowledge-based navigation of complex information spaces. In: Proceedings of the National Conference on Artificial Intelligence, vol. 462 (1996)

7. Drescher, L.S., Thiele, S., Mensink, G.B.: A new index to measure healthy food diversity better reflects a healthy diet than traditional measures. J. Nutr. **137**(3), 647–651 (2007)

8. Elsweiler, D., Harvey, M., Ludwig, B., Said, A.: Bringing the "healthy" into food recommenders. In: DMRS (2015)

9. Fleder, D.M., Hosanagar, K.: Recommender systems and their impact on sales diversity. In: Proceedings of the 8th ACM Conference on Electronic Commerce (2007)

10. Freyne, J., Berkovsky, S.: Intelligent food planning: personalized recipe recommendation. In: Proceedings of the 15th International Conference on Intelligent User Interfaces (2010)

11. Grace, K., Maher, M.L., Wilson, D., Najjar, N.: Personalised specific curiosity for computational design systems. In: Gero, J.S. (ed.) Design Computing and Cognition '16, pp. 593–610. Springer, Cham (2017). https://doi.org/10.1007/978-3-319-44989-0_32

12. Griffiths, E.J., Dooley, D.M., Buttigieg, P.L., Hoehndorf, R., Brinkman, F.S., Hsiao, W.W.: FoodON: a global farm-to-fork food ontology. In: ICBO/BioCreative (2016)

13. Hu, R., Pu, P.: Helping users perceive recommendation diversity. In: DiveRS@RecSys (2011)

14. Jannach, D., Manzoor, A., Cai, W., Chen, L.: A survey on conversational recommender systems. arXiv preprint arXiv:2004.00646 (2020)

15. Kahn, B.E., Wansink, B.: The influence of assortment structure on perceived variety and consumption quantities. J. Consum. Res. **30**(4), 519–533 (2004)

16. Kelly, J.P., Bridge, D.: Enhancing the diversity of conversational collaborative recommendations: a comparison. Artif. Intell. Rev. **25**(1–2), 79–95 (2006)

17. Knijnenburg, B.P., Sivakumar, S., Wilkinson, D.: Recommender systems for self-actualization. In: Proceedings of the 10th ACM Conference on Recommender Systems (2016)

18. McCarthy, K., Reilly, J., McGinty, L., Smyth, B.: On the dynamic generation of compound critiques in conversational recommender systems. In: De Bra, P.M.E., Nejdl, W. (eds.) AH 2004. LNCS, vol. 3137, pp. 176–184. Springer, Heidelberg (2004). https://doi.org/10.1007/978-3-540-27780-4_21

19. McCarthy, K., Reilly, J., McGinty, L., Smyth, B.: An analysis of critique diversity in case-based recommendation. In: FLAIRS Conference (2005)

20. McCarthy, K., Reilly, J., Smyth, B., Mcginty, L.: Generating diverse compound critiques. Artif. Intell. Rev. **24**(3), 339–357 (2005)

21. McGinty, L., Smyth, B.: On the role of diversity in conversational recommender systems. In: Ashley, K.D., Bridge, D.G. (eds.) ICCBR 2003. LNCS (LNAI), vol. 2689, pp. 276–290. Springer, Heidelberg (2003). https://doi.org/10.1007/3-540-45006-8_23

22. McSherry, D.: Diversity-conscious retrieval. In: Craw, S., Preece, A. (eds.) ECCBR 2002. LNCS (LNAI), vol. 2416, pp. 219–233. Springer, Heidelberg (2002). https://doi.org/10.1007/3-540-46119-1_17

23. van Pinxteren, Y., Geleijnse, G., Kamsteeg, P.: Deriving a recipe similarity measure for recommending healthful meals. In: Proceedings of the 16th International Conference on Intelligent User Interfaces (2011)

24. Reilly, J., McCarthy, K., McGinty, L., Smyth, B.: Dynamic critiquing. In: Funk, P., González Calero, P.A. (eds.) ECCBR 2004. LNCS (LNAI), vol. 3155, pp. 763–777. Springer, Heidelberg (2004). https://doi.org/10.1007/978-3-540-28631-8_55

25. Sajadmanesh, S., et al.: Kissing cuisines: exploring worldwide culinary habits on the web. In: Proceedings of the 26th International Conference on World Wide Web Companion (2017)

26. Smyth, B., McClave, P.: Similarity vs. diversity. In: Aha, D.W., Watson, I. (eds.) ICCBR 2001. LNCS (LNAI), vol. 2080, pp. 347–361. Springer, Heidelberg (2001). https://doi.org/10.1007/3-540-44593-5_25

27. Smyth, B., McGinty, L.: An analysis of feedback strategies in conversational recommenders. In: the Fourteenth Irish Artificial Intelligence and Cognitive Science Conference (AICS 2003). Citeseer (2003)

28. Vargas, S., Baltrunas, L., Karatzoglou, A., Castells, P.: Coverage, redundancy and size-awareness in genre diversity for recommender systems. In: Proceedings of the 8th ACM Conference on Recommender Systems (2014)

29. Zeyen, C., Hoffmann, M., Müller, G., Bergmann, R.: Considering nutrients during the generation of recipes by process-oriented case-based reasoning. In: Cox, M.T., Funk, P., Begum, S. (eds.) ICCBR 2018. LNCS (LNAI), vol. 11156, pp. 464–479. Springer, Cham (2018). https://doi.org/10.1007/978-3-030-01081-2_31

30. Zeyen, C., Müller, G., Bergmann, R.: Conversational retrieval of cooking recipes. In: ICCBR (Workshops) (2017)

# Evaluation of Similarity Measures for Flight Simulator Training Scenarios

Rubén Dapica[1]([✉]) and Federico Peinado[2]

[1] Departamento de Ingeniería Industrial y Aeroespacial,
Universidad Europea de Madrid, c/ Tajo s/n. Urb. El Bosque,
28670 Villaviciosa de Odón, Madrid, Spain
rubendap@ucm.es
[2] Departamento de Ingeniería del Software e Inteligencia Artificial,
Universidad Complutense de Madrid, c/ Profesor José García Santesmases 9,
28040 Madrid, Madrid, Spain
email@federicopeinado.com

**Abstract.** Flight simulator training is fundamental for the acquisition and maintenance of professional pilot skills. One of the key factors for the effectiveness of this type of training is the design of the scripts of the sessions, usually called "scenarios". Currently, civil aviation authorities are advocating a customization of the flight training scenarios based on the specific needs of each pilot, which makes their creation a very demanding task in time and resources. Automatic generation systems for these scenarios have been proposed in the scientific literature, but they have not been fully applied to commercial flight simulators yet.

In this paper, we review the most important advances in this field to date and introduce a first proposal of a case-based reasoning system for the generation of training scenarios for non-technical skills. Particularly, our goal is to evaluate a set of four different similarity measures for case retrieval of event sets found in these training scenarios, using the judgement of real experts in the field as validation method.

**Keywords:** Case retrieval · Scenario generation · Expert judgement · Simulation-based training · Flight simulation · Aerospace industry

## 1 Introduction

The use of simulators for flight crew training is a fundamental part of the aerospace sector today, because, among other advantages, it increases safety and reduces costs considerably [1]. Since flight skills, both cognitive and motor, that are not practiced on routine flights decay over time, for simulator training to be effective the sessions must be recurrent, varied, relevant and adapted to the pilots [2,4]. Scenario-Based Training (SBT) is a methodology that seeks to optimize such instruction that has been widely used not only in the aviation domain, but also in others where decision making is critical, including military [27] and medical applications [29]. The foundation of SBT is to expose trainees to

© Springer Nature Switzerland AG 2021
A. A. Sánchez-Ruiz and M. W. Floyd (Eds.): ICCBR 2021, LNAI 12877, pp. 17–31, 2021.
https://doi.org/10.1007/978-3-030-86957-1_2

realistic simulations of a context based on real-world tasks, specifically designed to elicit, practice and evaluate certain behaviors. These expected behaviors can be associated to the desired level of technical and non-technical performance in the real domain [11].

Flight simulator training sessions based on SBT mainly consist of the representation of a scenario, which basically consists of an operational context and a series of events distributed throughout the session. This scenario design approach is called event-based approach to training (EBAT). These events are unexpected occurrences like a system malfunction, a human error, a not predictable change in weather conditions or any other circumstances that set problems to be solved by the crew that demand the demonstration of relevant competencies. Some of these competencies are non-technical, like the ability to assess the situations correctly, make decisions, manage workload and communicate efficiently. The acquisition and maintenance of these non-technical skills has demonstrated being critical for flight safety [7].

In order to be efficient, EBAT application requires that the choice and distribution of scenario events be connected to the training objectives and performance evaluation criteria [8]. In recent years, the International Civil Aviation Organization[1] (ICAO) has fostered the implementation of evidence-based training (EBT) approach for recurring assessment and training sessions in flight simulators. EBT involves migration to an increasingly more dynamic system in the production of training scenarios, that requires their constant adaptation to the specific needs of the trainees taking into account their concrete competence profile and operational reality [9,10]. For all these reasons, the design of flight training scenarios is currently a complex process that must be carried out by experts, which makes it a considerable time and resource-consuming task.

Automated scenario generation is a possible solution to reduce costs. Many attempts to develop computer systems for generating scenarios for virtual training environments in several domains have been documented in the past. Some of the recent lines of research in this direction have included procedural methods [23], artificial neural networks [19–21] and reinforcement learning [26,31]. Although these lines of research are interesting, all of them have in common their heavy dependency on statistical correlations, which may be particularly problematic in flight training environments. The instructor who guides the training session in the simulator does not usually design that session, but he/she should be able to understand the relationship between the set of events chosen in the scenario and the skills to be trained. This connection between competences to develop and events that help to train them is essential to assess the behavior of the trainees and be able to give them appropriate feedback. In this sense, technologies that preserve traceability between problem features (pilot training needs) and their solutions (training scenarios) seem better approaches to customized aeronautical training than other methods.

In this work we present the general architecture for a case-based reasoning (CBR) system for scenario generation in the flight training domain and propose

---

[1] https://www.icao.int/.

a model to represent pilot training needs profiles based on the competence framework recommended by the aviation authorities. That model would constitute the base for storing, retrieving and adapting existing customized training scenario cases in the system's knowledge base. Once the model was defined, we tested four well known similarity operators to be used in the CBR retrieval stage. Flight simulation instructors were provided with a set of scenario chunks and asked to rate their usefulness one by one against randomly generated competency profiles. The questioned subjects were not aware that each of the pieces of scenario presented had been previously assessed (independently, by other experts) as suitable for a specific set of competencies that differed from those randomly assigned. The results showed that the values obtained by at least 3 of the considered similarity operators are consistent with the utility evaluations given by the experts.

The rest of the paper is structured as follows. Section 2 reviews relevant previous work in the field of flight simulator training. Section 3 defines the problem to be solved, analysing relationships between pilot competencies and scenario events. Section 4 proposes a model to represent these training cases and describes the architecture of the CBR system proposed as solution. Section 5 explains the evaluation method to validate the proposed similarity functions for case retrieval. Section 6 presents the data of the study conducted to test the proposed functions and elaborates a discussion on the obtained results. Finally, some conclusions based on the previous analysis are stated in Sect. 7, as well as the scope and limitations of the positive correlation found and future lines of work based on this result.

## 2    Related Work

Flight simulators based on high-fidelity and full-movement technologies developed between the late 40s and mid-60s allowed pilot training, previously conducted on real aircraft, to be transferred to a more efficient and safe environment [24]. These devices allowed first to develop important technical skills, such as executing emergency maneuvers that would be very dangerous or unfeasible in a real airplane, and later to incorporate non-technical skills, fundamental for solving real problems on board [14]. This integration of skills seeks to maximize realism and therefore transfer the complexity of the real world to the simulator. This methodology requires not only that the simulators have a high technical fidelity, but that the proposed scenarios also contain a high cognitive fidelity. This requirement forces the challenging task of constantly designing high-quality and varied simulator training scenarios, which is a challenging task. Hence the interest in the use of computer programs for their generation [6].

The first training scenario generator was designed in the 1980s for NASA space shuttle flight controller simulator [16–18]. But it was not until the 1990s, when the first attempt to solve the problem of the SBT in the field of commercial aviation was carried out. A tool called Rapidly Reconfigurable Line Operations Simulation (RRLOS) was launched in the framework of Advanced Qualification Programs of the Federal Aviation Administration [3,12]. RRLOS generated scenarios semi-automatically from event sets of other scenarios previously designed by experts. These event sets were stored in a database and they were recombined -randomly at first- in a new scenario. In order to provide realistic and useful scenarios, each event set was checked against a series of heuristics for each flight phase of the session. After an iterative process, when the heuristics constraints were satisfied in all the phases, the scenario was completed [23,25].

RRLOS intended to be a complete tool to reduce the cost of scenario development time, generating not only the scenario script (the event set distribution along the session) but all the related materials for the pilots and instructor (meteorological reports, navigation charts, evaluation sheets, etc.). However, it was a system difficult to update to the new training requirements that were arising, what caused many instructional organizations to gradually abandon their use [6]. This fact, together with the new EBT framework under implementation, highlights how important is for a flight training scenario generation tool being easy to update and maintain. In other words, the complexity of the system must be such that the time saved in the scenario development compensates for the time spent in updating the tool.

In the last few years, there have been several attempts to apply artificial intelligence techniques to the process of automated generation of training scenarios, especially for tactical domains. Procedural methods like L-systems were used by Martin [23]. Zook et al. [32] approached the problem by modeling it as a combinatorial optimization of events by means of genetic algorithms and applied it to decision-making training in the domain of first aid in combat situations. Luo et al. [19–21] used the same approach applied to a food distribution game, using artificial neural networks and an intelligent agent to evaluate the generated scenarios. Recent attempts in this line of research are also incorporating reinforcement learning [26].

Although previous approaches to combinatorial optimization may lead to useful and believable training scenarios, the process by which different solutions are reached is not easily explainable to users. In the context of aeronautical training, it is important that the instructor knows why some events are included in the scenario and not others, as well as what relationships these sets of events have with the training objectives. This knowledge is essential to be able to provide a comprehensible feedback to trainees at the end of the training session.

In contrast with previous efforts, our CBR approach would take advantage of expert knowledge, retrieving tailored scenarios that have previously shown their utility in real operation and training, keeping traceability between event sets in training scenarios and training needs and allowing continuous updating by adding new validated instances in the case base.

## 3    Problem Definition: Pilot Competencies and Event Sets

The goal of our research is to support the generation of flight simulator training scenarios adapted to pilot specific training needs. Concretely, for each specific set of competencies to be trained in a session, the system must be able to retrieve event sets that are useful to train those competencies, helping to build a useful and customized training scenario.

The proposed architecture is represented in Fig. 1. First of all, the input of the system is represented as a combination of the pilot's profile (obtained after an official assessment) and other training parameters such as the type of the aircraft involved. This information is considered the problem of the new case presented to the system, so the CBR cycle begins. In the case retrieval phase the system search for a similar problem in the case base, obtaining several cases with partial solutions (some event sets) to the global problem of generating a training scenario. The actual combination of these pieces in a complete scenario script is performed during the case reuse phase. After that, human trainers should use the proposed solution and validate its utility. After those empirical tests, it is possible to activate a partial retention case, breaking down the most relevant pieces of knowledge about use of event sets in one or more cases that should be added to the case base.

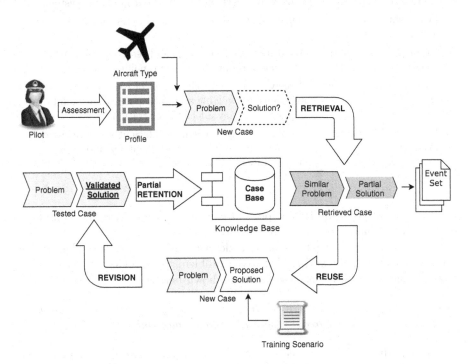

**Fig. 1.** General architecture of the proposed CBR Scenario Generator.

EBT framework established by ICAO defines eight basic competencies to be assessed and trained in recurrent flight simulator sessions [9,10], shown in Table 1. The problem of creating flight training scenarios can be modeled as follows. Given a pilot $p_i$ with a competency profile $p_i = \{c_1, c_2, ..., c_n\}$, being $c_i$ the measure of her a specific training scenario $S_i$ must be generated for that pilot. Assuming the ICAO pilot competency framework, vector $p_i$ will have eight dimensions ($n = 8$), according to the same table: $c_1 = APK$, $c_2 = COM$, $c_3 = FPA$, $c_4 = FPM$, $c_5 = LTW$, $c_6 = PSD$, $c_7 = SAW$ and $c_8 = WLM$.

Table 1. Summary of pilot competencies

| Competencies | Description |
|---|---|
| Application of Procedures (APK) | Identifies and applies procedures in accordance with published operating instructions and applicable regulations using the appropriate knowledge |
| Communication (COM) | Demonstrates effective oral, non-verbal and written communications in normal and non-normal situations |
| Aircraft Flight Management, automation (FPA) | Controls the aircraft flight path through automation, including appropriate use of flight management system(s) and guidance |
| Aircraft Flight Management, manual control (FPM) | Controls the aircraft flight path through manual flight, including appropriate use of flight management system(s) and flight guidance systems |
| Leadership and Teamwork (LTW) | Demonstrates effective leadership and team working |
| Problem Solving and Decision Making (PSD) | Accurately identifies risks and resolves problems. Uses the appropriate decision-making processes |
| Situational Awareness (SAW) | Perceives and comprehends all of the relevant information available and anticipates what could happen that may affect the operation |
| Workload Management (WLM) | Manages available resources efficiently to prioritize and perform tasks in a timely manner under all circumstances |

For the purposes of this paper, we are going to reduce the complexity of a flight scenario to an discrete array of $m$ (variable) events sets: $S_i = \{e_1, e_2, ..., e_m\}$. Each event set $e_j$ is a description of a situation that threats the safety of the flight and constitutes a problem to be solved by the trainees. An example of an event set description is shown below:

*Auto flight failure at decision altitude (DA) during a low visibility approach requiring a go-around flown manually.*

Indeed, these event sets could be decomposed in several elements, events with possible occurrences in a real flight, as in the case of this description: an aircraft equipment failure (the auto flight system), a specific aircraft location (the approach DA, where the pilots must decide if continue to the runway or abort landing), and a specific weather condition (low visibility under the limits of the approach procedure). On account of the fact that the current flight transport system is very complex, the diversity of events that can be part of an event set is very high. Some examples of event set elements are system failures, air traffic control errors, unexpected weather changes, critical aircraft positions, traffic conflicts, etc.

In our current model, an scenario is composed by one or more event sets assigned to each of the flight phases of a simulator session. The key to design a good training scenario is to choose a correct combination of event sets taking into account the pilot's competencies to train. Hence, the utility of a training scenario and therefore the efficiency of a simulator training session depends directly on the suitability of the selected event sets.

## 4    Case Retrieval of Scenario Event Sets

Generating a high quality event set is not a trivial task. Apart from the competency profile of the trainees, several variables must be taken into account, like the difficulty that emerge from the event set, the time needed to complete the associated tasks or the level of realism of the generated situation, among others. Due to this complexity, adding event sets to a training scenario is still an expert matter. Precisely, one of the key points of using a CBR approach in our system is taking advantage of that expertise to assist scenario designers and reduce time spent in building customized flight training scenarios. The main assumption of this work is that if some event sets succeeded training a specific set of pilot competencies, they must be useful for training new pilots with similar needs.

To reach our goals, in this first attempt we will rely on a case base of simple event sets, not complete scenarios, each one validated by experts for a specific pilot competence profile. Therefore, every case $C_{i,j}$ contains a single competence profile $p_i$ and a single event set $e_j$, as it is represented in Eq. 1.

$$C_{i,j} = (p_i, e_j) \tag{1}$$

The source for populating the case base in this fist attempt, when the system has not been implemented yet, is the recurrent assessment and training matrix for turbo-jet aeroplanes of the fourth generation provided by the International Air Transport Association (IATA) in its EBT Implementation Guide [9]. This matrix contains 150 validated event sets, each one designed to train a specific combination of pilot competencies. For instance, the event set example shown in Sect. 3 has been taken from that matrix, and could be called $e_x$. According to that matrix, that specific event set is useful to train APK, FPA, FPM and SAW competencies. In our model, we use a vector for representing the ideal pilot competence profile for that event set, with a value of 1 (or $true$) in the trainable competencies, and 0 (or $false$) in the rest. So the case will be represented in our case base as $C_x = (p_x, e_x)$ being $p_x = [1, 0, 1, 1, 0, 0, 1, 0]$, and $e_x$ the description of the event set.

According to previous studies, the stated flight competences are not independent, and they can be grouped in terms of their interrelationships [22]. In this work we are going to focus just on a subset of the flight competencies, specifically the non-technical ones (COM, LTW, PSD, SAW and WLM). The main reason to narrow down the study is that while non-technical skills are extensible to virtually all environments, technical skills are very dependent on the model of aircraft as well as the procedures of the specific flight company. Thus, in this study the vectors corresponding to the competence profiles have only five ($n = 5$) dimensions.

For the retrieval phase, we propose a k-nearest neighbors algorithm [13] using different similarity operators that compare stored scenario and trainee competency vectors. Given a pilot with a competence profile, her competence query vector $q$ is defined. The algorithm will compare $q$ with all the $p$ vectors in the case base, measuring similarity ($sim$) according to the chosen operator.

Once a minimum similarity threshold ($min$) is established, the system will retrieve the $k$ cases that meet $sim(q, p) \geq min$. The utility of the retrieved event cases for generating the training scenario is therefore dependant on the quality of the similarity function. In other words, the retrieved event sets will be useful to the extent that the similarity function captures the relevant connection between competence profiles. These similarity measures should therefore be exposed to the judgement of real experts in order to determine if they are capable of matching cases with similar characteristics in terms of perceived opportunities for the development of flight competencies.

## 5    Evaluation of Similarity Measures

From the abundant number of similarity measures available, four functions widely accepted for use in binary vectors were chosen [15]. The functions and their corresponding definitions are shown below. The first one, Cosine similarity or $sim_{cos}$ (Eq. 2), has proved being very successful in other CBR applications like item-based collaborative filtering recommendation [30].

$$sim_{cos} = \frac{\sum_{i=1}^{n} q_i \cdot p_i}{\sqrt{\sum_{i=1}^{n} q_i^2} + \sqrt{\sum_{i=1}^{n} p_i^2}} \tag{2}$$

$$sim_T = \frac{\sum_{i=1}^{n} q_i \cdot p_i}{\sum_{i=1}^{n} q_i^2 + \sum_{i=1}^{n} p_i^2 - \sum_{i=1}^{n} q_i \cdot p_i} \tag{3}$$

$$sim_{S-D} = \frac{2 \cdot \sum_{i=1}^{n} q_i \cdot p_i}{\sum_{i=1}^{n} q_i^2 + \sum_{i=1}^{n} p_i^2} \tag{4}$$

$$sim_{SM} = \frac{M_{00} + M_{11}}{M_{00} + M_{01} + M_{10} + M_{11}} \tag{5}$$

One of the reasons for choosing this similarity measure is that its result only depends on the elements present in both vectors, and not on the absent ones. The hypothesis maintained in this paper is that the similarity of two flight training scenarios in terms of their competence profile must be sought at the intersection of competencies that both train, rather than in their similarity in what they do not train.

The validity of the cosine function in this context will be contrasted, via the experts' judgment, with the values given by three other similarity measures. The Tanimoto operator, $sim_T$ (Eq. 3) is a version for binary vectors of the widely known Jaccard similarity measure. The Sorensen-Dice similarity function, $sim_{S-D}$ (Eq. 4) is a variation of $sim_{cos}$ by means of a multiplicative coefficient. Any of these three functions are linear combinations of the other two, that is, they differ in the relative importance of the common and distinct elements of the two vectors [15]. These three operators were contrasted with a fourth measure, the Simple Matching similarity, $sim_{SM}$ (Eq. 5), in which the common absent elements in both vectors also increase similarity. In this last similarity operator, the number $M$ of positive matching in each coordinate of the vectors is added, both by presence and absence of the characteristic (competency), and divided by the total number of coincidences and non-coincidences. This measure is different from the other three, including negative matching as a value for similarity [5]. This measure is different from the other three in that it gives weight to the negative matching in the similarity value. This contrast allows us to verify whether measures focus on positive matching are more appropriate in this context, as predicted by our hypothesis.

The procedure followed to contrast similarity functions with the experts' criteria is detailed in the next section.

## 5.1   Experimental Setup

We now set the procedure to evaluate the chosen similarity functions with the help of expert judgement. The objective was to determine whether there is a correlation between the value of the proposed similarity functions and a measure of the perceived utility of the retrieved cases given by experts. Provided with random pairs of competence profiles and event sets, the task of the experts was to fill an online survey assessing the utility of each event set for training the specific competencies required by each corresponding profile. Finally the central tendency measures of the assessed utility for all the pairs was compared with the corresponding similarity values given by the equations for that competency profile and the original competency profile attributed in the knowledge base (the EBT Implementation Guide Matrix, see Sect. 4) for that event set.

We recruited 11 active flight instructors approved by the European Union Aviation Safety Agency for this study. The experts were provided with a questionnaire in which they had to specify their credentials as a flight instructor as well as the number of years of experience in that activity. They were asked 10 questions, in each of which they were shown an event set (from the knowledge base) and a competence profile, both chosen randomly, and were asked to evaluate the suitability of said event set for the profile provided. Each question was then formed by a query profile and an event set $(q_i, e_i)$.

The way to generate those pairs was as follows. The first array was created with a 5-digit binary random number generator, discarding the zero-vector ($[0, 0, 0, 0, 0]$). The second item, the event set, was randomly selected from the case base with a random number generator, dismissing repetitions. In each question, experts had to answer, according to his free judgment, on a 5-point Likert scale to the question: *"What level of adequacy do you think exists between the profile of this pilot and the corresponding training scenario element?"*. Possible answers were *"Very good"* (VG), *"Good"* (G), *"Acceptable"* (A), *"Poor"* (P) and *"Very poor"* (VP).

In the last part of the questionnaire, experts were asked to make any suggestions regarding the experiment, as well as providing contact information to inform them of the result of this research.

### 5.2   Analysis

After the surveys were answered and before collecting the data, the similarity operators were applied to all the pairs $(q_i, p_j)$, being $p_j$ the competency vector associated to the event set $e_j$ in the knowledge base. The pairs of profiles were then ordered from least to greatest similarity according to $sim_{cos}$, naming them P1, P2, P3, P4, P5, P6, P7, P8, P9 and P10. Each of these pairs were then evaluated by the rest of similarity operators. In the end, each pair had 4 similarity values and 11 utility values given by the human experts. The median of the expert evaluations for each pair was calculated and then the Spearman and Kruskall correlation tests were conducted to compare it with the computed similarities.

## 6   Results and Discussion

The data is summarized in Table 2. Each line in the table corresponds to the 10 pairs of competency profiles ordered from least to greatest, and named in the first column from P1 to P10. The second column details the competency profiles to be compared with the event sets supplied to the experts. The third column shows the competence profiles associated with these events. This information was not available to the experts when they were evaluating. In the fourth column the values of the similarity function for each pair of profiles are shown. Finally, in the fifth column the medians of the answers given by the judges for each pair are calculated.

The distributions of utility values per pair are shown in form of box plots in Fig. 2. As can be easily seen from both data and distributions, there is an increasing trend in evaluations as the value of similarity increases.

When conducting the Kruskal-Wallis test to check differences in the evaluations between experts, the result obtained is statistically significant ($H = 24.951; p - value < 0.01$). From this result it follows that between two experts in the sample there may be considerable differences in criteria in the evaluation of an event set based on their suitability to train a specific competency profile. However, these differences tend to disappear when the number of experts increases, as occurs when seen in the measure of central tendency of the total sample. This is so even though the sample of experts has been taken among instructors of the same approved training organization. This indicates that the diversity in the scenario evaluation criteria can be considerable even among experts with similar training and professional environment. In view of this result, broad consensus seems necessary for the evaluation of specific scenarios.

**Table 2.** Comparison of similarity operators and utility values assigned by the experts. The pairs were initially set randomly and then named in order according to the value of the cosine similarity operator.

| Pair | $q$ | $p$ | $sim_{cos}$ | $sim_T$ | $sim_{S-D}$ | $sim_{SM}$ | Median of utility values |
|------|-----|-----|-------------|---------|-------------|------------|--------------------------|
| P1   | $[1,1,0,0,0]$ | $[0,0,0,1,0]$ | 0 | 0 | 0 | 0.4 | *Acceptable* |
| P2   | $[0,1,1,0,1]$ | $[0,0,0,1,0]$ | 0 | 0 | 0 | 0.2 | *Good* |
| P3   | $[0,0,0,1,1]$ | $[1,0,1,0,0]$ | 0 | 0 | 0 | 0.2 | *Acceptable* |
| P4   | $[1,1,0,1,0]$ | $[0,0,1,1,1]$ | 0.3333 | 0.2 | 0.3333 | 0.2 | *Good* |
| P5   | $[0,1,1,1,0]$ | $[1,0,0,1,0]$ | 0.4082 | 0.25 | 0.4 | 0.4 | *Good* |
| P6   | $[0,1,1,0,0]$ | $[1,0,1,0,0]$ | 0.5000 | 0.3333 | 0.5 | 0.6 | *Good* |
| P7   | $[1,1,1,0,1]$ | $[0,0,1,1,1]$ | 0.5774 | 0.4 | 0.5714 | 0.4 | *Good* |
| P8   | $[0,0,0,1,0]$ | $[1,0,0,1,0]$ | 0.7071 | 0.5 | 0.6667 | 0.8 | *Good* |
| P9   | $[0,1,1,1,1]$ | $[0,0,0,1,1]$ | 0.7071 | 0.5 | 0.6667 | 0.6 | *Very Good* |
| P10  | $[1,1,1,0,0]$ | $[1,1,1,1,0]$ | 0.8660 | 0.75 | 0.8571 | 0.8 | *Very Good* |

Nevertheless, the data obtained showed a strong correlation between the expert evaluation median and three of the similarity functions. The Spearman's correlation test for Cosine, Tanimoto and Sorensen-Dice similarity measures with the median of expert evaluation gave a value of $r = 0.8106$ ($p < 0.01$), but could not find an acceptable value of correlation in the case of Simple Matching similarity function ($r = 0.5680; p < 0.08$).

The Kendall's $\tau_b$ test was also conducted for all the similarity functions, in order to check bias due to the small size of the Likert scale ordinal variable. The correlation was confirmed in the four measures, being stronger in Sorensen-Dice ($\tau = 0.7664; p < 0.01$), Cosine, and Tanimoto measures ($\tau = 0.7378; p < 0.01$) than in Simple Matching ($\tau = 0.5281; p < 0.01$).

In Fig. 3 it can be seen that the cosine and Sorensen-Dice similarity functions give very similar results, the values given by Tanimoto being approximately 20% lower when pairs of scenario-profile competence vectors of medium to high utility are evaluated, according to the experts (pairs 4 to 10 on the graph). This result suggests that both Cosine and Sorensen-Dice better capture the concept of utility as perceived by experts in this context, given that the median assessment in these pairs were between *"Good"* and *"Very good"*.

Although there is a clear correlation between considered similarity operators and the expert assessment, it is not perfect, especially when small utility values are considered. If the utility values given by the experts in pairs 1 to 3 are compared with the similarity values given by the three most reliable similarity operators (Cosine, Tanimoto and Sorensen-Dice) a considerable difference emerge. For example, in cases P1, P2 and P3, whose similarity is 0 in all three cases,

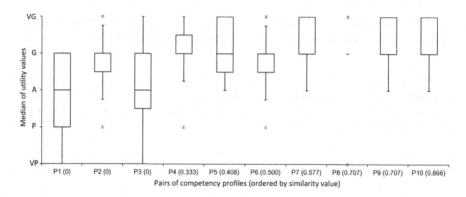

**Fig. 2.** Results of the judges evaluation for each pair of profile-scenario competency vectors.

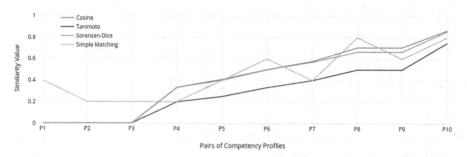

**Fig. 3.** Comparison between 4 well known similarity measures for binary data in the evaluation of a set of 10 pairs profile-scenario competency vectors.

the medians of the expert evaluation are *"Acceptable"*, *"Good"*, and *"Acceptable"*, respectively. The possible reason for this is that the zeros in the profile vectors are not "nonexistent" training necessities. Although the competency profiles associated with the event sets in the case base were also assigned by experts, in that matrix there are no gradations of the more realistic involvement of all the competencies in each of them. All the information that they could register -and is therefore available- is only the set of competencies that are most developed. This does not mean that there is no presence of the others. As mentioned above, strong relationships have been identified between competencies that prevent a small set of them from being completely isolated in a flight situation [22].

Take P1 as an example. The pair in this case is shown in Eq. 6

$$P1 = (q_1, p_1) = ([1, 1, 0, 0, 0], [0, 0, 0, 1, 0]))$$  (6)

The vector $q_1$ defines the skills to be trained by the pilot and the vector $p_1$ the competency profile associated to the event set randomly recovered from the case base. Although the pilot must train COM and LTW competencies, the recovered event was evaluated as optimal only for developing Situational

Awareness competence, but that does not mean that the event truly does not need some Communication and Teamwork skills to be overcome successfully. In fact, Situational Awareness is strongly linked to communication and cooperation with the rest of the team in order to check whether pilot's own interpretation of the situation is correct [28]. Therefore, although the cosine similarity function may give a value of 0 for a pair, as it is the case of P1, it is difficult to find flight training situations with absolute no necessities in any non-technical competences.

Another problem of taking measures of central tendency in the evaluation is the ambiguity that is appreciated at medium-high values of similarity. As can be seen in Table 2, the median of the good evaluation ranges from 0 to 0.7071. A possible solution to partially reduce this may be the refinement of the evaluation system, for example by means of a 7-point Likert scale. Such a system would increase the distance on the evaluation scale of some cases of medium-high utility with respect to the optimal cases. For example, if an event set is evaluated as position 4 on a 5-point scale, on the 7-point scale it could correspond to position 5 or 6. In this way, the evaluation system would be more sensitive to possible small differences between experts in the evaluation of certain event sets, discriminating them more clearly from those of higher quality.

# 7  Conclusions

In our research, we are dealing with the problem of generating training scenarios for adapted flight simulator sessions using a CBR approach based on the current standard pilot competence framework. In this paper, we have managed to support its viability by evaluating four similarity functions for the retrieval of event sets for these scenarios through expert judgment.

Specifically, the study was successful in providing a high correlation between the Cosine, Tanimoto and Sorensen-Dice similarity functions and the human expert assessment of flight training scenario event sets, with a better fitting in the first two measures. Some limitations of this study include, on the one hand, the number of expert evaluations, and on the other hand, the consideration only of non-technical competencies. In addition, the nature of simulator training scenarios, which constitute natural decision-making environments, makes it impossible to perfectly isolate competencies within each event set. The use of a more refined assessment of the event sets, could minimize this problem.

Assuming the postulates of CBR, and that the scenarios constituted by event sets adjusted to specific competence profiles are themselves adjusted to said profile, the solution for generating training scenarios through a CBR tool using the proposed similarity function seems to be a promising option.

As future lines of work we propose the extension of the study of the similarity functions to the rest of the flight competences, the use of more refined evaluation scales in order to better define utility, and the design and testing of an algorithm that allows choosing between different scenarios with identical similarities based on other factors such as aircraft systems that must be trained in the session, historical record of the pilot and the diversity of the event sets of the generated scenarios.

**Acknowledgments.** This work has been possible thanks to the collaboration of the EASA approved training organization *Global Training and Aviation, S.L.*

# References

1. Allerton, D.: The impact of flight simulation in aerospace. Aeronaut. J. **114**(1162), 747–756 (2010)
2. Arthur, W., Jr., Bennett, W., Jr., Stanush, P.L., McNelly, T.L.: Factors that influence skill decay and retention: a quantitative review and analysis. Hum. Perform. **11**(1), 57–101 (1998)
3. Bowers, C., Jentsch, F., Baker, D., Prince, C., Salas, E.: Rapidly reconfigurable event-set based line operational evaluation scenarios. In: Proceedings of the Human Factors and Ergonomics Society Annual Meeting, vol. 41, no. 2, pp. 912–915 (1997)
4. Childs, J.M., Spears, W.D.: Flight-skill decay and recurrent training. Percept. Mot. Skills **62**(1), 235–242 (1986)
5. Choi, S.S., Cha, S.H., Tappert, C.C.: A survey of binary similarity and distance measures. J. Syst. Cybern. Inform. **8**(1), 43–48 (2010)
6. Curtis, M., Jentsch, F.: Line operations simulation development tools. In: Crew Resource Management, pp. 323–341. Elsevier (2019)
7. Flin, R., O'Connor, P.: Safety at the Sharp End: A Guide to Non-technical Skills. CRC Press, Boco Raton (2017)
8. Fowlkes, J., Dwyer, D.J., Oser, R.L., Salas, E.: Event-based approach to training (EBAT). Int. J. Aviat. Psychol. **8**(3), 209–221 (1998)
9. IATA: Evidence-based training implementation guide July 2013. International Air Transport Association, Montreal (2013)
10. ICAO: Manual of evidence-based training. International Civil Aviation Organization, Montreal (2013)
11. ICAO: Doc 9868: Procedures for air navigation services. Training. Second Edition. International Civil Aviation Organization, Montreal (2015)
12. Jentsch, F., Bowers, C., Berry, D., Dougherty, W., Hitt, J.M.: Generating line-oriented flight simulation scenarios with the RRLOE computerized tool set. In: Proceedings of the Human Factors and Ergonomics Society Annual Meeting, vol. 45, no. 8, pp. 749–749 (2001)
13. Kataria, A., Singh, M.: A review of data classification using k-nearest neighbour algorithm. Int. J. Emerg. Technol. Adv. Eng. **3**(6), 354–360 (2013)
14. Koteskey, R.W., Hagan, C., Lish, E.T.: Line oriented flight training: a practical guide for developers. In: Crew Resource Management, pp. 283–322. Elsevier (2019)
15. Lesot, M.J., Rifqi, M., Benhadda, H.: Similarity measures for binary and numerical data: a survey. Int. J. Knowl. Eng. Soft Data Paradigms **1**(1), 63–84 (2009)
16. Loftin, R.B., Wang, L., Baffes, P., Hua, G.: An intelligent training system for space shuttle flight controllers. Telematics Inform. **5**(3), 151–161 (1988)
17. Loftin, R.B., Wang, L., Baffes, P., Rua, M.: An intelligent training system for payload-assist module deploys. In: Space Station Automation III, vol. 851, pp. 83–91. International Society for Optics and Photonics (1987)
18. Loftin, R., Wang, L., Baffes, P.: Intelligent scenario generation for simulation-based training. In: 7th Computers in Aerospace Conference, p. 3054 (1989)
19. Luo, L., Yin, H., Cai, W., Lees, M., Othman, N.B., Zhou, S.: Towards a data-driven approach to scenario generation for serious games. Comput. Anim. Vir. Worlds **25**(3–4), 393–402 (2014)

20. Luo, L., Yin, H., Cai, W., Lees, M., Zhou, S.: Interactive scenario generation for mission-based virtual training. Comput. Anim. Vir. Worlds **24**(3–4), 345–354 (2013)
21. Luo, L., Yin, H., Cai, W., Zhong, J., Lees, M.: Design and evaluation of a data-driven scenario generation framework for game-based training. IEEE Trans. Comput. Intell. AI Games **9**(3), 213–226 (2016)
22. Mansikka, H., Harris, D., Virtanen, K.: An input-process-output model of pilot core competencies. Aviation Psychology and Applied Human Factors (2017)
23. Martin, G.A.: Automatic scenario generation using procedural modeling techniques. Doctoral Dissertation. Ph.D. thesis, University of Central Florida (2012)
24. Page, R.L.: Brief history of flight simulation. In: SimTecT 2000 Proceedings, pp. 11–17 (2000)
25. Prince, C., Jentsch, F.: Aviation crew resource management training with low-fidelity devices. In: Improving Teamwork in Organizations: Applications of Resource Management Training, pp. 147–164 (2001)
26. Rowe, J., Smith, A., Spain, R., Lester, J.: Understanding novelty in reinforcement learning-based automated scenario generation. In: Proceedings of the 7th Annual GIFT Users Symposium, p. 76. US Army Combat Capabilities Development Command-Soldier Center (2019)
27. Salas, E., Priest, H.A., Wilson, K.A., Burke, C.S.: Scenario-based training: improving military mission performance and adaptability. Mi. Life: Psychol. Serv. Peace Combat **2**, 32–53 (2006)
28. Salas, E., Prince, C., Baker, D.P., Shrestha, L.: Situation awareness in team performance: implications for measurement and training. Hum. Factors **37**(1), 123–136 (1995)
29. Salas, E., et al.: Simulation-based training for patient safety: 10 principles that matter. J. Patient Saf. **4**(1), 3–8 (2008)
30. Sarwar, B., Karypis, G., Konstan, J., Riedl, J.: Item-based collaborative filtering recommendation algorithms. In: Proceedings of the 10th international conference on World Wide Web, pp. 285–295 (2001)
31. Sottilare, R.: A hybrid machine learning approach to automated scenario generation (ASG) to support adaptive instruction in virtual simulations and games. In: The International Defense & Homeland Security Simulation Workshop of the I3M Conference, Budapest, Hungary (2018)
32. Zook, A., Lee-Urban, S., Riedl, M.O., Holden, H.K., Sottilare, R.A., Brawner, K.W.: Automated scenario generation: toward tailored and optimized military training in virtual environments. In: Proceedings of the International Conference on the Foundations of Digital Games, pp. 164–171. ACM (2012)

# Instance-Based Counterfactual Explanations for Time Series Classification

Eoin Delaney[1,2,3]([✉]), Derek Greene[1,2,3], and Mark T. Keane[1,2,3]

[1] School of Computer Science, University College Dublin, Dublin, Ireland
{eoin.delaney,derek.greene,mark.keane}@insight-centre.org
[2] Insight Centre for Data Analytics, University College Dublin, Dublin, Ireland
[3] VistaMilk SFI Research Centre, Dublin, Ireland

**Abstract.** In recent years, there has been a rapidly expanding focus on explaining the predictions made by black-box AI systems that handle image and tabular data. However, considerably less attention has been paid to explaining the predictions of opaque AI systems handling *time series* data. In this paper, we advance a novel model-agnostic, case-based technique – *Native Guide* – that generates counterfactual explanations for time series classifiers. Given a query time series, $T_q$, for which a black-box classification system predicts class, $c$, a counterfactual time series explanation shows how $T_q$ could change, such that the system predicts an alternative class, $c'$. The proposed instance-based technique adapts existing counterfactual instances in the case-base by highlighting and modifying discriminative areas of the time series that underlie the classification. Quantitative and qualitative results from two comparative experiments indicate that Native Guide generates plausible, proximal, sparse and diverse explanations that are better than those produced by key benchmark counterfactual methods.

**Keywords:** Counterfactual explanation · XCBR · Time series

## 1 Introduction

In recent years, the predictive success of machine learning systems has been undermined by their lack of interpretability and beset by growing public disquiet about the fairness, accountability, and transparency of intelligent systems [1,19]. These challenges have led to major efforts in Explainable AI (XAI), where a raft of techniques has been developed to shed light on opaque predictions. Most of this research focuses on image and tabular data, with less attention being given to the explanation of time series data [39]. Explaining time series predictions, arguably, presents a whole new set of issues for XAI, due to the multi-dimensional nature of the data, strong feature dependencies, and the need to define the contexts where explanations could be used. In this paper, we advance an explainable case-based reasoning (XCBR) solution to this XAI problem.

© Springer Nature Switzerland AG 2021
A. A. Sánchez-Ruiz and M. W. Floyd (Eds.): ICCBR 2021, LNAI 12877, pp. 32–47, 2021.
https://doi.org/10.1007/978-3-030-86957-1_3

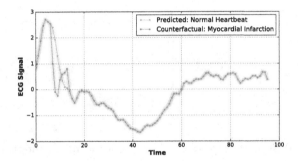

**Fig. 1.** A counterfactual instance explains the classification of an ECG signal. Here, a black-box's classification of a normal heartbeat is explained with a counterfactual, from *Native Guide*, showing an abnormal, heart-attack signal.

Recently, a variety of CBR methods for XAI has been proposed (see [51] for a review). For image and tabular data, these XCBR techniques provide factual, example-based explanations (e.g., [23,53]), feature-importance explanations (CBR-LIME; [45]), and counterfactual explanations [25]. In particular, counterfactual explanations have become popular as a post-hoc explanation technique, with over 100 distinct methods being proposed [24]. However, few of these methods consider the explanation of time series [3,8,18,22]. Hence, we advance *Native Guide*, a novel model-agnostic explanation technique for time series classification (TSC) systems that provides counterfactual explanations for their predictions.

**XAI's Promise for time Series Classification (TSC).** TSC has demonstrated significant promise in a variety of domains, including healthcare and food spectroscopy. However, there is a requirement to explain these decisions to end users. In healthcare, one practical application involves the classification of electrocardiogram signals, where explainable insights can aid medical practitioners in determining what portions of the time series are most informative for detecting abnormalities [42] (e.g. myocardial infarction). Figure 1 shows one such example, where a cardiologist might be shown the normal heartbeat of a patient along with a counterfactual signal as an explanation, basically saying "for this patient, their normal profile looks like this (purple-blue line), but if it changes to this counterfactual profile (purple-pink line), then they are experiencing an infarction" (see also Fig. 2). Similar examples can be found in spectroscopy analyses when determining the provenance of different foods. For example, near-infrared spectrographs can distinguish between Arabica and Robusta coffee beans or honey from different regions [5] (see also Fig. 7). By identifying portions of the time series that are discriminative for classification, cheaper sensors can be designed that only consider a small portion of the wider spectra. Similarly, in deep learning systems, explainable insights can uncover the portions of a time series that may be most prone to adversarial attacks and show them to model developers [11].

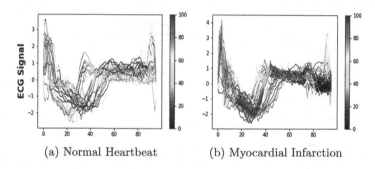

(a) Normal Heartbeat                    (b) Myocardial Infarction

**Fig. 2.** Class Activation Maps (CAM) generated for the ECG200 dataset highlighting (in red) those areas of the time series which are most discriminative for a CNN Classifier. Here, the initial portion of the time series is most discriminative for both classes. (Color figure online)

**Outline of Paper.** In the remainder of this paper we first review the related work on time series XAI (Sect. 2). We then discuss the untapped promise of counterfactual explanations in TSC and the potential properties of good counterfactual explanations in this context (see Sect. 3). Next we describe the proposed technique (Sect. 4) and conduct comparative experiments to evaluate the quality of the explanations produced before discussing our results and suggesting promising avenues for future work (see Sects. 5 and 6).

## 2  Related Work

The XAI literature on explaining time series classification has progressed along similar lines to XAI, in general; initial techniques focused on explanation through visualization and feature-importance, rather than on instance-based methods, such as factual or counterfactual explanations [26,33].

*Saliency* methods typically visualize an extracted explanation weight vector $\omega$ that captures discriminative areas of a time series for classification [39]. For example, Class Activation Maps (CAMs) [60] utilize these weight vectors to highlight areas of a time series that are most informative for classification decisions of deep neural networks (DNNs) [12,57] (see Fig. 2). Similarly, *shapelets* can also find discriminative subsequences of a time series that can either be directly extracted from a set of time series [58] or learned by minimizing an objective function [16]. Shapelets can capture relationships between features and are closely related to saliency maps as both techniques offer visual explanations for classification tasks. Some have considered using shapelets for contrastive explanation [18]. However, concerns have been raised about the interpretability of shapelets produced by the deployed *learning-shapelets* algorithm [56].

*Feature-importance analyses* are another method used to find relevant portions of a time series for use in explanation. Many state-of-the-art time series classifiers (e.g. Mr-SEQL [30]) transform the input data and deploy a linear model for classification, where $\omega$ can be directly extracted from the regression coefficients of the classifier. Indeed, model-agnostic techniques such as LIME [46] and SHAP [35], can be used to compute $\omega$ if it is not readily provided by the base classifier [39]. However, concerns about the stability of these methods have been raised through examining how small perturbations can change the explanation [2,39]. Schlegel et al. [50] tested the informativeness and robustness of different feature-importance techniques in time series classification. LIME was found to produce poor results across all evaluated datasets (a problem attributed to the high dimensionality of the data); in contrast, saliency-based approaches and SHAP were found to be more robust across different architectures.

More recently, a handful of *instance-based techniques* have been proposed to explain time series classification. *Prototypes* are instances that are maximally representative of a class and have demonstrated promise in producing global insights for time series classification in the healthcare domain [14] but they do not provide insights into the most discriminative areas of the time series. Case-based approaches using twin systems [23,27,49] have also been extended to time series data; Leonardi et al. [32] suggested mapping features from a DNN to a CBR system for interpretable haemodialysis classification. However, these techniques do not consider very popular counterfactual explanations. In an earlier unpublished version of the present paper [8], we considered how instances from the case-base could be retrieved for counterfactual explanation. However, we did not retrieve and integrate discriminative feature information in counterfactual generation, a significant novelty in the current method. Here, we advance a new XCBR method for generating good explanatory counterfactuals for any black-box time series classifier.

## 3    Good Counterfactuals for Time Series: Key Properties

There is a growing consensus that counterfactual explanations are causally informative [33,43], psychologically effective [6,9,25,36,37], and legally compliant with respect to GDPR [55]. Arguably, counterfactuals provide more robust and informative explanations than feature-importance methods, such as LIME or SHAP [17]. Although it can be difficult to visualize counterfactual explanations for tabular data [38], in the time series domain their visualization is more straight-forward (see Fig. 1). However, counterfactual XAI solutions for time series classification are rare (see e.g. [3,18,22] for closest works) and we know of no existing XCBR solutions. Indeed, it is unclear if (i) existing counterfactual techniques for tabular/image data can be applied to time series data, (ii) the properties of good counterfactuals from tabular and image data transfer to the time series domain. In Sect. 4, we present the details of our novel XCBR method, but before that we first consider four potential properties of good counterfactual explanations for time series: namely, proximity, sparsity, plausibility, and diversity.

**Proximity.** Proximity refers to how close the to-be-explained query is to the generated counterfactual instance. Typically, closeness is measured using predefined distance metrics; close counterfactuals measured using Manhattan distance have been found to be informative [25,38]. Following recent recommendations on evaluation [10,21,24], we use several different distance metrics and a relative counterfactual distance measure, to monitor the proximity of the generated counterfactual with respect to existing in-sample counterfactual solutions [25].

**Sparsity.** As noted by [38], counterfactual instances that change fewer features are preferred for informative explanations. Keane and Smyth [25] suggested that a sparsity of $\leq 2$ feature differences was preferable for tabular data, on psychological grounds (that have been confirmed in recent user studies). However, the multi-dimensional nature of time series data means that a simple application of this idea is untenable. For image data, it has been argued that counterfactuals need to modify "semantically-meaningful" features instead of small, pixel-level features that may not be humanly-perceptible [28,48]. For time series data, it has been proposed that semantically-meaningful/discriminative information is contained in contiguous subsequences of the series [30,58]. So, by analogy, we argue that "good", sparse counterfactuals need to modify a single discriminative portion of the time series (i.e., a contiguous subsequence), rather than distributed, discrete time-points in series (e.g., see Figs. 1 and 7).

**Plausibility.** Informative counterfactual explanations also need to be plausible [36]. Many suggest that proximity is a good proxy for plausibility [37], though others argue that falling within the data distribution is a better proxy [28,29,44,54]. Poyiadzi et al. [44] argue that plausible counterfactuals are representative of the underlying data distribution. Figure 6 shows some examples of implausible counterfactuals that are out-of-distribution, even though they have high proximity. Hence, in our evaluations we explore several novelty-detection algorithms to find better measures of plausibility/implausibility (see e.g. [20]).

**Diversity.** Mothal et al. [38] advanced the idea that a system should be able to produce multiple *diverse* explanations for a single query case. One advantage of this is that different users may find different explanations helpful [51]. So, our proposed method generates multiple explanations for a single test instance. However, we explicitly ensure that diversity should not come at the cost of either (i) plausibility or (ii) the loss of semantically meaningful information.

In the next section, we describe the proposed Native Guide method, before we consider a series of tests of it on several different datasets.

## 4    Native Guide: Counterfactual XAI for Time Series

Like other case-based XAI methods [25,27,31,41], at its core Native Guide relies upon existing instances in the training data, so-called *native guides* or nearest unlike neighbors (NUNs), that it retrieves and adapts to generate counterfactual explanations (see Fig. 3). In this section, we outline the two main steps in the algorithm, after first describing the notation adopted.

**Notation.** Staying consistent with the notation of [15,18], a time series $T = \{< t_1, t_2, ..., t_m >\}$ is an ordered set of real values, where $m$ is the length. A

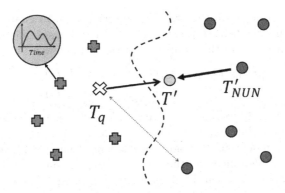

**Fig. 3.** A query time series $T_q$ (X with solid arrow) and a nearest-unlike neighbor, $T'_{NUN}$ (red circle with solid arrow) are used to guide the generation of counterfactual $T'$ (see yellow circle) in a binary classification task. Another in-sample counterfactual (i.e., the *next* NUN; other red circle with dashed arrow) could also be used to generate another counterfactual for diverse explanations. (Color figure online)

time series data set $\mathbf{T} = \{T_1, T_2 ..., T_n\} \in \mathbb{R}^{n \times m}$ is a collection of such time series where each time series has a class label $c$ forming a vector of class labels $\mathbf{Y} \in \mathbb{Z}$. Consider a black-box classifier $b(T)$ that takes a time series $T$ as an input and predicts a probability output $P(\mathbf{Y}|T)$ over the label output space. Given a to-be-explained query time series $T_q$, with predicted label $c$ from the black-box classifier (formally $b(T_q) = c$), a counterfactual explanation aims to find how $T_q$ needs to change for the system to classify it alternatively, as $c'$. We refer to $T'$ as a counterfactual explanation for $T_q$ such that $b(T') = c'$. Although there are many candidate solutions for $T'$, the method prioritizes those that meet the four key properties of proximity, sparsity, plausibility and diversity.

**Step 1: Retrieve Native Guide.** Given a query time series, $T_q$, find a counterfactual instance, $T'_{Native}$, that exists in the case-base. An example of one such instance is the query's nearest unlike neighbor ($T'_{NUN}$). In using these "native counterfactual" cases the method guarantees the explanation's *plausibility* as it is, by definition, within the distribution. However, such instances are not guaranteed to be sufficiently proximate to the query or, indeed, sparse, so an adaption step is necessary to generate the "explanatory counterfactual", $T'$ (see Fig. 3).

**Step 2: Adapt Native Guide to generate Counterfactual.** To produce a more proximate explanatory counterfactual, $T'$, the native guide, $T'_{Native}$ is perturbed towards the to-be-explained query-case, $T_q$ (see Fig. 3). Typically, counterfactual methods use some $L_p$ distance metric to guide this perturbation (such as Manhattan distance, [55]) and in time series where dynamic time warping (DTW) distance is often more appropriate an analogous averaging technique known as weighted dynamic barycentre averaging can be used [13]. In cases where we are explaining a deep-learner's predictions, the feature-weight vectors of the classifier, $\omega$, can be used to perturb "semantically-meaningful" features of the time series, rather than the "raw" time series data, to guarantee sparsity[1]. Accordingly, using the feature-weights, the method seeks to modify contiguous, subsequences, rather than the whole time series, as follows:

$$T_q = \{< t_1, t_2, t_3, t_4, t_5 ..., tn >\} \text{ s.t. } b(T_q) = c$$
$$T' = \{< t_1, t_2', t_3', t_4', t_5 ..., tn >\} \text{ s.t. } b(T') = c'$$

Specifically, the feature-weight vector, $\omega$, can be extracted using techniques such as Class Activation Mapping in the case of DNNs (see e.g. Fig. 2). Given $T'_{Native}$ and $\omega$, the most influential contiguous subsequence (measured by the magnitude of weights in $\omega$) is identified and the corresponding region in $T_q$ is replaced with these values. This process can be initialized using a small subsequence and the length of this subsequence can be iteratively incremented until $b(T') = c'$. In the very worst case scenario, the size of the subsequence will be equal to the length of $T_q$ and the native counterfactual $T'_{Native}$ is returned. This adaptation step improves the *proximity* and *plausibility* of the generated counterfactuals. Finally, *diversity* can also be met, as other in-sample instances can be used as guides (e.g., the *next* nearest unlike neighbor), to produce alternative counterfactual explanations for the original query (see Fig. 3).

**Table 1.** Summary of TSC datasets used to evaluate counterfactual explanations

| Dataset | Train size | Test size | Length | Type | No. classes |
|---|---|---|---|---|---|
| CBF | 30 | 900 | 128 | Simulated | 3 |
| Chinatown | 20 | 343 | 24 | Traffic | 2 |
| Coffee | 28 | 28 | 286 | Spectro | 2 |
| ECG200 | 100 | 100 | 96 | ECG | 2 |
| GunPoint | 50 | 150 | 150 | Motion | 2 |

## 5    Testing Native Guide: Two Comparative Experiments

We test the Native Guide counterfactual method in two experiments evaluating how it meets the properties of good explanatory counterfactuals relative to two benchmark methods on 5 representative datasets. Our focus is on explaining a black-box fully convolutional neural network classifier (FCN). Experiment 1 assesses the proximity and sparseness of the counterfactuals generated. Experiment 2 examines the plausibility and diversity of the counterfactuals generated. Here, we describe the setup for these experiments in terms of the datasets, comparative benchmark methods and black-box classification system.

**(I) Datasets.** Five diverse datasets (binary and multiclass) from the UCR archive [7] (see Table 1) were used for the classification task. To encourage reproducibility we use the default train-test splits provided by the archive and provide all experimental code, fully detailing hyper-parameters[2].

---

[1] Note, SHAP can also be used to generate such vectors, if we are directly explaining any given model, rather than twinning.

[2] https://github.com/e-delaney/Instance-Based_CFE_TSC.

**(II) Baseline Models.** The performance of Native Guide was compared to two baseline models: the $w$-counterfactual and NUN-CF methods. The *w-counterfactual method (w-CF)* proposed by Wachter et al., [55] is a key benchmark method; it is the most cited counterfactual XAI method in the literature and many other methods are variants of it[3] [24]. It proposes that that counterfactuals can be generated by minimizing a loss function;

$$L(x, x', y', \lambda) = \lambda(b(x') - c')^2 + d(x, x') \tag{1}$$

$$\underset{x'}{argmin} \underset{\lambda}{max} L(x, x', c', \lambda) \tag{2}$$

The first collection of terms in this loss function encourage the output of the classifier $b$, to be close to the desired class $c'$. The $\lambda$ parameter acts as a balancing term. The distance metric $d(x, x')$ measures the amount of change between the to-be explained instance $x$ and the counterfactual candidate $x'$. A Manhattan distance weighted feature-wise with the inverse median absolute deviation (MAD) is typically used here in order to ensure the generation of sparse solutions that are robust to outliers [55]. One noted weakness of the $\lambda$ parameter is that it tends to infinity raising stability issues in counterfactual generation [47]. The second method used, the *NUN-CF method*, can be viewed as a simplified variant of Native Guide; it simply uses the NUN for the query case directly, without any adaptation of deep or discriminative features (e.g., see [41]). This model represents a good comparison point as it allows us to see the contributions of the adaptation steps in Native Guide.

**(III) Time Series Classifier.** The black-box classifier used was a fully convolutional neural network (FCN)[4], by [57], a state of the art DNN architecture for time series classification (Fig. 4). Notably, the Global Average Pooling (GAP) layer reduces the number of parameters in a neural network while enabling the use of the Class Activation Map (CAM) [60]; the latter highlights parts of the input time series that contribute the most to a given classification, enabling the extraction of $\omega$. For each test/query instance, counterfactuals were generated using (i) *NUN-CF*, where the NUN, using a Euclidean distance measure, was selected [40], (ii) *w-Counterfactual* method (w-CF), as proposed by Wachter et al. [55], initializing it with $\lambda = 0.1$ and termination condition $P(c'|T) \geq 0.5$, minimizing the loss function in (see EQ.1) with adaptive Nelder-Mead optimization (iii) *Native Guide*, using the closest in-sample counterfactual and the feature-importance vector, $\omega$, given by the Class Activation Map (CAM) [39].

---

[3] We tried and failed in these tests, to use DiCE [38], a variant of $w$-CF with added constraints for diversity. We found that DiCE did not generate diverse counterfactuals within reasonable time-limits, suggesting that it is not well suited to high-dimensional time series data (even for shallower ANNs).

[4] Counterfactuals for other classifiers, such as MR-SEQL, were found but not reported.

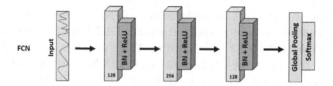

**Fig. 4.** A fully convolutional neural network (FCN) with three convolutional layers, batch normalization, ReLu activations and global average pooling preceding the final softmax layer enabling the use of a Class Activation Map (CAM) [57].

### 5.1 Experiment 1: Probing Proximity and Sparsity

This experiment compares the three counterfactual techniques on the five datasets, evaluating the counterfactuals produced in terms of proximity and sparsity. Proximity was evaluated using the relative counterfactual distance ($RCF = \frac{d(T_q, T')}{d(T_q, T'_{NUN})}$) enabling explicit comparisons to in-sample counterfactual instances (as suggested by [24,25]). Basically, this measure determines whether the distance between the query and the generated counterfactual is closer than that between the query and its "naturally-occurring" NUN. As in some other studies [21], three distance metrics were used; (i) Manhattan Distance ($\ell_1$ norm), (ii) Euclidean Distance ($\ell_2$ norm), and (iii) Chebyshev Distance ($\ell_\infty$ norm) (Fig. 5).

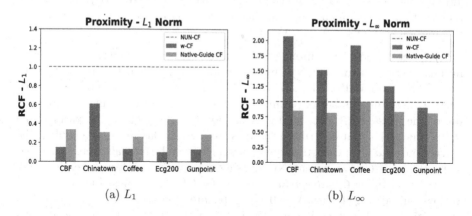

(a) $L_1$                               (b) $L_\infty$

**Fig. 5.** A comparison of the proximity of query-counterfactual pairs relative to query-NUN pairs for five datasets. In (a) the generated counterfactual explanations are closer to the query compared to the in-sample NUNs, in terms of $\ell_1$ distance. Perhaps more interesting is the fact that the $w$-counterfactuals are consistently less close than the NUNs, in terms of $\ell_\infty$ norm. This effect may be due to erroneous spikes in the counterfactual explanations generated by this method.

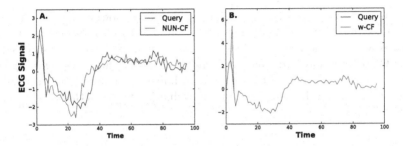

**Fig. 6.** Comparing counterfactuals (red line) for an ECG200 classification (blue line) generated by (a) NUN-CF and (b) $w$-CF. Here, NUN-CF fails to generate a proximate/sparse solution and $w$-CF's erratic spikes raise concerns about whether the counterfactual is out-of-distribution (see Fig. 1 for comparison). (Color figure online)

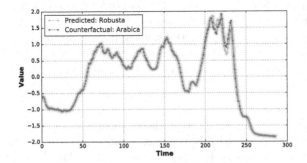

**Fig. 7.** A Native Guide counterfactual explanation for the coffee dataset. The method perturbs a contiguous subsequence corresponding to a semantically-meaningful and discriminative area of the spectrograph; this area provides information about the caffeine content of the coffee beans. Arabica coffee beans have a lower caffeine and chlorogenic acid content contributing to their finer taste and higher market value [5].

**Results and Discussion.** Both Native Guide and the $w$-CF counterfactual explanations produce proximate explanations that are significantly closer to the query instance compared to the existing NUNS, on both $\ell_1$ and $\ell_2$ norms (Wilcoxon, $p < 0.01$). In the case of $w$-CF this is somewhat unsurprising as it minimizes an $\ell_1$-based distance-metric optimizing to generate close, sparse counterfactuals. $w$-CF is known to sometimes produce implausible counterfactual explanations [25,28]. It is interesting to find that many of the perturbed features in its counterfactuals can be erratic spikes in the time series, reflecting out-of-distribution occurrences (see Fig. 6b). Moreover, the explanations produced by $w$-CF often perturb several different features in non-contiguous locations of the time series, considering these values to be independent. Conversely, Native Guide constrains its perturbations to selected, contiguous subsequences producing counterfactual explanations that are more plausible and more meaningful (see e.g. Figs. 1 and 7). These results indicate that good counterfactual explanations in the time series domain are not necessarily instances that are closest to the query, reflecting previous findings by Downs *et al.* for tabular data [10]. Notably, the $\ell_\infty$ norm seems to be able to diagnose counterfactual instances with erratic feature values. Counterfactual explanations produced by Native Guide are more proximate in terms of $\ell_\infty$ norm, further suggesting that the $w$-counterfactuals may not be realistic. Admittedly, a more robust evaluation of plausibility should be considered when evaluating these methods. We turn to this issue in Expt. 2.

### 5.2   Experiment 2: Exploring Plausibility and Diversity

In this experiment, we aim to evaluate the *plausibility* of generated counterfactual explanations in time series using novelty detection algorithms to detect out-of-distribution (OOD) explanations. We implement the Local Outlier Factor Method [4,20], Isolation Forest (IF) [34] and OC-SVM [52] (on both raw time and matrix profile [59] representations of the time series). We test if Native Guide can generate *diverse* explanations, when it uses alternative counterfactual instances as guides (see the other red instances shown in Fig. 3). The datasets, classifier, and methods tested are identical to those in Experiment 1.

**Results and Discussion.** The counterfactuals produced by the Native Guide are consistently more plausible than those generated by the benchmark, $w$-counterfactual method ($w$-CF; see Table 2). Results also confirm the hypothesis that proximity to the query is a poor heuristic for plausibility. One possible reason why case-based solutions produce more plausible counterfactual explanations is that they are grounded in the training data echoing previous findings by Laugel *et al.* [29]. Unlike Native Guide, $w$-CF fails to perturb discriminative, meaningful subsequences [30,58]. It is also interesting to note that different novelty detection algorithms produce very different results. For example the local outlier factor method was considerably less sensitive than the kernel-based

**Table 2.** Comparing the Native Guide (NG-CF) and $w$-Counterfactual ($w$-CF) models on plausibility using four OOD metrics (IF, LOF, OC-SVM, OC-SVM MP). Results indicate the percentage of generated counterfactuals that are out-of-distribution (n.b., lower scores are better and the best are highlighted in bold).

| | Fully convolutional neural network | | | | | | | |
| | IF | | LOF | | OC-SVM | | OC-SVM MP | |
| Dataset | w-CF | NG-CF | w-CF | NG-CF | w-CF | NG-CF | w-CF | NG-CF |
|---|---|---|---|---|---|---|---|---|
| CBF | 0.15 | **0.09** | 0.09 | **0.00** | 0.69 | **0.50** | 0.61 | **0.34** |
| Chinatown | 0.48 | **0.37** | 0.11 | **0.00** | 0.44 | **0.07** | 0.87 | **0.22** |
| Coffee | 0.41 | **0.37** | 0.04 | 0.04 | 0.25 | **0.14** | 0.43 | **0.21** |
| ECG200 | 0.28 | **0.26** | 0.22 | **0.02** | 0.50 | **0.16** | 0.44 | **0.13** |
| Gunpoint | 0.23 | **0.20** | **0.19** | 0.23 | 0.18 | **0.11** | 0.57 | **0.3** |

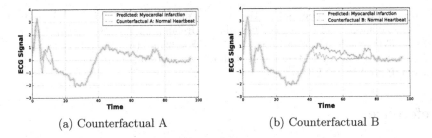

(a) Counterfactual A                          (b) Counterfactual B

**Fig. 8.** Two diverse counterfactual explanations, generated by Native Guide, for the same query case based on perturbing different in-sample counterfactual cases.

techniques in detecting OOD explanations (see Table 2). Unlike many blind perturbation techniques, Native Guide has the ability to generate diverse counterfactual explanations (see Fig. 8). This is particularly useful because (i) different users may prefer different explanations [51] and (ii) counterfactual explanations can also help humans to identify meaningful regions for classification (of which there may be many) [15]. For example, in the electrocardiogram domain we hypothesize that retaining counterfactual cases could help cardiologists to identify abnormalities that are useful for future problem scenarios. While one can evaluate diversity by monitoring feature wise distances between counterfactuals [38], the generated explanations may fail to satisfy domain constraints. Indeed, extensive user testing with experts will be an important avenue for future evaluation as novelty detection can be an imperfect proxy for plausibility.

## 6  Conclusion and Future Directions

In this paper a novel case-based technique, *Native Guide*, was proposed to provide proximate, sparse, plausible, and diverse counterfactual explanations for time series classification tasks. The method uses existing instances in the casebase to generate better counterfactual candidates. The technique is grounded in relevant evidence from the psychological and social sciences [6,36] and can integrate explanation weight-vectors extracted from techniques such as Class Activation Mapping [60]. Comparative tests on diverse datasets from the UCR archive using a fully convolutional neural network, demonstrate that the explanatory counterfactuals produced by Native Guide are significantly better than (i) explanations that already existed in the case-base (from NUN-CF) and (ii) explanations produced by constraint-based optimisation techniques (from $w$-CF). The experiments also indicated that techniques designed for tabular data often failed to produce meaningful explanations in the time series domain. Native Guide generates new time series data which holds promise for data augmentation purposes [13]. Given the ubiquitous nature of time series data and the frequent requirement for explanation, it is clear that experiments with human users and CBR solutions have much to offer in future work.

**Acknowledgments.** This publication has emanated from research conducted with the financial support of (i) Science Foundation Ireland (SFI) to the Insight Centre for Data Analytics under Grant Number 12/RC/2289_P2 and (ii) SFI and the Department of Agriculture, Food and Marine on behalf of the Government of Ireland under Grant Number 16/RC/3835 (VistaMilk).

# References

1. Adadi, A., Berrada, M.: Peeking inside the black-box: a survey on explainable artificial intelligence (XAI). IEEE Access **6**, 52138–52160 (2018)
2. Adebayo, J., Gilmer, J., Muelly, M., Goodfellow, I., Hardt, M., Kim, B.: Sanity checks for saliency maps. In: NeurIPS, pp. 9505–9515 (2018)
3. Ates, E., Aksar, B., Leung, V.J., Coskun, A.K.: Counterfactual explanations for machine learning on multivariate time series data. arXiv preprint arXiv:2008.10781 (2020)
4. Breunig, M.M., Kriegel, H.P., Ng, R.T., Sander, J.: Lof: identifying density-based local outliers. In: ACM SIGMOD, pp. 93–104 (2000)
5. Briandet, R., Kemsley, E.K., Wilson, R.H.: Discrimination of arabica and Robusta in instant coffee by Fourier transform infrared spectroscopy and chemometrics. J. Agric. Food Chem. **44**(1), 170–174 (1996)
6. Byrne, R.M.: Counterfactuals in explainable artificial intelligence (XAI): evidence from human reasoning. In: IJCAI-19, pp. 6276–6282 (2019)
7. Dau, H.A., et al.: The UCR time series archive. IEEE/CAA J. Automatica Sinica **6**(6), 1293–1305 (2019)
8. Delaney, E., Greene, D., Keane, M.T.: Instance-based counterfactual explanations for time series classification. arXiv preprint arXiv:2009.13211 (2020)
9. Dodge, J., Liao, Q.V., Zhang, Y., Bellamy, R.K., Dugan, C.: Explaining models: an empirical study of how explanations impact fairness judgment. In: International Conference on Intelligent User Interfaces, pp. 275–285 (2019)
10. Downs, M., Chu, J.L., Yacoby, Y., Doshi-Velez, F., Pan, W.: Cruds: counterfactual recourse using disentangled subspaces. In: ICML Workshop Proceedings (2020)
11. Fawaz, H.I., Forestier, G., Weber, J., Idoumghar, L., Muller, P.A.: Adversarial attacks on deep neural networks for time series classification. In: 2019 International Joint Conference on Neural Networks (IJCNN), pp. 1–8. IEEE (2019)
12. Ismail Fawaz, H., Forestier, G., Weber, J., Idoumghar, L., Muller, P.-A.: Deep learning for time series classification: a review. Data Min. Knowl. Disc. **33**(4), 917–963 (2019). https://doi.org/10.1007/s10618-019-00619-1
13. Forestier, G., Petitjean, F., Dau, H.A., Webb, G.I., Keogh, E.: Generating synthetic time series to augment sparse datasets. In: ICDM, pp. 865–870. IEEE (2017)
14. Gee, A.H., Garcia-Olano, D., Ghosh, J., Paydarfar, D.: Explaining deep classification of time-series data with learned prototypes. In: CEUR Workshop Proceedings, vol. 2429, pp. 15–22 (2019)
15. Goyal, Y., Wu, Z., Ernst, J., Batra, D., Parikh, D., Lee, S.: Counterfactual visual explanations. In: ICML, pp. 2376–2384. PMLR (2019)
16. Grabocka, J., Schilling, N., Wistuba, M., Schmidt-Thieme, L.: Learning time-series shapelets. In: ACM SIGKDD, pp. 392–401 (2014)
17. Guidotti, R., Monreale, A., Giannotti, F., Pedreschi, D., Ruggieri, S., Turini, F.: Factual and counterfactual explanations for black box decision making. IEEE Intell. Syst. **34**(6), 14–23 (2019)

18. Guidotti, R., Monreale, A., Spinnato, F., Pedreschi, D., Giannotti, F.: Explaining any time series classifier. In: CogMI 2020, pp. 167–176. IEEE (2020)
19. Gunning, D., Aha, D.: Darpa's explainable artificial intelligence (XAI) program. AI Mag. **40**(2), 44–58 (2019)
20. Kanamori, K., Takagi, T., Kobayashi, K., Arimura, H.: Dace: distribution-aware counterfactual explanation by mixed-integer linear optimization. In: IJCAI-20, pp. 2855–2862 (2020)
21. Karimi, A.H., Barthe, G., Balle, B., Valera, I.: Model-agnostic counterfactual explanations for consequential decisions. In: AISTATS, pp. 895–905 (2020)
22. Karlsson, I., Rebane, J., Papapetrou, P., Gionis, A.: Explainable time series tweaking via irreversible and reversible temporal transformations. In: ICDM (2018)
23. Keane, M.T., Kenny, E.M.: How case-based reasoning explains neural networks: a theoretical analysis of XAI using *Post-Hoc* explanation-by-example from a survey of ANN-CBR twin-systems. In: Bach, K., Marling, C. (eds.) ICCBR 2019. LNCS (LNAI), vol. 11680, pp. 155–171. Springer, Cham (2019). https://doi.org/10.1007/978-3-030-29249-2_11
24. Keane, M.T., Kenny, E.M., Delaney, E., Smyth, B.: If only we had better counterfactual explanations: five key deficits to rectify in the evaluation of counterfactual XAI techniques. In: IJCAI-21 (2021)
25. Keane, M.T., Smyth, B.: Good counterfactuals and where to find them: a case-based technique for generating counterfactuals for explainable AI (XAI). In: Watson, I., Weber, R. (eds.) ICCBR 2020. LNCS (LNAI), vol. 12311, pp. 163–178. Springer, Cham (2020). https://doi.org/10.1007/978-3-030-58342-2_11
26. Kenny, E.M., Delaney, E.D., Greene, D., Keane, M.T.: Post-hoc explanation options for XAI in deep learning: the *Insight centre for data analytics* perspective. In: Del Bimbo, A., et al. (eds.) ICPR 2021. LNCS, vol. 12663, pp. 20–34. Springer, Cham (2021). https://doi.org/10.1007/978-3-030-68796-0_2
27. Kenny, E.M., Keane, M.T.: Twin-systems to explain artificial neural networks using case-based reasoning: comparative tests of feature-weighting methods in ANN-CBR twins for XAI. In: IJCAI-19, pp. 2708–2715 (2019)
28. Kenny, E.M., Keane, M.T.: On generating plausible counterfactual and semi-factual explanations for deep learning. In: AAAI-21, pp. 11575–11585 (2021)
29. Laugel, T., Lesot, M.J., Marsala, C., Renard, X., Detyniecki, M.: The dangers of post-hoc interpretability: unjustified counterfactual explanations. In: Proceedings of IJCAI-19, pp. 2801–2807 (2019)
30. Le Nguyen, T., Gsponer, S., Ilie, I., O'Reilly, M., Ifrim, G.: Interpretable time series classification using linear models and multi-resolution multi-domain symbolic representations. Data Min. Knowl. Disc. **33**(4), 1183–1222 (2019). https://doi.org/10.1007/s10618-019-00633-3
31. Leake, D., Mcsherry, D.: Introduction to the special issue on explanation in case-based reasoning. Artif. Intell. Rev. **24**(2), 103 (2005)
32. Leonardi, G., Montani, S., Striani, M.: Deep feature extraction for representing and classifying time series cases: towards an interpretable approach in haemodialysis. In: Flairs-2020. AAAI Press (2020)
33. Lipton, Z.C.: The mythos of model interpretability. Queue **16**(3), 30 (2018)
34. Liu, F.T., Ting, K.M., Zhou, Z.H.: Isolation forest. In: ICDM, pp. 413–422 (2008)
35. Lundberg, S.M., Lee, S.I.: A unified approach to interpreting model predictions. In: Advances in Neural Information Processing Systems, pp. 4765–4774 (2017)
36. Miller, T.: Explanation in artificial intelligence: insights from the social sciences. Artif. Intell. **267**, 1–38 (2019)

37. Molnar, C.: Interpretable machine learning. Lulu.com (2020)
38. Mothilal, R.K., Sharma, A., Tan, C.: Explaining machine learning classifiers through diverse counterfactual explanations. In: ACM FAccT, pp. 607–617 (2020)
39. Nguyen, T.T., Le Nguyen, T., Ifrim, G.: A model-agnostic approach to quantifying the informativeness of explanation methods for time series classification. In: Lemaire, V., Malinowski, S., Bagnall, A., Guyet, T., Tavenard, R., Ifrim, G. (eds.) AALTD 2020. LNCS (LNAI), vol. 12588, pp. 77–94. Springer, Cham (2020). https://doi.org/10.1007/978-3-030-65742-0_6
40. Nugent, C., Cunningham, P.: A case-based explanation system for black-box systems. Artif. Intell. Rev. 24(2), 163–178 (2005)
41. Nugent, C., Doyle, D., Cunningham, P.: Gaining insight through case-based explanation. J. Intell. Inf. Syst. 32(3), 267–295 (2009). https://doi.org/10.1007/s10844-008-0069-0
42. Olszewski, R.T.: Generalized feature extraction for structural pattern recognition in time-series data, Technical report. Carnegie-Mellon Univ, Pittsburgh (2001)
43. Pearl, J., Mackenzie, D.: The Book of Why. Basic Books, New York (2018)
44. Poyiadzi, R., Sokol, K., Santos-Rodriguez, R., De Bie, T., Flach, P.: FACE: feasible and actionable counterfactual explanations. In: AIES, pp. 344–350 (2020)
45. Recio-Garcia, J.A., Diaz-Agudo, B., Pino-Castilla, V.: CBR-LIME: a case-based reasoning approach to provide specific local interpretable model-agnostic explanations. In: Watson, I., Weber, R. (eds.) ICCBR 2020. LNCS (LNAI), vol. 12311, pp. 179–194. Springer, Cham (2020). https://doi.org/10.1007/978-3-030-58342-2_12
46. Ribeiro, M.T., Singh, S., Guestrin, C.: Why should i trust you?: Explaining the predictions of any classifier. In: Proceedings of SIGKDD'16, pp. 1135–1144. ACM (2016)
47. Russell, C.: Efficient search for diverse coherent explanations. In: Conference on Fairness, Accountability, and Transparency, pp. 20–28 (2019)
48. Samangouei, P., Saeedi, A., Nakagawa, L., Silberman, N.: ExplainGAN: model explanation via decision boundary crossing transformations. In: Ferrari, V., Hebert, M., Sminchisescu, C., Weiss, Y. (eds.) ECCV 2018. LNCS, vol. 11214, pp. 681–696. Springer, Cham (2018). https://doi.org/10.1007/978-3-030-01249-6_41
49. Sani, S., Wiratunga, N., Massie, S.: Learning deep features for kNN-based Human Activity Recognition. In: Proceedings of the International Conference on Case-Based Reasoning Workshops, pp. 95–103. CEUR Workshop Proceedings, Trondheim (2017). https://rgu-repository.worktribe.com/output/246837/learning-deep-features-for-knn-based-human-activity-recognition
50. Schlegel, U., Arnout, H., El-Assady, M., Oelke, D., Keim, D.A.: Towards a rigorous evaluation of xai methods on time series. arXiv preprint arXiv:1909.07082 (2019)
51. Schoenborn, J.M., Weber, R.O., Aha, D.W., Cassens, J., Althoff, K.D.: Explainable case-based reasoning: a survey. In: AAAI-21 Workshop Proceedings (2021)
52. Schölkopf, B., Platt, J.C., Shawe-Taylor, J., Smola, A.J., Williamson, R.C.: Estimating the support of a high-dimensional distribution. Neural Comput. 13(7), 1443–1471 (2001)
53. Sørmo, F., Cassens, J., Aamodt, A.: Explanation in case-based reasoning-perspectives and goals. Artif. Intell. Rev. 24(2), 109–143 (2005). https://doi.org/10.1007/s10462-005-4607-7
54. Van Looveren, A., Klaise, J.: Interpretable counterfactual explanations guided by prototypes. arXiv preprint arXiv:1907.02584 (2019)
55. Wachter, S., Mittelstadt, B., Russell, C.: Counterfactual explanations without opening the black box: automated decisions and the GDPR. Harv. J. Law Tech. 31, 841 (2017)

56. Wang, Y., et al.: Learning interpretable shapelets for time series classification through adversarial regularization. arXiv preprint arXiv:1906.00917 (2019)
57. Wang, Z., Yan, W., Oates, T.: Time series classification from scratch with deep neural networks: a strong baseline. In: IJCNN, pp. 1578–1585. IEEE (2017)
58. Ye, L., Keogh, E.: Time series shapelets: a novel technique that allows accurate, interpretable and fast classification. Data Min. Knowl. Disc. **22**(1–2), 149–182 (2011). https://doi.org/10.1007/s10618-010-0179-5
59. Yeh, C.C.M., et al.: Matrix profile i: all pairs similarity joins for time series: a unifying view that includes motifs, discords and shapelets. In: ICDM (2016)
60. Zhou, B., Khosla, A., Lapedriza, A., Oliva, A., Torralba, A.: Learning deep features for discriminative localization. In: IEEE CVPR, pp. 2921–2929 (2016)

# User Evaluation to Measure the Perception of Similarity Measures in Artworks

Belén Díaz-Agudo$^{(\boxtimes)}$ , Guillermo Jimenez-Diaz ,
and Jose Luis Jorro-Aragoneses

Department of Software Engineering and Artificial Intelligence,
Instituto de Tecnologías del Conocimiento, Universidad Complutense de Madrid,
Madrid, Spain
{belend,gjimenez,jljorro}@ucm.es

**Abstract.** Similarity measures do not typically capture subjective elements of perception of similarity. Our research contributes an experimental methodology for validating and learning similarity computation algorithms against human perceptions of similarity in subjective domains like Art and emotions. In this paper, we explain the first experiment to check our hypotheses and methodology. We have obtained promising results that explain the differences between users profiles and their perception of similarity between artworks and how to combine local similarity functions to be able to compute similarity measures reflecting users' perception.

**Keywords:** Similarity measures · Arts · Similarity perception

## 1 Introduction

The well-known main assumption in case-based reasoning (CBR) relies on the hypothesis that similar problems should have similar solutions. No need to say that similarity is a core concept for different processes of the CBR cycle. Similarity between the query and the cases is typically computed using the description features represented using attribute-value pairs. These features can be simple, textual or, in some applications, it may be necessary to use derived features obtained by inference based on domain knowledge. In yet other applications, cases are represented by complex structures (such as graphs or first-order terms) and retrieval requires an assessment of their structural similarity. As might be expected, the use of deep features or structural similarity is computationally expensive; however, the advantage is that relevant cases are more likely to be retrieved [14]. Our research group has previous works on semantic and structured semantic similarity with ontological an taxonomic knowledge [9,20,21].

In this paper we highlight that *similarity measures* on the structures that represent the cases (either attribute-value or graphs) does not typically capture

© Springer Nature Switzerland AG 2021
A. A. Sánchez-Ruiz and M. W. Floyd (Eds.): ICCBR 2021, LNAI 12877, pp. 48–63, 2021.
https://doi.org/10.1007/978-3-030-86957-1_4

*subjective* elements of perception of similarity. For example, when comparing two menus that are similar for me, the similarity can be due to the fact that both include my favourite meals and not because of the similarity between their ingredients. There are other examples, like recommender systems for songs or movies, where retrieval performance is affected when similarity between items depends on subjective criteria for different users. There are many different application domains where human subjectivity is an issue. However, perception of similarity is difficult to measure, even if the item's descriptions is structured and includes semantic domain knowledge. In this paper we present a case study in the Art domain where similarity perception is clearly a subjective criteria.

The context of the research conducted in this paper is the SPICE project[1]. The overall aim of SPICE is to develop tools and methods to support Citizen Curation [7], in which citizens actively engage in curatorial activities in order to learn more about themselves and develop a better understanding of, and empathy for, other communities.

One challenge is to be able to identify communities of citizens that allow the reflection processes inside and between communities. This is a two way process. On one side, community detection relies heavily on the definition and use of semantic similarity measures over complex graph structures representing citizens, opinions, artworks, contributions, reflections and emotions[2]. On the other side, communities of users can be seen as useful resources to identify common profiles from the similarity perception point of view.

Our research contributes an experimental method for validating similarity computation algorithms against human perceptions of similarity. Such validation enables researchers to ground their similarity methods in context of intended use instead of relying on assumptions of fit. In addition to the methodology, this paper presents the results of experimentation using real data with artworks from the Prado Museum. We also present some analysis of potential causes of differences between the compared cases in which this model matches human perceptions of similarity. This method will allow to personalise a similarity measure to compare items using subjective perception criteria adapted to the user who compares the items. She is the user that retrieves the case in CBR systems, or the user that get the recommendation in a recommender system or for any other applications relying on similarity computation where similarity perception is an issue.

The paper runs as follows. Section 2 reviews some related work about similarity. Section 3 describes out methodology for capturing and learning knowledge reflecting human perceptions of similarity. Section 4 describes an experiment associated to the step 1 of our methodology. In the experiment with Art data from the Prado Museum we define local similarity measures for comparing attributes of artworks and validate them regarding user perception of similarity. Section 5 concludes the paper and review some lines of future work.

---

[1] Social cohesion, Participation, and Inclusion through Cultural Engagement - Horizon 2020 programme https://spice-h2020.eu/.

[2] SPICE relies in Linked Data technologies that include a huge mass of interlinked knowledge.

## 2   Related Work About Similarity

CBR relies strongly in similarity computation. However, Similarity measures are considered also essential tools to solve problems in a broad range of AI domains and applications, specially when semantic matters. For example semantic web and linked data [29], recommender systems [11], Natural Language Processing [22], Information Retrieval, Knowledge Engineering [3], and many others. There are several similarity measures that have been used in CBR systems, and some comparison studies and frameworks exist [14,18]. The results obtained in these studies show that the different similarity measures have a performance strongly related to the type of attributes representing the case and to the importance of each attribute. Thus, it is very different to deal with only continuous data, with ordered discrete data or non-ordered discrete data. In [15] authors distinguish between case similarity measures that are learnt from data and those that are typically modelled by experts with the relevant domain knowledge together with CBR experts, who know how to encode this domain knowledge into the similarity measures by selecting what are the properties of the case descriptions have more impact in the similarity of the solutions. For example, in a cars for sale application the amount of miles driven has a greater importance than the color of the car [6].

When dealing with conceptual background domain models, like graphs, networks or taxonomies, another possibility is the representational approach that assigns similarity meaning to the path joining two individuals. In general, a graph-based semantic similarity measure is a mathematical tool used to estimate the strength of the semantic interaction between entities (concepts or instances) based on the analysis of ontologies [12]. Similarity is computed for a given pair of individuals. An individual is defined in terms of the concepts of which is an instance and the properties asserted for it, which are represented as relations connecting the individual to other individuals or primitive values (fillers). In [9] we have described a similarity framework where we distinguished between the *structural similarity* that will be computed based on the composition relations (part-of, has-part), the *semantics similarity* is due to all the concepts and relations describing the meaning of the case, and the *contextual similarity* that depends on the case context relations and the *adaptation similarity* that will use the adaptation related knowledge (also used in previous approaches like [23]). Note that the application of this measure is strongly dependent on the availability of an ontology or conceptual model that represents the application domain. The work in [17] classifies the distance and similarity functions on graph-based representations in four types: (1) graph matching, (2) based on edit distances, (3) based on the types of relationships and refinement operators and (4) based on kernels.

Regarding perception of similarity related work exists in the field of human psychology, where similarity is defined a relationship that holds between two perceptual or conceptual objects and serves to classify objects, form concepts and make generalizations [26]. As it is noted in [5], similarity between objects is not solely dependent on the characteristics of those objects. It is also affected by the context, and by other present and immediately past stimuli, as well as long-term experience with related objects. A well-known example is that humans have the effect of experience on similarity among phonemes. Native English speakers find spoken "L" and "R" quite distinct, whereas to native speakers of Japanese they sound extremely similar.

Perception of similarity is also relevant when dealing with textual representations. For example, in [25] authors deal with the problem of navigation in large text collections (blogs, forums, idea management systems, online deliberation platforms...), and analyse how the algorithmic similarity measures being used match up with human perceptions. They found out that in favourable conditions human similarity judgements and algorithmic similarity measurement often (75%) agree. However, that agreement is not so good (66%) when documents are selected more generally. Other previous studies have also examined the match between typical algorithmic similarity based approaches (such LDA, or cosine similarity) and human perceptions of text document similarity. For example, in [27] authors compare the relevance (according to human judges) of the results of the retrieval task on an abstracted document collection given an information-need query. Also related is the work of [10] that refers to the individual word level. They compared the computed cosine similarity between feature vectors that incorporated information from lexicons and large corpora, against benchmark datasets containing pairs of English words that had been assigned similarity ratings by humans, finding out discrepancies between the perceived and the computed similarity.

In [4] authors describe a CBR system that helps the users make online privacy decisions by identifying similar situations from the past. They calculate the similarity between privacy policies and provide results from a focus group study on the perceived similarity of data items and data handling purposes from a privacy point of view. Particular attention has been also placed by the similarity perceived by experts on the use of analogical reasoning [13, 28].

## 3   Methodology for Learning Similarity Measures Reflecting Human Perceptions

In this section, we propose a methodology for the construction of similarity measures that reflect the perception of similarity. The challenge is that similarity perception is different for different people, so it can not be computed with a common similarity measure that is shared for all the users.

Our proposal aims at configuring different similarity measures for different users and being able to generalise these measures for users of the same *profile*, supposing that users who belong to the same profile have similar perception of similarity. Our methodology relies on the following hypothesis:

1. **Hypothesis 1.** Different users have different perceptions of similarity and consider differently the attributes describing the items. In this paper, we study how local similarity measures on the individual attributes relate with the perceived similarity in each one of these attributes.

2. **Hypothesis 2.** Users can be grouped together using profiles and users from the same profile have similar perceptions of similarity. That means that a similarity measure can be learnt for each profile group.
3. **Hypothesis 3.** Profiles can be learnt from the common properties of users of the same community. Our proposal consist on applying community detection algorithms and use the communities as the profiles to construct a similarity measure between items.

These hypothesis needs to be proven by cross-validation and it is very dependent on how the profiles are defined.

### 3.1 Methodology for Learning Perception Aware Similarity Measures

Our methodology aims to build improved measures to compare both similarity between items and between users reflecting perception:

1. $SIM'_x Item$: similarity measure **between items** that is a computable model that is adjusted either to a particular user $u$ ($SIM'_u Item$) or to the users of the same profile $p$ ($SIM'_p Item$).
2. $SIM'User$: similarity measure **between users** that reflects shared perceptions regarding items. Similar users will be those that have similar emotions regarding similar items.

The process starts with the following **input requirements**:

- Set of *Items* defined by the set of descriptive attributes ($atr_j$)
- Set of *Users* defined by the set of descriptive attributes ($UserAtr_j$)
- Basic similarity measure between users ($SimUsers$) defined as a linear combination of the user descriptive attributes $UserAtr_j$.
- Set of Profiles to classify users. To simplify we assume from now that each user belong to exactly one profile (see Sect. 3.2).

We propose a methodology organised in three steps:

1. **Step 1.** Definition and validation of the Local Similarity measures for each individual attribute $atr_j$. We define local similarity measures associated to each attribute describing the items, so $SimAtr_k(I_{ik}, I_{jk})$ is the local similarity between the value of attribute $k$ in items $i$ and $j$. We assume that, in this way, we can define weighted similarity measures $SIM(I_i, I_j) = \sum_k w_k \cdot SimAtr_k(I_{ik}, I_{jk})$, where $I_i$ and $I_j$ are two items; $w_k$ is the weight or importance assigned to attribute $k$. Local similarity measures can be complex (graph based) or simple depending on the domain background knowledge. It is necessary to study how each local measure affect the perception of global similarity and study the correlation within the different user profiles.
2. **Step 2.** Construction of $SIM'_x Item$ as a computable model to calculate the similarity between items. As this measure should reflect perception it will

reflect either perception of one specific user $u$, or more interestingly perception of a group of users $p$ sharing a common profile. Because this measure is a weighted similarity measures $SIM(I_i, I_j)$, we address the fundamental problem of learning a weight model for features. i.e., it is necessary to give a greater similarity contribution to an important attribute than to other less important ones regarding perception of similarity:

- $SIM'_u Item$ with $u \in Users$. The weight for each attribute is adjusted to reflect the perception of the specific user $u$ using the results of the experiment for *user u.*
- $SIM'_p Item$ The weight for each attribute is adjusted to reflect the perception of the users of the *profile p. p $\in$ Profiles.*

We plan to use an approach similar to the one described in [24] to learn weights of the local similarity measures through a genetic algorithm.

3. **Step 3.** Construct a similarity measure $Sim'User$ that combines $SimUser$ with the polarity of the compared users with items. The similarity measure $Sim'User$ should reflect that users with similar emotions on similar items are similar (using the perceived measure $SIM'_p Items$).

One advantage of this approach is that it is scalable. New users and new items can be included in the system. New users benefit of personalised similarity measures reflecting perceptions of similar users. A key aspect of this methodology is the definition of user profiles that is described next.

## 3.2   Profile Definition

As our methodology depends on the existence of profiles we consider two options:

- Manual definition of simple profiles at-hand reflecting the knowledge about the domain. This option is used in this paper using the knowledge in the Art domain, where there is a dependency between the level of expertise with the perception of similarity between artworks (see Sect. 4).
- Use community detection algorithms [1] and define the *profile* as the common features for the users in this community. Again, community detection algorithms rely on similarity measures between users. As future work we will explore an iterative process to improve Community Detection processes by improving the similarity measure between users, as follows:
  1. Initial Community detection using basic $SimUser$
  2. Use each community in the communities set $c \in C$ as profiles for steps 1,2 and 3 of the methodology to learn $Sim'Users$ and $Sim'_c Items$.
  3. Recalculate Communities using the improved similarity measure $Sim'Users$ and study community model adequacy.

# 4   Experiment on the Perception of Similarity for Artworks

In this paper we propose to validate similarity computation algorithms against human perceptions of similarity. This section describes an experiment associated to the Step 1 of our methodology (described in Sect. 3): defining local similarity measures for comparing attributes of items and validating them regarding user perception of similarity.

Our experiment uses an artwork dataset. The set of Items is a set of Artworks from the Prado Museum described by four attributes $atr_j$: the dominant colour, the motion evoked to the users, the content depicted in the artwork and the domain knowledge that the user has about the artwork (like the painter or its art movement). Our first experimental goal is to validate the general acceptance of these aspects by real users and the difference of criteria in the perception of similarity between artworks in different user profiles.

The set of Items (artworks) employed in the experiments come from a dataset created as an excerpt from Wikiart Emotion Dataset [16]. We use this dataset because it includes data about the emotions that the artworks evoked to different users, so they will be employed to compute the local similarity concerning to evoked emotions. This dataset contains 30 artworks from Prado Museum and 1760 annotations of emotions from 171 different users. We limited the number of emotions to the set of emotions described by Plutchik Emotion Theory [19] (anger, anticipation, joy, trust, fear, surprise, sadness, and disgust), so the dataset is reduced to 1040 annotations from 168 unique users. The original artwork dataset has also been enriched with the Wikidata URLs of the paintings and artists, as long as the entity identifier in Wikidata[3] in order to compute the local similarity measure concerning the content depicted in the artwork.

According to our methodology, we first define local similarity measures associated to each attribute describing the items. This will be employed to exemplify the validation of our similarity measures against the user perceptions. Additionally, we will check if the combination of these local similarities can enhance the precision of a similarity measure according to the perceived similarity by users.

The experiment is divided into two steps: the implementation of local similarity measures and the gathering of user data about perceived similarity. These steps will be described in following subsections.

### 4.1 Definition of Local Similarity Measures

The four attributes selected as the ones that support the similarity between two artworks $I_i$ and $I_j$ has been converted into four local similarity measures $(SimAtr_k(I_{ik}, I_{jk}))$:

1. **Colour similarity:** This measure uses the weighted euclidean distance between the dominant colour of each painting in HSV space [8]. The dominant colour is the center of the biggest cluster when applying k-means on the artwork image pixels in RGB space.

$$SimAttr_{col}(I_i, I_j) = 1 - dist(hsv_i, hsv_j)$$

---

[3] Wikidata: https://www.wikidata.org.

$$dist(hsv_i, hsv_j) = \sqrt{(v_i - v_j)^2 + s_i{}^2 + s_j{}^2 + 2s_i s_j(h_i - h_j)}$$

where $hsv_i$ is the dominant colour or artwork $I_i$ in HSV space.

2. **Content similarity:** This measure employs the knowledge about the elements depicted in an artwork stored in Wikidata. For each artwork, we created a list of contents collecting the values for the "depict" property in Wikidata. A first test over these lists highlighted that common contents were not frequent so we enlarge the list of contents using the concept hierarchy defined in Wikidata with the properties "instance of" and "subclass". The final list is computed traversing these hierarchy up to 2 levels. Finally, the similarity measure is computed using Jaccard over the list of contents.

$$SimAttr_{con}(I_i, I_j) = Jaccard(C_i, C_j) = \frac{C_i \cap C_j}{C_i \cup C_j}$$

where $C_i$ is the list of contents in artwork $a_i$.

3. **Emotion similarity:** This similarity uses the annotations in Wikiart Emotion Dataset about the emotions evoked by the artworks in different users. It is computed using the 3 most popular emotions and calculating the distance between emotions according to the Plutchik wheel of emotions –that places similar emotions close together and opposites 180° apart, like complementary colours (see Fig. 1).

$$SimAttr_{emo}(I_i, I_j) = 1 - \frac{1}{3}\sum_{k=1}^{3} dist(e_{ik}, e_{jk})$$

$$dist(e_i, e_j) = min_{dist}(e_i, e_j)/4$$

where $e_{ik}$ is the k-th most popular emotion in artwork $a_i$.

4. **Knowledge similarity:** This similarity uses the information about the artist and the art movement that the artworks belong to. These information is extracted from the WikiArt Emotion Dataset.

$$SimAttr_{kno}(I_i, I_j) = \begin{cases} \alpha & \text{if } author(a_i) = author(a_j) \\ \beta & \text{if } artMov(a_i) = artMov(a_j) \\ 0 & \text{otherwise} \end{cases}$$

where $author(a_i)$ is the artist who painted $a_i$; $artMov(a_i)$ is the art movement that $a_i$ belongs to and $\alpha$ and $\beta$ are constants in $[0, 1]$ and $\alpha > \beta$.

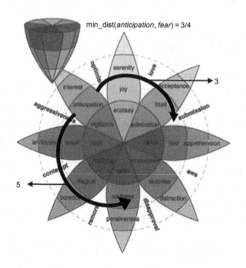

**Fig. 1.** Emotion wheel conceptualised by Plutchik [19]

## 4.2  Data Gathering of Perceived Similarity

We have collected user perceptions of similarity between different artworks through an online questionnaire[4]. This questionnaire also collects user information in order to sketch some initial profiles that will be employed to study how the perception relates with the different user profiles and learn similarity measures that reflect the common perceptions (see Step 2 in Sect. 3).

Figure 2 (left) shows the first part of the questionnaire. In this part, we collect the information used to create user profiles. A tentative profile for this experiment is based on demographic aspects (age and gender), the user expertise or knowledge about art (professional, amateur, a fan of an artist or not interested in art), and the user habits on how often they visit museums (rarely, sporadic or often). This information allowed us to manually define different user profiles and evaluate the perception of similarity among them.

The next step of the questionnaire aims to gather the perception of similarity between different artworks. In this step, the application shows two different artworks (Fig. 2, right) and users should select a value of similarity between 1 –artworks are very different– and 5 –artworks are very similar. The artworks are extracted from the Wikiart Emotion Dataset, described above. Although the pair of artworks is randomly chosen, the experiment is designed in a way that the randomisation process tries to balance the number of times that every pair is presented in the questionnaire.

To understand the reasons behind similarity perception, we include another question where users choose which criteria they applied to rate this similarity between both artworks. The criteria categories are the colour, the content represented, the user knowledge about these artworks (like the author, style, etc.)

---

[4] The questionnaire is available at https://tinyurl.com/2dn7wey4.

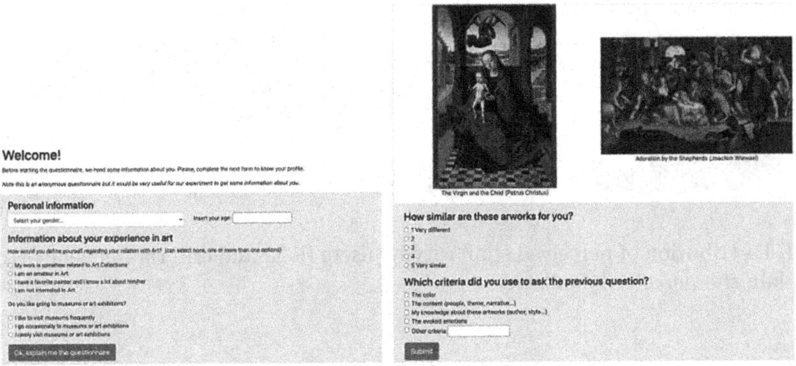

**Fig. 2.** Web application for collecting perceived similarity. On the left, questions about user profile information. On the right, the interface for assessing the perceived similarity between two artworks for choosing the criteria applied to explain the similarity perceived

and evoked emotions by these artworks. Users can select more than one criteria category and they can add any other criteria not included in the questionnaire.

## 4.3  Experimental Results

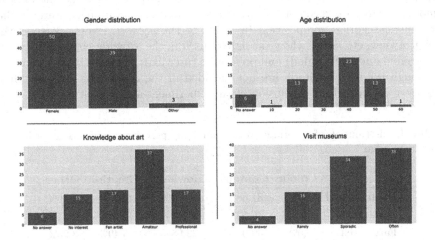

**Fig. 3.** Distribution of users by age, gender, their knowledge about art and how often they visit a museum.

During the experiment, 92 unique users filled the questionnaire for assessing the perceived similarity. Figure 3 shows the users distribution based on each profile variable described in Sect. 4.

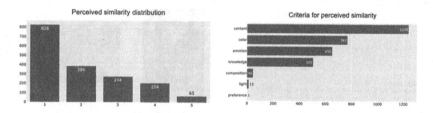

**Fig. 4.** Distribution of perceived similarity by users (left) and criteria frequency (right) to explain the perceived similarity

These users generated a total of 1792 answers about the similarity perceived for a pair of artworks. All artworks received at least 2 answers, and most of them received 3 or 4 answers. The left graph included in Fig. 4 shows the distribution of the perceived similarity values provided by users. It is worth noting that most of the answers correspond to perceiving dissimilarity (value equal to 1) between the artworks shown.

Next, we analysed the criteria employed by users to explain the similarity value provided. The right graph in Fig. 4 shows the frequency of each criterion employed by users. It is important to remember that users can choose several criteria to explain the perceived similarity value. This graph shows that content is the most criteria employed, followed by the colour and the emotion evoked by artworks. In addition, we obtained 95 answers that considered other criteria out of the initial categories provided (i.e. colour, content, knowledge and evoked emotion). After a revision of these answers, we added 3 additional criteria to the previous categories: composition (it refers to the artwork layout, perspective, point of view, etc.), light (how the light is used in the artwork, contrast between foreground and background) and preference (user likes or dislike both artworks). Although these new criteria are not included in the rest of the analysis, they represent an important conclusion of our experiment and our future work.

**Table 1.** Distribution of criteria used for explaining perceived similarity according to user knowledge

|              | Colour  | Content | Knowledge | Emotion | Other  |
|--------------|---------|---------|-----------|---------|--------|
| **Professional** | 28,03%  | 31,68%  | 18,83%    | 20,00%  | 1,46%  |
| **Amateur**      | 24,54%  | 36,53%  | 14,72%    | 22,20%  | 2,01%  |
| **Fan artist**   | 9,09%   | 59,09%  | 22,08%    | 6,49%   | 3,25%  |
| **No interest**  | 16,37%  | 50,44%  | 15,49%    | 16,81%  | 0,88%  |
| **No answer**    | 33,49%  | 33,97%  | 4,31%     | 27,27%  | 0,96%  |

When we correlate these criteria with the different user profiles, we can see some dependencies. In Table 1, we see that content criterion is more employed by users categorised in a lower knowledge level profile (fan of an artist/no interest) than amateur and professional profiles. Professional and amateur profiles use more Colour and Emotion criteria, as long as the composition and light criteria (other criteria) discovered during the questionnaire data analysis. This fact supports our Hypothesis 2 that different profiles are affected by different aspects in the perception of similarity between artworks.

**Table 2.** Most frequent criteria combinations to explain the perceived similarity

| Criteria combination | Frequency |
|---|---|
| content | 439 |
| colour-content-knowledge-emotion | 177 |
| colour-content | 172 |
| colour | 162 |
| colour-content-emotion | 136 |
| content-emotion | 115 |
| knowledge | 81 |
| emotion | 81 |

Although a single criterion is, in general, widely used to explain the perceived similarity, users also have employed combinations of these criteria. Table 2 summarises the top selected combinations of criteria. The combinations of colour, content and emotion are the most popular criteria to explain the perceived similarity.

**Table 3.** Results of Mean Absolute Error (MAE) and Mean Squared Error (MSE) between similarity measures and user similarity perception

|  | Colour | Content | Knowledge | Emotion |
|---|---|---|---|---|
| **N** | 366 | 401 | 314 | 336 |
| **MAE** | 0.271 | 0.246 | 0.236 | 0.337 |
| **MSE** | 0.105 | 0.087 | 0.091 | 0.153 |

The next step in our analysis is to determine if the two first steps of our methodology can be applied to this problem. In Step-1, we define and validate local similarity measure for each attribute. To do that, we calculated the Mean Absolute Error (MAE) and the Main Squared Error (MSE) between the local similarity measures explained in Sect. 3 and each corresponding values of humans perceptions of similarity. Table 3 shows the results, and we can observe that the *emotion similarity measure* is the least accurate comparing with the users'

**Table 4.** Results of Mean Absolute Error (MAE) and Mean Squared Error (MSE) between combination of local similarity measures and user similarity perceptions

|       | Col-Con-K-E | Col-Con | Col-Con-E | Con-E |
|-------|-------------|---------|-----------|-------|
| **N**   | 247         | 362     | 305       | 333   |
| **MAE** | 0.140       | 0.148   | 0.166     | 0.155 |
| **MSE** | 0.031       | 0.037   | 0.040     | 0.037 |

perceptions. On the other hand, the *knowledge similarity function* has the most accurate results regarding similarity perceptions.

We have additionally calculated the corresponding error values combining the top selected criteria. To do that, we compare the perception of users that combines these criteria with a simple weighted similarity function that uses the average of the local similarity measure based on these criteria. Table 4 presents the result of this analysis. Although, in the current state of this work, we have applied the same weight for each local similarity value, results show that the combination of similarity functions increases the accuracy. These are preliminary promising results and, in the next step (Step-2) of our methodology, we will apply learning algorithms to better adjust these weights to users' perceptions. In summary, in the experiment conducted in this paper we have validated Step 1 (correlation between local similarity measures and perception of similarity), we have observed that our Hypotheses 1 and 2 work on this experiment, and we have obtained promising informal results for Step 2 (learning and validating similarity measure $SIM'Item$ reflecting perception).

## 5    Conclusions and Future Work

Defining similarity measures is a requirement for some AI methods including CBR. Typically most of the approaches capture and define similarity measures analytically. However, research about automatically learning similarity measures has also been an active area of research in CBR. In this paper we have considered the problem of similarity measures definition for tasks where subjectivity in the perception of similarity is an issue. We have proposed a methodology for the definition of similarity measures that reflects the perception of similarity and applied it to the domain of Art.

In the experiment described in this paper we have captured datasets from the similarity perception between artworks. This dataset contains the knowledge to construct or learn such a similarity measure. We have proposed a methodology, acquired the dataset, and validated Hypothesis 1 –different users have different perceptions of similarity and consider differently the attributes describing the items; and Hypothesis 2 –users can be grouped together using profiles and users from the same profile have similar perceptions of similarity. This means that a similarity measure can be learnt for each profile group. This will be done as future work in the Step 2 of the methodology that aims at the construction of a computable model to calculate the similarity between items reflecting the perception data acquired during step 1. We will automate the construction of similarity measures using machine learning from the acquired data. We address the fundamental problem of weight model for features. i.e., it is necessary to give a greater similarity contribution to an important attribute than to other less important ones regarding perception of similarity. We will explore different approaches [15] to automate the construction of similarity measures using machine learning algorithms. We also need to deal with the heterogeneity problem that arises when different attributes are used to describe different cases. We plan to use an approach similar to the one described in [24] to learn weights of the local similarity measures through an evolutionary algorithm (EA) and apply different solutions in the metric learning research area [2].

Also as future work in Step 3 we will construct a similarity measure to compare users ($Sim'User$) by combining a simple measure $SimUser$ with the polarity and emotions of the compared users regarding the domain items. The similarity measure $Sim'User$ should reflect that users with similar emotions on similar items are similar regarding the measure $SIM'_pItems$. This methodology will be validated and applied in the SPICE project. Our proposal is applying community detection algorithms and use the communities as the profiles to learn improved similarity measures.

**Acknowledgments.** The research leading to this publication has received funding from the European Union's Horizon 2020 research and innovation programme under grant agreement SPICE No 870811

# References

1. Arinik, N., Labatut, V., Figueiredo, R.: Characterizing and comparing external measures for the assessment of cluster analysis and community detection. IEEE Access **9**, 20255–20276 (2021). https://doi.org/10.1109/ACCESS.2021.3054621
2. Bellet, A., Habrard, A., Sebban, M.: A survey on metric learning for feature vectors and structured data. Research report, Laboratoire Hubert Curien UMR 5516 (2013). https://hal.inria.fr/hal-01666935
3. Bergmann, R., Müller, G.: Similarity-based retrieval and automatic adaptation of semantic workflows. In: Nalepa, G.J., Baumeister, J. (eds.) Synergies Between Knowledge Engineering and Software Engineering. AISC, vol. 626, pp. 31–54. Springer, Cham (2018). https://doi.org/10.1007/978-3-319-64161-4_2
4. Bernsmed, K., Tøndel, I.A., Nyre, Å.A.: Design and implementation of a CBR-based privacy agent. In: 2012 Seventh International Conference on Availability, Reliability and Security, pp. 317–326. IEEE (2012)
5. Blough, D.S.: The perception of similarity. In: Edited and Published by Dr. Robert G. Cook Department of Psychology Tufts University In cooperation with Comparative Cognition Press, vol. Avian Visual Recognition. Department of Psychology, Brown University (2001). http://www.pigeon.psy.tufts.edu/avc/dblough/
6. Cunningham, P.: A taxonomy of similarity mechanisms for case-based reasoning. IEEE Trans. Knowl. Data Eng. **21**(11), 1532–1543 (2009)
7. Daga, E., et al.: Enabling multiple voices in the museum: challenges and approaches. Digit. Cult. Soc. **6**, 259–266 (2020)

8. Dasari, H., Bhagvati, C., Jain, R.K.: Distance measures in RGB and HSV color spaces. In: Hu, G. (ed.) 20th International Conference on Computers and Their Applications, CATA 2005, Proceedings, pp. 333–338. ISCA (2005)

9. Díaz-Agudo, B., González-Calero, P.A.: A declarative similarity framework for knowledge intensive CBR. In: Aha, D.W., Watson, I. (eds.) ICCBR 2001. LNCS (LNAI), vol. 2080, pp. 158–172. Springer, Heidelberg (2001). https://doi.org/10.1007/3-540-44593-5_12

10. Faruqui, M., Dodge, J., Jauhar, S.K., Dyer, C., Hovy, E.H., Smith, N.A.: Retrofitting word vectors to semantic lexicons. In: Mihalcea, R., Chai, J.Y., Sarkar, A. (eds.) The 2015 Conference of the North American Chapter of the Association for Computational Linguistics: Human Language Technologies, pp. 1606–1615 (2015). https://doi.org/10.3115/v1/n15-1184

11. Gazdar, A., Hidri, L.: A new similarity measure for collaborative filtering based recommender systems. Knowl. Based Syst. **188**, 105058 (2020). https://doi.org/10.1016/j.knosys.2019.105058

12. Harispe, S., Ranwez, S., Janaqi, S., Montmain, J.: Semantic similarity from natural language and ontology analysis. CoRR abs/1704.05295 (2017). http://arxiv.org/abs/1704.05295

13. Holyoak, K.J., Koh, K.: Surface and structural similarity in analogical transfer. Mem. Cogn. **15**(4), 332–340 (1987). https://doi.org/10.3758/BF03197035

14. de Mántaras, R.L., et al.: Retrieval, reuse, revision and retention in case-based reasoning. Knowl. Eng. Rev. **20**(3), 215–240 (2005)

15. Mathisen, B.M., Aamodt, A., Bach, K., Langseth, H.: Learning similarity measures from data. Progr. Artif. Intell. **9**(2), 129–143 (2019). https://doi.org/10.1007/s13748-019-00201-2

16. Mohammad, S., Kiritchenko, S.: WikiArt emotions: an annotated dataset of emotions evoked by art. In: Proceedings of the 11th International Conference on Language Resources and Evaluation (LREC 2018). European Language Resources Association (ELRA) (2018)

17. Ontañón, S.: An overview of distance and similarity functions for structured data. Artif. Intell. Rev. **53**(7), 5309–5351 (2020). https://doi.org/10.1007/s10462-020-09821-w

18. Osborne, H.R., Bridge, D.G.: A case base similarity framework. In: Smith, I., Faltings, B. (eds.) EWCBR 1996. LNCS, vol. 1168, pp. 309–323. Springer, Heidelberg (1996). https://doi.org/10.1007/BFb0020619

19. Plutchik, R.: The nature of emotions: human emotions have deep evolutionary roots, a fact that may explain their complexity and provide tools for clinical practice. Am. Sci. **89**(4), 344–350 (2001)

20. Puga, G.F., Díaz-Agudo, B., González-Calero, P.A.: Similarity measures in hierarchical behaviours from a structural point of view. In: Guesgen, H.W., Murray, R.C. (eds.) Proceedings of the Twenty-Third International Florida Artificial Intelligence Research Society Conference. AAAI Press (2010)

21. Sánchez-Ruiz, A.A., Ontañón, S.: Structural plan similarity based on refinements in the space of partial plans. Comput. Intell. **33**(4), 926–947 (2017)

22. Shanavas, N., Wang, H., Lin, Z., Hawe, G.: Knowledge-driven graph similarity for text classification. Int. J. Mach. Learn. Cybern. **12**(4), 1067–1081 (2020). https://doi.org/10.1007/s13042-020-01221-4

23. Smyth, B., T.Keane, M.: Adaptation-guided retrieval: questioning the similarity assumption in reasoning (1998). https://doi.org/10.1016/S0004-3702(98)00059-9

24. Stahl, A., Gabel, T.: Using evolution programs to learn local similarity measures. In: Ashley, K.D., Bridge, D.G. (eds.) ICCBR 2003. LNCS (LNAI), vol. 2689, pp. 537–551. Springer, Heidelberg (2003). https://doi.org/10.1007/3-540-45006-8_41
25. Towne, W.B., Rosé, C.P., Herbsleb, J.D.: Measuring similarity similarly: LDA and human perception. ACM Trans. Intell. Syst. Technol. 8(1), 7:1-7:28 (2016). https://doi.org/10.1145/2890510
26. Tversky, A.: Features of similarity. Psychol. Rev. 84(4), 327–352 (1977). https://doi.org/10.1037/0033-295X.84.4.327
27. Voorhees, E.M.: TREC: continuing information retrieval's tradition of experimentation. Commun. ACM 50(11), 51–54 (2007). https://doi.org/10.1145/1297797.1297822
28. Voskoglou, M.G., Salem, A.M.: Analogy-based and case-based reasoning: two sides of the same coin. CoRR abs/1405.7567 (2014). http://arxiv.org/abs/1405.7567
29. Wenige, L., Ruhland, J.: Similarity-based knowledge graph queries for recommendation retrieval. Semant. Web 10(6), 1007–1037 (2019). https://doi.org/10.3233/SW-190353

# Measuring Financial Time Series Similarity with a View to Identifying Profitable Stock Market Opportunities

Rian Dolphin[1]($\boxtimes$) (ID), Barry Smyth[1,2] (ID), Yang Xu[3] (ID), and Ruihai Dong[1,2] (ID)

[1] School of Computer Science, University College Dublin, Dublin, Ireland
rian.dolphin@ucdconnect.ie, {barry.smyth,ruihai.dong}@ucd.ie
[2] Insight Centre for Data Analytics, University College Dublin, Dublin, Ireland
[3] School of Economics and Management, Beihang University, Beijing, China
yang_xu@buaa.edu.cn

**Abstract.** Forecasting stock returns is a challenging problem due to the highly stochastic nature of the market and the vast array of factors and events that can influence trading volume and prices. Nevertheless it has proven to be an attractive target for machine learning research because of the potential for even modest levels of prediction accuracy to deliver significant benefits. In this paper, we describe a case-based reasoning approach to predicting stock market returns using only historical pricing data. We argue that one of the impediments for case-based stock prediction has been the lack of a suitable similarity metric when it comes to identifying similar pricing histories as the basis for a future prediction—traditional Euclidean and correlation based approaches are not effective for a variety of reasons—and in this regard, a key contribution of this work is the development of a novel similarity metric for comparing historical pricing data. We demonstrate the benefits of this metric and the case-based approach in a real-world application in comparison to a variety of conventional benchmarks.

**Keywords:** Case-based reasoning · Financial time series · Stock market · Similarity metric

## 1 Introduction

The stock market represents a challenging target when it comes to analysis and prediction [6,16,30]. The stochastic nature of stock prices reflects a complex network of interactions involving a web of hidden factors and unpredictable events. At the same time, the potential to identify even fleeting patterns in market data promises tremendous rewards and in a world where nanoseconds count even a modest degree of prediction accuracy can provide traders with a valuable edge over the competition.

It is not surprising therefore that many researchers have attempted to use a variety of data analysis and machine learning techniques [19] to extract meaningful patterns from market data whether attempting to determine the fair value

© Springer Nature Switzerland AG 2021
A. A. Sánchez-Ruiz and M. W. Floyd (Eds.): ICCBR 2021, LNAI 12877, pp. 64–78, 2021.
https://doi.org/10.1007/978-3-030-86957-1_5

of a stock (so-called *fundamental analysis* [18]) or predicting its future trajectory (so-called *technical analysis* [25]). For example, traditionally, stock returns prediction has been tackled using statistical techniques such as autoregressive integrated moving average models [5], but with recent advances in machine learning and deep learning, research applying advanced computational techniques to the problem of stock market prediction has become increasingly popular [32].

Indeed, the potential role for case-based reasoning (CBR) in financial domains was discussed early on in the CBR literature [36] and there have been numerous attempts to apply case-based ideas to a variety of financial decision making and prediction tasks over the years [10,13,24,27,31,37] with varying degrees of success. However, one of the problems facing similarity-based methods concerns the challenge of developing a suitable similarity metric with which to assess the similarity of price-based time series. For example, conventional Euclidean distance and correlation based metrics have typically fallen short, leading some researchers to explore alternatives; see for example, Chun and Ko's [13] shape-based distance metric.

In this paper we apply case-based reasoning techniques to stock selection based on the prediction of future returns, using only historical pricing data; in Sect. 3 we describe the basic case representation. The main contribution is the development of a novel hybrid similarity metric combining information about price deviations and trends into a single metric; see Sect. 4. Then, in Sect. 5 we present the results of a comprehensive offline evaluation of this approach by evaluating the returns produced by trading strategies using this approach, and in comparison to a variety of alternative benchmarks, to demonstrate significant benefits due to our approach across a range of suitable evaluation metrics.

## 2   Related Work

As a reuse-based problem solving method, guided by similarity [1], case-based reasoning is an appealing paradigm when it comes to a variety of decision problems in financial domains. Intuitively, the idea of basing current decisions on the outcomes of similar decisions that have been made in the past—the core of CBR—seems like an excellent fit in many financial settings. Even though historical patterns will not always prove to be a reliable guide to the future, markets are often driven by cyclical patterns and seasonal trends, which can be exploited to good effect. Indeed case-based reasoning has had a long history when it comes to tackling a range of important problems in financial domains, with applications spanning several distinct topics such as bond rating prediction [34,35], bankruptcy [2,3,21], financial risk assessment [22], real estate valuation [39] and stock market prediction [8,10–15,17,20].

While recent work on the application of case-based reasoning to stock market prediction have been somewhat scarce [13], previous efforts have explored a variety of approaches in terms of their case representations and similarity metrics. Often cases are represented as (multivariate) time series [12,14,15] but sometimes more conventional feature-based approaches are used; [20] selects twelve fundamental and technical indicators as predictor variables. In this paper, cases are represented by a simplified univariate time series using historical monthly returns.

When it comes to case similarity, the literature discusses a variety of options including conventional approaches such as Euclidean, Manhattan and Gaussian distance metrics [12,14,15]; [20] proposes the use of genetic algorithms to determine the feature weights in a Euclidean distance metric. One of the problems with such metrics is that they fail to adequately account for the temporal nature of time-series data such as pricing data [13]. This has motivated recent work by Chun and Ko [13]to develop a more geometrically inspired approach to time-series similarity. Their *shape-based* similarity metric focuses on the patterns of rising and falling price-data, between two time series, rather than on the differences between prices at a given point in time. The work presented in this paper is similarly motivated and we too propose a new similarity metric as the centrepiece of our CBR approach to stock selection and returns prediction.

## 3    From Prices to Cases

The dataset used in this work spans the fifteen-year period from 01/01/2005 to 01/01/2021. Assets were selected from a range of international markets with the inclusion criterion being: (i) the availability pricing data spanning the period in question and (ii) their inclusion in the Nasdaq 100, EURO STOXX 50 or FTSE 100 indices. The resulting dataset was downloaded from Yahoo! Finance and contained 160 unique stock/asset tickers from six stock exchanges.

The resulting raw data consisted of daily adjusted closing prices for each stock. When considering the problem of stock price prediction, a common approach in the literature has been next-day price prediction [29,38], with some considering even shorter time spans [4,33]. However, stock market returns are notoriously hard to predict, especially for shorter time spans due to the increased influence of market noise on price movements [23]. Thus we first convert the raw daily data into monthly price data with each $p$ indicating the price of a stock at the end of a given month. Then we transform the monthly pricing data into monthly returns data in order to extract a more reliable signal (see Eq. 2).

$$prices(a_i) = \{p_1^{a_i}, ..., p_n^{a_i}\} \tag{1}$$

$$r_t^{a_i} = \frac{p_t^{a_i} - p_{t-1}^{a_i}}{p_{t-1}^{a_i}} \tag{2}$$

Accordingly, each case, for asset $a_i$ at time $t$ ($c(a_i, t)$) consists of a sequence of monthly returns for the previous twelve months (the *problem description* part of the case) and a corresponding return for the single next month (the *solution* part of the case); see Eq. 3.

$$c(a_i, t) = \{r_{t-12}^{a_i}, r_{t-11}^{a_i}, ..., r_{t-1}^{a_i} \mid r_t^{a_i}\} \tag{3}$$

Obviously, this case structure is a very simple one, purposely so. It has been chosen for two main reasons. First, it is a good fit for the type of similarity metric that we develop in the following section. Second, by simplifying our case structure in this way we can avoid the many additional factors that may complicate performance analysis and obscure the reason for a particular evaluation outcome, not to mention limiting the explainability of this approach. Indeed, we suggest that if we can generate good predictions using this simple case structure then it suggests an effective performance baseline and a platform for future enhancements.

This case structure was used to build a case base as follows. First, for reasons of computational efficiency, we limited our data to the period between January 2005 and December 2020 (180 months in total). Then, for each of the 160 companies/stocks in our dataset, we constructed cases during this period, with each case containing the past returns for the preceding 12 months and the return for the current (13th) month leading to 28,880 (180 × 160) unique cases. Later we will discuss how this case base was used during our evaluation.

## 4  Similarity in Financial Time Series

Similarity is obviously central to case-based reasoning but conventional similarity metrics such as Euclidean distance or cosine similarity tend not to fare well when it comes to assessing time-series similarity because they ignore the temporal relationship that exists between the different feature values, or monthly returns in this case. In this section, we propose a new similarity metric that emphasises two aspects of similarity that are important in a financial setting: (i) the correlation between time-series returns; and (ii) similar cumulative returns at the end of an investment period. In other words, given a target query case $q$, we wish to identify a set of similar cases whose monthly returns behave in a manner that is similar to the monthly returns of $q$ and whose cumulative return is similar to $q$'s cumulative return.

As an aside, at this stage it is worth highlighting a somewhat unusual and counter-intuitive feature of similarity assessment in a stock-trading setting. Namely, it is not only important to be able to identify a set of similar stocks, but also a set of dissimilar stocks that are expected to behave in opposition to the similar stocks. This is because, in a trading context, traders will often need to offset or hedge their positions in selected stocks by also trading in maximally dissimilar stocks; the idea being that under-performance in a selected (similar) stock can be offset by gains in a dissimilar stock, thereby allowing a trader to limit their overall risk. While we do not consider this aspect in more detail in this paper it is nevertheless an important consideration when selecting a suitable similarity metric and we will comment on this further below.

### 4.1  The Problem with Correlation

It is a commonly held belief, by investors, and even some academics, that a positive correlation between two stock-price time-series indicates that the stocks move in the same direction at the same time, while a large negative correlation indicates that the asset tends to move in opposite directions [26]. In fact, a correlation-based metric, such as Pearson's, actually tends to measure the degree to which the returns deviate above or below their mean at the same time. This distinction is significant in the financial domain and will be highlighted below through an example.

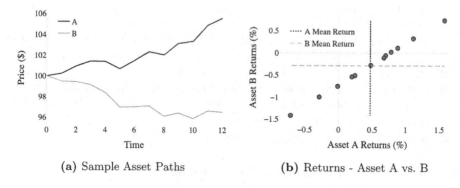

(a) Sample Asset Paths          (b) Returns - Asset A vs. B

**Fig. 1.** Correlation example

Consider the price evolution of two hypothetical assets A and B in Fig. 1(a), but note that correlation is calculated based on *returns* (differences in prices) rather than prices, in order to discount the underlying trends that would otherwise overly influence the correlation. In other words, the correlation between A and B is based on the sequence and magnitude of their price changes rather than the actual prices themselves. The point is that an investment in asset A performed well over the period, with a consistent positive return, while an investment in asset B lost money. Despite this, a traditional correlation metric such as Pearson's correlation coefficient (see Eq. 4) determines that they are perfectly positively correlated; Pearson's returns a near perfect correlation value of 0.99 in this case. This is illustrated in Fig. 1(b) which shows each pair of monthly returns from Fig. 1(a) as a set of points with a clear linear correlation.

$$\rho(x, y) = \frac{\sum_{i=1}^{n}(x_i - \bar{x})(y_i - \bar{y})}{\sqrt{\sum_{i=1}^{n}(x_i - \bar{x})^2 \sum_{i=1}^{n}(y_i - \bar{y})^2}} \qquad (4)$$

### 4.2   An Adjusted Correlation Metric

This correlation phenomenon is particularly problematic in a financial setting and it is a known problem with conventional correlation, such as Pearson's correlation coefficient [26] but few practical solutions have been proposed. One actionable diagnosis of the problem is that it occurs because individual monthly returns are assessed relative to *mean* returns ($\bar{x}$ and $\bar{y}$ in Eq. 4) [28]. Asset A has an overall positive mean return, compared with an overall negative mean return for Asset B, leading to the positive correlation. Thus, a straightforward solution is to derive a modified correlation by simply eliminating the dependency on the means and instead shifting the point of reference to zero. This modification is shown in Eq. 5

$$\tau(x,y) = \frac{\sum_{i=1}^{n} x_i \cdot y_i}{\sqrt{\sum_{i=1}^{n} x_i^2 \cdot \sum_{i=1}^{n} y_i^2}} \tag{5}$$

In what follows, we will refer to this as the *adjusted* correlation metric. Similar to Pearson's correlation metric, this adjusted metric returns values in the interval $[-1, +1]$ and in the case of the data shown in Fig. 1(a), this adjusted metric returns a value of 0.425.

### 4.3   A Novel Similarity Metric for Returns-Based Time-Series

We mentioned earlier that it is desirable for our metric to measure similarity in terms of the tendency for a pair of stock cases to deliver similar returns at similar times – the adjusted correlation metric provides for this – but also to ensure that their cumulative returns are similar. To address the latter requirement we propose using Eq. 6 which calculates the relative difference between two cases, $c(a_i, t)$ and $c(a_j, s)$, based on the product of their monthly returns; this product of monthly returns is mathematically equivalent to the relative difference between the start and end price of each stock over their 12 month periods, but since cases are represented using returns data, rather than price data, we calculate the cumulative return in this way.

$$e(c_{a_i,t}, c_{a_j,s}) = \sqrt{\left( \prod_{\hat{t}=t-1}^{t-12} (1 + r_{\hat{t}}^{a_i}) - \prod_{\hat{s}=s-1}^{s-12} (1 + r_{\hat{s}}^{a_j}) \right)^2} \tag{6}$$

Then, we present our overall similarity metric as Eq. 7, which calculates the cumulative returns and adjusted correlation metric; note that the cumulative returns metric has been incorporated in Eq. 7 in such a way that it serves as a true similarity metric, rather than a distance metric.

$$sim(c_{a_i,t}, c_{a_j,s}) = \frac{w}{1 + e(c_{a_i,t}, c_{a_j,s})} + (1 - w) \cdot \tau(c_{a_i,t}, c_{a_j,s}) \tag{7}$$

Obviously, the relative importance of the cumulative returns and correlation components can be adjusted by varying $w$; when $w = 0$ the similarity equation is based solely on the adjusted correlation metric and when $w = 1$ it resorts to euclidean distance between cumulative returns only. In order to evaluate the impact of $w$ on similarity, using each case in our case base as a *query* we calculate the top-20 most similar cases using the above metric with different values of $w$ ($0 \le w \le 1$) and then compare the next-month returns for the similar cases to the actual next-month return for the corresponding query cases. The absolute relative difference between the return of the similar cases and the query case serves as an error score and the mean error score by $w$ is shown in Fig. 2(a). We can see that the optimal error occurs for values of $w$ between 0.4 and 0.5 and for the remainder of this study we use $w = 0.5$; obviously, this optimal weight may be sensitive to different case bases and case structures. Figure 2(b) shows a histogram of the similarity values obtained during this analysis; the results suggest that the metric behaves as expected as a similarity metric.

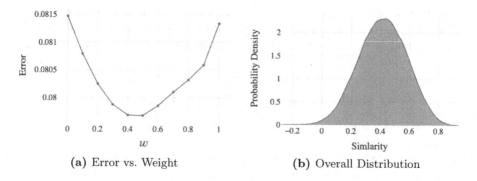

(a) Error vs. Weight          (b) Overall Distribution

Fig. 2. Analysis of proposed metric

### 4.4   Most and Least Similar Cases

Visualising the most and least similar cases for a randomly selected query case is a simple way of verifying the efficacy of the proposed similarity metric and we do this for three separate examples in Fig. 3. Taking Fig. 3(a) as an example, it illustrates the two most and least similar cases for the target query defined by asset Engie SA over the time period 11/2018 to 11/2019. It is seen that the price evolution of the two most similar cases track that of the query case very closely. Not only do the high similarity cases exhibit similar cumulative returns (end up very close), but they also tend to rise and fall at the same points in time. Conversely, the two least similar cases almost mirror the query case over the x-axis. Firstly, their cumulative returns are highly negative in contrast with the strong positive cumulative return of the query case. Secondly, the deviations at each point in time tend to be opposite in sign but similar in magnitude to that of the query case, as we would hope. This is particularly evident at time 3, for example, where the query has a strong positive return while both low similarity cases have very large negative returns for that month.

In conventional approaches, the tendency is to focus on either the deviations at each individual point in time *or* the overall trend. For example, both Pearson's correlation and Chun's [13] recent geometrical similarity metric focus solely on the rises and falls at each individual time period but disregard the overall trend. The novel formulation proposed in Eq. 7 allows us to capture both of these components simultaneously as the examples in Fig. 3 illustrate.

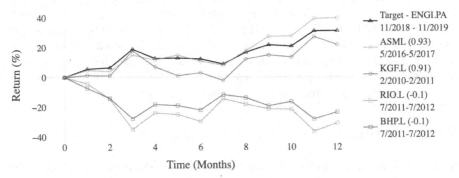

(a) The query case represents the 12 month price evolution of Engie SA.

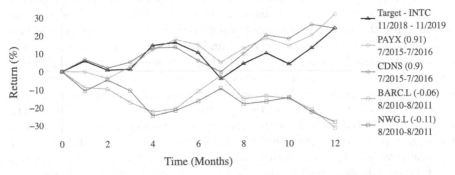

(b) The query case represents the 12 month price evolution of Intel Corporation.

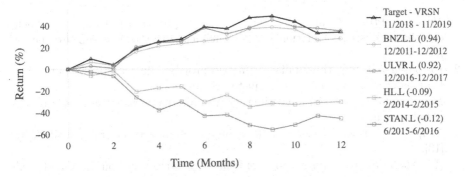

(c) The query case represents the 12 month price evolution of Verisign.

**Fig. 3.** Example of the two most and least similar cases for randomly chosen query cases in the time period 11/2018–11/2019. The two most similar cases are plotted in light blue with circle symbols while the two least similar cases are plotted in red with square symbols. The asset ticker, similarity and time period for each similar case is given in the legend.

Interestingly, the most and least similar cases in Fig. 3(a), for example, are all from different markets to the target. The target, Engie SA (ENGI.PA) is a French company listed on the Euronext stock exchange in Paris while its most similar case ASML Holdings (ASML) is traded on the NASDAQ exchange in the USA. The second most similar case and the two least similar cases are all listed on the London Stock Exchange. Additionally, we note that the most and least similar cases occur up to eight years prior to the target case with both low similarity cases coming from the same 12-month period.

## 5    Evaluation

So far we have described a case representation for encoding the relationship between the previous 12 months of returns for a given stock and the next month of returns, and we have presented a novel similarity metric, which we believe can provide a better sense of similarity in this task domain. In this section, we will describe the results of an evaluation to compare the performance of this new metric to a variety of alternatives in an investment setting[1]. In fact, we will conduct two related evaluations: (1) predicting next-month returns; and (2) using predicted next-month return to inform stock selection as part of an extended investment strategy. In the former we will compare our proposed metric to a variety of alternatives in terms of their ability to accurately predict next-month returns. In the latter we will use these predictions to select the top-5 stocks with the highest predicted returns each month, over a 172 month period, to compare different strategies in terms of the compounded, cumulative investment gains.

In both evaluations we compare the results obtained using the following different similarity variations:

1. *ProposedAdjusted*, the main similarity metric proposed in this paper which combines adjusted correlation and the Euclidean distance between cumulative returns.
2. *ProposedPearson*, the similarity metric proposed in this paper but using Pearson correlation instead of adjusted correlation.
3. *PearsonOnly*, a conventional time-series similarity measure using Pearson's correlation metric.
4. *Shape*, the authors' version of the geometric similarity metric described by [13].
5. *AdjustedOnly*, the adjusted form of Pearson's correlation from Eq. 5 and used in *Proposed*.
6. *CumulativeOnly*, the Euclidean distance between cumulative returns metric from Eq. 6.

---

[1] The relevant code can be found at https://github.com/rian-dolphin/ICCBR2021-Financial-TS-Similarity.

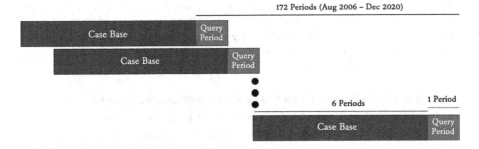

**Fig. 4.** Rolling window layout

## 5.1 Predicting Monthly Returns

In this part of the evaluation, the goal is to predict the next-month returns for a stock based on its previous 12 months of returns. Due to the temporal nature of the data, care must be taken to ensure that only cases that refer to periods prior to the query case period are considered during retrieval; thus if we wish to predict the next-month return for March 2019 then we can only draw on cases whose next-month returns occur prior to March 2019. As a result, a simple leave-one-out strategy cannot be directly implemented, and so, we employ a rolling window approach inspired by [7,9], but tailored to a CBR framework. This approach, illustrated in Fig. 4, allows us to utilise as many query cases as possible in our evaluation but has the effect that the case base depends on the query case. In particular, the case base is defined to contain all cases from the previous six time periods, with six being chosen due to computational limitations. Equation 8 formalises the case base, $\mathcal{C}(c(a_i, t))$, for a general query case $c(a_i, t)$.

$$\mathcal{C}\Big(c(a_i, t)\Big) = \left\{ c(a_i, t - j) \,\middle|\, \begin{matrix} i \in \{1, 2, ..., 160\} \\ j \in \{1, 2, ..., 6\} \end{matrix} \right\} \tag{8}$$

For each query case $q$, the task is to predict $q$'s next-month returns based on a similarity-weighted mean of the next-month returns for the $k$ most similar cases to $q$. Each prediction is compared to the actual next-month returns for $q$ via an absolute difference, giving us an error measurement. We repeat this for different values of $k$ from 1 to 50.

The mean error results presented in Table 1 show how the proposed metric tends to produce predictions with lower error rates than all of the other variations considered. Although the differences are small, it must be remembered that this reflects the errors associated with a single monthly prediction and obviously these error have the potential to compound and accumulate if their corresponding metrics are used to inform an extended trading strategy over time. We will return to this in the section that follows.

Post-hoc Tukey HSD tests confirm that there are significant differences between the pairs of techniques shown in Table 1. Although the

**Table 1.** Mean errors in next-month returns for varied $k$ and similarity metrics. $H_0$ refers to the result of a Tukey HSD test with null hypothesis that the mean of the *ProposedAdjusted* metric is not significantly different from the given baseline metric with $k = 10$. Rejection of the null hypothesis at $\alpha = 0.01$ is indicated by ✓.

|                    | 1      | 5      | 10     | 25     | 50     | $H_0$ |
|--------------------|--------|--------|--------|--------|--------|-------|
| *ProposedAdjusted* | **0.0887** | **0.0882** | **0.0883** | **0.0882** | **0.0883** | –     |
| *ProposedPearson*  | 0.0893 | 0.0885 | 0.0884 | 0.0883 | 0.0884 | ✗     |
| *Shape*            | 0.0888 | 0.0889 | 0.0889 | 0.0891 | 0.0893 | ✓     |
| *PearsonOnly*      | 0.0910 | 0.0909 | 0.0909 | 0.0909 | 0.0908 | ✓     |
| *CumulativeOnly*   | 0.0911 | 0.0908 | 0.0907 | 0.0904 | 0.0903 | ✓     |
| *AdjustedOnly*     | 0.0911 | 0.0908 | 0.0908 | 0.0907 | 0.0905 | ✓     |

*ProposedAdjusted* technique shows significant improvement with respect to the *PearsonOnly*, *AdjustedOnly*, *Shape*, and *CumulativeOnly* metrics at $k = 10$ it is not significantly better than *ProposedPearson*, at least in terms of the error associated with a single monthly returns prediction. It is worth noting, however, that *ProposedPearson* is the only other strategy, in addition to *ProposedAdjusted*, which uses the novel formulation proposed in Eq. 7.

## 5.2   Comparing Trading Strategies

As mentioned above, the previous experiment focused on a single next-month returns prediction, but in practice trading performance is measured over an extended period of time based on the cumulative returns obtained during many buy-sell cycles. In order to evaluate this, in this section we consider a simple trading scenario in which a trader begins with a $1,000 float and invests this uniformly in the stocks with the top-5 highest predicted returns each month, selling these stocks at the end of the month, and rolling-up any profits/losses into their next month investment. This continues for a period of 172 months, as outlined in Fig. 4, and the cumulative gains are calculated at the end of this period. We use the six different similarity metrics, as before, to generate the monthly returns predictions, and various values of $k$ are used, also as before.

Since we are simulating buying the top-5 assets in terms of the highest predicted return each month, the strategies will execute 860 (5 trades × 172 months) buy orders in total over the course of the experiment. Though a sizable number of trades, running the simulation only once would mean the evaluation has the potential to be influenced by a small number of 'lucky' trades. To prevent this, we ran the simulation one hundred times, each time randomly removing 20% of the assets from the dataset.

The results are presented in Table 2 for each strategy with $k = 10$. Under the *ProposedAdjusted* strategy the initial float of $1000 accumulates to $8305.23, corresponding to an annualised return of 15.9%, the highest of all the strategies. A Tukey test indicates the mean return for *ProposedAdjusted* is significantly

**Table 2.** Results of a trading simulation spanning mid 2006–end 2020 with $k = 10$ and initial capital of \$1000. $H_0$ refers to the result of a Tukey HSD test with null hypothesis that the mean annualised return of the *ProposedAdjusted* metric is not significantly different from the given baseline metric with $k = 10$. Rejection of the null hypothesis at $\alpha = 0.01$ is indicated by ✓.

| | Accumulated Value | Annualised Return | Annualised Volatility | $H_0$ |
|---|---|---|---|---|
| *ProposedAdjusted* | **\$8,305.23** | **15.9%** | 22.3% | – |
| *ProposedPearson* | \$6,551.87 | 14.0% | 22.2% | ✓ |
| *Shape* | \$7,797.25 | 15.4% | **19.7%** | ✗ |
| *PearsonOnly* | \$6,233.49 | 13.6% | 21.3% | ✓ |
| *CumulativeOnly* | \$6,527.48 | 14.0% | 21.6% | ✓ |
| *AdjustedOnly* | \$7,883.33 | 15.5% | 21.1% | ✗ |

higher than that of *ProposedPearson*, *PearsonOnly* and *CumulativeOnly*. Though a higher mean return is seen, the Tukey test does not confirm statistical significance over *Shape* and *AdjustedOnly* at $\alpha = 0.01$.

It is notable too that the *AdjustedOnly* strategy beats the *PearsonOnly* strategy (significantly at $\alpha = 0.01$) indicating that the modified correlation metric described in Sect. 4.2 is outperforming the more conventional Pearson correlation metric when applied in a trading simulation. In fact, Pearson correlation alone performs worse than all other strategies. Chun's [13] more recent geometrical similarity metric (*Shape*) performs well in this trading evaluation, producing annualised returns that are better than most of the other strategies, although not the proposed strategy. Its performance is very similar to that of the *AdjustedOnly* similarity metric which is unsurprising since the adjusted correlation can, in some sense, be thought of as a continuous version of the geometric metric as both use 0 as a reference point.

As predicted, although the individual monthly gains in prediction accuracy are small, when compounded as part of a selective investment strategy, then even modest gains can accumulate to offer significant differences in annualised returns. Obviously, this experiment represents a very simplified trading scenario that is limited by factors such as the number of stocks selected for investment each month and how current funds are shared among the selected stocks. In reality, one would expect more sophisticated trading strategies to be used, which vary the number of stocks selected each month and how funds are divided up for the purpose of investment. It may be prudent to include other indicators to aid the trade selection process and it may also be appropriate to consider inter-stock similarity when selecting stocks to provide some level of hedging/diversification as part of an investment strategy. All of these factors will further influence the returns obtained and none have been considered in this initial evaluation.

# 6    Conclusion and Future Work

This work has focused on the problem of measuring similarity between financial time series with a particular focus on stock market pricing and returns data. Our proposed metric combines an adjusted correlation coefficient with a Euclidean metric to simultaneously identify similarity from two angles which, to the best of our knowledge, has not been explored before.

In addition, we have applied our novel similarity metric to the problem of predicting stock market returns and using this to inform a trading strategy. Although this is no doubt a challenging problem, it is motivated by the knowledge that even modest returns and improvements can prove to be extremely useful in the high-stakes world of finance.

We have described a straightforward approach to using ideas from case-based reasoning for this task, including a simple returns-based case representation and a novel approach to measuring the similarity between stock time-series. The results of an initial evaluation demonstrate strong results in terms of returns-based prediction accuracy which in turn lead to significant benefits in terms of annualised returns when used as part of a stock trading strategy. Moreover, the results reported for our novel similarity-metric are superior to those for a variety of alternative including conventional and state-of-the-art baselines.

There is substantial scope for future work with the approach described in this paper. The trading strategy used during the evaluation is likely too simple to be useful in practice and can be enhanced in a number of ways to more reliably evaluate the benefits of the new similarity metric. Comparing our results to state-of-the-art non-CBR baselines such as long short-term memory (LSTM) networks as well as testing varied case lengths are other planned areas of future work. Moreover, modern portfolio theory is based heavily on the use of Pearson correlation to ensure diversification and there is an obvious opportunity to evaluate our revised similarity metric and the adjusted correlation coefficient in the portfolio optimisation domain.

**Acknowledgments.** This publication has emanated from research conducted with the financial support of Science Foundation Ireland under Grant number 18/CRT/6183.

# References

1. Aamodt, A., Plaza, E.: Case-based reasoning: foundational issues, methodological variations, and system approaches. AI Commun. **7**(1), 39–59 (1994)
2. Ahn, H., Kim, K.J.: Bankruptcy prediction modeling with hybrid case-based reasoning and genetic algorithms approach. Appl. Soft Comput. **9**(2), 599–607 (2009)
3. Alaka, H.A., et al.: Systematic review of bankruptcy prediction models: Towards a framework for tool selection. Expert Syst. Appl. **94**, 164–184 (2018)
4. Alostad, H., Davulcu, H.: Directional prediction of stock prices using breaking news on twitter. In: 2015 IEEE/WIC/ACM International Conference on Web Intelligence and Intelligent Agent Technology (WI-IAT), vol. 1, pp. 523–530. IEEE (2015)

5. Ariyo, A.A., Adewumi, A.O., Ayo, C.K.: Stock price prediction using the ARIMA model. In: 2014 UKSim-AMSS 16th International Conference on Computer Modelling and Simulation, pp. 106–112. IEEE (2014)
6. Bachelier, L.: Théorie de la spéculation. In: Annales scientifiques de l'École normale supérieure, vol. 17, pp. 21–86 (1900)
7. Bao, W., Yue, J., Rao, Y.: A deep learning framework for financial time series using stacked autoencoders and long-short term memory. PLoS One **12**(7), e0180944 (2017)
8. Bedo, M.V.N., dos Santos, D.P., Kaster, D.S., Traina, C.: A similarity-based approach for financial time series analysis and forecasting. In: Decker, H., Lhotská, L., Link, S., Basl, J., Tjoa, A.M. (eds.) DEXA 2013. LNCS, vol. 8056, pp. 94–108. Springer, Heidelberg (2013). https://doi.org/10.1007/978-3-642-40173-2_11
9. Chan Phooi M'ng, J., Mehralizadeh, M.: Forecasting east Asian indices futures via a novel hybrid of wavelet-PCA denoising and artificial neural network models. PloS One **11**(6), e0156338 (2016)
10. Chang, P.C., Fan, C.Y., Lin, J.L.: Trend discovery in financial time series data using a case based fuzzy decision tree. Expert Syst. Appl. **38**, 6070–6080 (2011). https://doi.org/10.1016/j.eswa.2010.11.006
11. Chang, P.C., Liu, C.H., Lin, J.L., Fan, C.Y., Ng, C.S.: A neural network with a case based dynamic window for stock trading prediction. Expert Syst. Appl. **36**, 6889–6898 (2009). https://doi.org/10.1016/j.eswa.2008.08.077
12. Chun, S.H., Kim, S.H.: Data mining for financial prediction and trading: application to single and multiple markets. Expert Syst. Appl. **26**(2), 131–139 (2004)
13. Chun, S.H., Ko, Y.W.: Geometric case based reasoning for stock market prediction. Sustainability (Switzerland) **12** (2020). https://doi.org/10.3390/su12177124
14. Chun, S.H., Park, Y.J.: Dynamic adaptive ensemble case-based reasoning: application to stock market prediction. Expert Syst. Appl. **28**(3), 435–443 (2005)
15. Chun, S.H., Park, Y.J.: A new hybrid data mining technique using a regression case based reasoning: application to financial forecasting. Expert Syst. Appl. **31**(2), 329–336 (2006)
16. Fama, E.F.: The behavior of stock-market prices. J. Bus. **38**(1), 34–105 (1965)
17. Goswami, M.M., Bhensdadia, C.K., Ganatra, A.: Candlestick analysis based short term prediction of stock price fluctuation using SOM-CBR. In: 2009 IEEE International Advance Computing Conference, pp. 1448–1452. IEEE (2009)
18. Haque, S., Faruquee, M.: Impact of fundamental factors on stock price: a case based approach on pharmaceutical companies listed with Dhaka stock exchange (2013)
19. Hu, Z., Zhao, Y., Khushi, M.: A survey of forex and stock price prediction using deep learning. Appl. Syst. Innov. **4**(1), 9 (2021)
20. Ince, H.: Short term stock selection with case-based reasoning technique. Appl. Soft Comput. J. **22**, 205–212 (2014). https://doi.org/10.1016/j.asoc.2014.05.017
21. Jo, H., Han, I., Lee, H.: Bankruptcy prediction using case-based reasoning, neural networks, and discriminant analysis. Expert Syst. Appl. **13**(2), 97–108 (1997)
22. Kapdan, F., Aktaş, M.G., Aktaş, M.S.: Financial risk prediction based on case based reasoning methodology. In: 2019 Innovations in Intelligent Systems and Applications Conference (ASYU), pp. 1–6. IEEE (2019)
23. Kenton, W.: Noise and time frames (2020). https://www.investopedia.com/terms/n/noise.asp. Accessed 21 Mar 2021
24. Kim, K.J.: Toward global optimization of case-based reasoning systems for financial forecasting. Appl. Intell. **21**(3), 239–249 (2004)

25. Kumar, G., Jain, S., Singh, U.P.: Stock market forecasting using computational intelligence: a survey. Arch. Comput. Methods Eng. **28**(3), 1069–1101 (2020). https://doi.org/10.1007/s11831-020-09413-5
26. Lhabitant, F.S.: Correlation vs. trends: a common misinterpretation (2020)
27. Li, S.T., Ho, H.F.: Predicting financial activity with evolutionary fuzzy case-based reasoning. Expert Syst. Appl. **36**(1), 411–422 (2009)
28. Libesa: Correlation with prices or returns: that is the question. https://quantdare.com/correlation-prices-returns/. Accessed 03 Apr 2020
29. Long, J., Chen, Z., He, W., Wu, T., Ren, J.: An integrated framework of deep learning and knowledge graph for prediction of stock price trend: an application in Chinese stock exchange market. Appl. Soft Comput. **91**, 106205 (2020)
30. Malkiel, B.G., Fama, E.F.: Efficient capital markets: a review of theory and empirical work. J. Finance **25**(2), 383–417 (1970)
31. Oh, K.J., Kim, T.Y.: Financial market monitoring by case-based reasoning. Expert Syst. Appl. **32**(3), 789–800 (2007)
32. Ozbayoglu, A.M., Gudelek, M.U., Sezer, O.B.: Deep learning for financial applications: a survey. Appl. Soft Comput. **93**, 106384 (2020)
33. Selvin, S., Vinayakumar, R., Gopalakrishnan, E., Menon, V.K., Soman, K.: Stock price prediction using LSTM, RNN and CNN-sliding window model. In: 2017 International Conference on Advances in Computing, Communications and Informatics (ICACCI), pp. 1643–1647. IEEE (2017)
34. Shin, K.S., Han, I.: Case-based reasoning supported by genetic algorithms for corporate bond rating. Expert Syst. Appl. **16**(2), 85–95 (1999)
35. Shin, K.S., Han, I.: A case-based approach using inductive indexing for corporate bond rating. Decis. Support Syst. **32**(1), 41–52 (2001)
36. Slade, S.: Case-based reasoning for financial decision making. In: Proceedings of the First International Conference on Artificial Intelligence Applications on Wall Street, New York, NY. IEEE Computer Society (1991)
37. Wang, Y., Wang, Y.: A case-based reasoning-decision tree hybrid system for stock selection. Int. J. Comput. Inf. Eng. **10**(6), 1223–1229 (2016)
38. Yang, L., et al.: Explainable text-driven neural network for stock prediction. In: 2018 5th IEEE International Conference on Cloud Computing and Intelligence Systems (CCIS), pp. 441–445. IEEE (2018)
39. Yeh, I.C., Hsu, T.K.: Building real estate valuation models with comparative approach through case-based reasoning. Appl. Soft Comput. **65**, 260–271 (2018)

# A Case-Based Reasoning Approach to Predicting and Explaining Running Related Injuries

Ciara Feely[1(✉)], Brian Caulfield[2], Aonghus Lawlor[2], and Barry Smyth[2]

[1] ML Labs, University College Dublin, Dublin, Ireland
Ciara.Feely@ucdconnect.ie
[2] Insight Centre for Data Analytics, University College Dublin, Dublin, Ireland
{brian.caulfield,aonghus.lawlor,barry.smyth}@ucd.ie

**Abstract.** When training for endurance activities, such as the marathon, the risk of injury is ever-present, especially for first-time or inexperienced athletes. And because injuries depend on various factors, there is an opportunity to provide athletes with greater levels of support and guidance when it comes to the risks associated with their training. Hence, in this work we propose a case-based reasoning approach to predict injury risk for marathoners and provide actionable explanations so that runners can understand this risk and potentially reduce it. We do this using the type of activity data collected by common training apps, with extended training breaks used as a proxy for injury (in the absence of explicit injury data), and we show how future breaks can be predicted based on the training patterns of similar runners. Furthermore, we demonstrate how counterfactual explanations can be used to highlight those features that are unique to injured runners (those suffering from training breaks) to emphasise training behaviours that may be responsible for higher levels of injury risk for the target runner. We evaluate our work with a dataset of real-world training data by more than 5,000 real marathon runners.

**Keywords:** CBR for health and exercise · Marathon running · Injury prediction · Counterfactual explanations

## 1 Introduction

Running is one of the most popular forms of personal exercise and endurance events such as the marathon are becoming increasingly popular among recreational runners. But running puts considerable stress on the body—each stride delivers a force equivalent to 2.5 times our bodyweight [1]—and while experienced runners can often deal with this stress, the risk of impact and overuse injuries remains ever-present, but especially among novices and less experienced runners [2]. However, understanding the risk factors associated with running related injuries (RRIs) has proven to be a complex undertaking, as exemplified

© Springer Nature Switzerland AG 2021
A. A. Sánchez-Ruiz and M. W. Floyd (Eds.): ICCBR 2021, LNAI 12877, pp. 79–93, 2021.
https://doi.org/10.1007/978-3-030-86957-1_6

by the considerable research into the various factors that are likely to determine injury risk [3–9]. Yet the scientific jury remains out when it comes to a definitive determination about how and why some runners become injured while others do not; in fact, the only reliable risk factor appears to be a history of a previous injury [10,11]. It is not surprising, therefore, to find that many runners remain somewhat in the dark when it comes to what they can and should do to lower their personal risk of injury [12].

Thus, this work is motivated by the desire to provide more targeted feedback to runners, to highlight aspects of their training that may be associated with higher rates of RRIs. Using a dataset of more than 5,000 past marathon training histories (the 16 week training histories of Strava runners who competed in the Dublin marathon between 2014 and 2017) we build a case-based reasoning (CBR) system in order to estimate the level of injury risk for a given runner, based on their recent training and the (injury) experiences of runners with similar training patterns. In addition to presenting an injury risk score we also use ideas from counterfactual reasoning to explain the nature of the predicted risk in terms of those features of training that appear to be linked to similar runners who have become injured in the past. It is worth noting that our expectations for success in this research were modest, given the difficulty of predicting RRIs reported in the scientific literature. That being said, given that runners are not well-served, if at all, by targeted injury advice currently, there is reason to be optimistic that even modest success would have the potential to help runners complete their training more effectively and more safely.

In the next section we summarise some of the related work on understanding and predicting RRIs and how it connects with recent work on using machine learning methods to support recreational running. Then, in Sect. 3 we describe our CBR approach, focusing on: (a) how suitable cases can be extracted from raw activity/training data; (b) how these cases can be used to predict future injury or estimate the risk of injury; and (c) how these cases can be used to generate counterfactual explanations to help runners understand the nature of this risk and potential actions that they can take to reduce the risk. Before concluding, in Sect. 4, we present the results of an initial evaluation to demonstrate the potential benefits of this approach to supporting marathon runners.

## 2   Related Work

Numerous studies have been conducted to better understand the risk factors for RRIs. These have considered a variety of variables from physiological features related to the mechanics of running gait [3,4] to the impact of training load [5,6], and from a runner's level of experience [7–9] to their specific personality traits [13,14]. Often conducted using cohorts of very experienced, even elite runners, these studies have failed to provide a definitive account of the factors that are predictive of many common forms of RRI, beyond the tendency of runners who have been injured in the past to become injured again in the future [10,11]. That being said, it is commonly held that training load—an estimate of an athlete's training volume and intensity—is likely to be an important factor in several common RRIs such as so-called overuse injuries [15–19]; although even this assertion is not without a level of controversy [20].

Training activity data provides a rich source of training data for machine learning, by integrating fitness and physiology data with training volumes, and even user-provided training assessments, e.g. by logging effort perceptions, documenting injury and illness; see also related work with GPS sensors and soccer players [21]. In due course, it may be possible to identify novel patterns linking fitness, training, recovery and injury and so develop effective early-warning systems for athletes, to alert them to changes in their performance, which may be a precursor to the onset of illness of injury [22–24], although as has been highlighted already, predicting whether a runner will become injured, or is at greater risk of injury, is an extremely challenging task [25–27] and will likely remain so for the foreseeable future. Beyond marathon running, in soccer machine learning has been utilized to predict the necessary amount of recovery time needed by soccer players following an injury [28].

This work aims to link training behaviour with injury risk. While previous efforts have not been successful, we argue that this is likely due to small sample sizes of homogeneous runners (e.g. 10 s of elite or experienced runners). In contrast, this work focuses on a much larger cohort of mostly recreational runners with a wide variety of experience levels and abilities. It also builds on recent work in the area of smart approaches to endurance training [29–31] and particularly work involving case-based reasoning methods [32–35] which has mostly focused on predicting race-times or recommending training plans. Thus far, there has been limited work on injury prediction such as the work of [36], which although it failed to predict the future incidence of injuries, was able to generate an injury risk score that was correlated with future injury rates. However, the work of [36] didn't explain this injury-risk score to runners leaving them unable to take action to improve their injury prospects. In the present work, we will employ recent ideas from counterfactual reasoning [37] and explainable AI (XAI) [38] to explore the potential for using counterfactual cases to distinguish between the training patterns of injured and non-injured runners.

## 3  CBR for Injury Prediction and Explanation

During a typical 12–16 week training plan, most marathon runners will gradually increase their weekly distance and improve their pace as they progress through a variety of carefully coordinated sessions designed to build endurance and strength. Most runners will train 3–6 times per week and most plans will punctuate (typically every 4 weeks) an increasing training load with periodic "down" weeks so that runners can recover from weeks of heavy training load before starting a new training block. These days, most runners record the details of these training sessions (distance, time, pace, heart-rate, cadence, etc.) using smartwatches, sensors, and apps and such activity traces provide a valuable source of raw data that is used here as the basis of a novel CBR system for injury risk assessment.

## 3.1  Representing Training Load

For the purpose of this work we use anonymous activity data collected by the Strava[1] fitness app and made available to the authors via a data sharing agreement with Strava. This data includes distance, time, and elevation data. Some activities also include heart-rate and cadence data but these are not used here. The time and distance data is converted into pacing data (mins/km) for 100 m intervals and pacing data can be further converted into *grade-adjusted pacing* using the elevation data.

More formally for a given runner $r$, we can represent their entire training programme as a time-ordered sequence of activities where each activity $A_i(r) = (d, P)$ and $d$ is the start date and time of the activity and $P$ is a set of paces for the activity:

$$T(r) = \big\{ A_1(r), A_2(r), \ldots, A_n(r) \big\} \tag{1}$$

Then, a runner's training load during a given week $w$ can be characterised in terms of the following distance and pace features; the runner's sex is also included in $F(r, w)$.

1. Total Weekly Distance (m): the total distance during a given week.
2. Longest Run Distance (m): the distance of the single longest run in a given week.
3. Number of Active Days: the number of days with training sessions in a given week.
4. Mean Grade Adjusted Pace (mins/km): the average grade-adjusted pace for the week.
5. Fastest 10 km pace (mins/km): the fastest grade-adjusted pace over a 10 km segment during the week.

## 3.2  Representing Injury Cases

So far we have been talking about injuries and yet the activity data we have access to does not contain explicit injury data. However, we can detect *likely* injuries among these data by identifying extended periods without activities. Specifically, in this work a training break of 14 days or more is considered a likely injury candidate. This is an imperfect heuristic because there may be other reasons for a $\geq$14-day training break (illness, travel, etc.), but we believe that such long breaks are likely to reasonable proxies for injury. Thus for the purpose of this work, predicting whether a runner will become injured amounts to predicting whether they will have a $\geq$14-day training break at some point beyond the current week and before race-day.

We convert the above training load into *injury* cases. Each case is made up of a set of features, derived from the training load features from the previous 4 weeks of training, including:

---

[1] www.strava.com.

1. Average: the average training load (total weekly distance, longest run distance, number of active days etc.) over the previous 4 weeks, $F(r, w + 3)$, ..., $F(r, w)$.
2. Standard Deviation: the standard deviation of the training load features over the past 4 weeks.
3. Relative Change: The average week-on-week relative change for each training load feature over the past 4 weeks.

In what follows, we will sometimes refer to *total weekly distance* as the feature *type* and the average, standard deviation, and relative change forms of this feature type as the actual features.

Then, a *positive* injury case $C^+(r, w)$ denotes a runner $r$ who experiences a $\geq$14-day training break after week $w$ and it is associated with two further features, $BD(r, w)$, the date of this training break and $BL(r, w)$ the length of this training break in days; see Eq. 2. Alternatively, a *negative* case, $C^-(r, w)$, denotes a runner who does not experience such a break before race-day; its $BD$ and $BL$ features are both null. Note, for convenience we refer to $F(r, w)$ as $F_w$ without loss of generality.

$$C^+(r, w) = \{F_{w-3}, F_{w-2}, F_{w-1}, F_w\} \rightarrow injury, BD_w, BL_w \qquad (2)$$

$$C^-(r, w) = \{F_{w-3}, F_{w-2}, F_{w-1}, F_w\} \rightarrow healthy \qquad (3)$$

### 3.3 Balancing the Case Base

In this way, each runner can be associated with a number of injury cases based on different training weeks and their future long-break status. However, such a case base is not balanced as it contains far more negative (non-injury) cases than positive (injury) cases.

To address this we balance the case base by undersampling from the negative cases to select the same number of negative cases as there are positive cases. In this work we consider two sampling strategies:

1. *Random Sampling*: select a random subset of negative cases to match the number of positive cases.
2. *Nearest Unlike Neighbour Sampling*: for each positive case select the most similar negative case.

However, the random strategy was found to be more effective during evaluation and so in what follows we will focus on the use of this approach.

### 3.4 Task 1: Predicting Training Breaks

Now that we have a suitable case representation we can turn our attention to the prediction of future training breaks, as proxies for injury, see Algorithm 1;

---

**Algorithm 1.** Predicting training disruptions as a proxy for injury

---

**Input: q**, the query case for runner **r** in week **w**; **CB** the balanced case base; **k** the number of similar cases to retrieve during prediction.

1: $C \leftarrow filter(CB, week = w)$
2: $C' \leftarrow sort(C, sim(q, c))$
3: $C_k \leftarrow C'.head(k)$
4: $p \leftarrow majority(C_k.class)$
5: $r \leftarrow |C_l[C_k.class == p]|/k$
6: **return** $p, r$

---

**Algorithm 2.** Explaining training disruption predictions

---

**Input: q**, the query case for runner **r** in week **w**; **CB**, the balanced case base; **p** the predicted training disruption class (positive/injured or negative/healthy; **n**, the number of supporting and counterfactual cases to use; **m**, the number of significant features to return).

1: $C \leftarrow filter(CB, week = q.week)$
2: $C' \leftarrow sort(C, sim(q, c))$
3: $S \leftarrow C'[C'.class == p].head(n)$
4: $CF \leftarrow C'[C'.class! = p].head(n)$
5: $sig \leftarrow []$
6: **for** $f$ in $CB.features$ **do**
7:     $sig.append(f)$ **if** $ttest(S.f, CF.f) < 0.05$
8: **end for**
9: **return** $sig, S, CF$

---

note that for the purpose of similarity assessment all case features are scaled using a standard minmax scaler.

Given a query case $q$ for a runner $r$ at week $w$, we first select only those cases in the (balanced) case base that are also $w$ weeks form race-day (line 1) and identify the $k$ most similar cases using a standard cosine distance metric based on their 4-week features $F'(r, w), ..., F'(r, w + 3)$ (lines 2–3). Then, the injury prediction is based on the majority class (positive or negative) and an injury risk score is based on the number of positive cases as a fraction of $k$ (lines 4–5).

### 3.5   Task 2: Explaining Training Breaks

In this work we argue that presenting a runner with a risk assessment when it comes to the injury prospects is not enough because it does not, on its own, help the runner to understand the reason for this assessment – what it is about their recent training that is associated with a given risk level – and nor does it provide them with the means to modulate their risk. This is especially true in the context of first-time or inexperienced marathoners.

With this in mind we propose an approach to explaining positive (injury) and negative (non-injury) prediction outcomes by using a combination of factual and counterfactual cases; see Algorithm 2. Consider a positive recommendation for a given query case, that is the runner is deemed to be at risk of injury: using $q$ we identify the $n$ most-similar *supporting* cases, $S$, (positive cases in this example) and a corresponding set of $n$ *counterfactual* cases, $CF$, (negative cases); lines 1–4. Thus, $S$ corresponds to a set of runners with similar training patterns to the query runner and who have $\geq$14-day extended breaks in their training just as the query runner is predicted to experience, while $CF$ corresponds to a set of runners with similar training patterns but who have not experienced $\geq$14-day future training breaks. The assumption is that any significant differences between the $S$ and $CF$ may usefully explain the reason for their different outcomes. Thus, for each case feature we determine whether its mean values from $S$ and $CF$ are significantly different using a $t$ test with $p<0.05$; lines 5–8.

The end result, for a given query $q$ and prediction, is a short-list of features (*sig*) whose values in the supporting cases are different from their corresponding values in the counterfactual cases. From among these *explanatory* features we can select the top $k$ features to use in constructing a final explanation for the runner. There are a number of options available when it comes to this selection process. One option is to select those features with the greatest relative difference between their $S^+$ and $CF^-$ means; an alternative could be to use a more statistically robust *effect-size* metric, such as Cohen's $d$, to select the $k$ features with the largest effect-size. In what follows we adopt a third approach by scoring each feature *type* (e.g. total distance vs active days vs mean pace etc.) based on the sum of the absolute values of the $t$ statistic for their significant features to provide a standardized difference score across various features; a larger score means that a given feature type is more important when it comes to distinguishing between the two injured and healthy cases.

For example, in the case of a query runner who is predicted to be at risk of a future $\geq$14-day break (as a proxy for injury), this approach may identify a feature such as the average weekly distance or their average weekly pace as the most discriminating features supporting this prediction. In other words, the fact that the query runner's weekly distance is significantly *higher* among supporting cases or that the runner's average weekly pace is *lower* (faster) among supporting cases might suggest that the query runner should reduce their average weekly pace and slow their pace in the coming weeks; the mean weekly distance and pace from the supporting and counterfactual groups can provide guidance on how much to change their weekly distance and pace.

## 4   Evaluation

So far we have described an approach to predicting and explaining the future injury status of a runner using the type of raw activity data that is commonly available from fitness apps such as Strava, Runkeeper, MapMyRun, and others. In this section we describe the results of a preliminary evaluation of these approaches using real-world data. The evaluation is *preliminary* in the sense that it is retrospective rather than prospective. It speaks to what happened in the past but does not tell us about what might happen if such approaches were deployed for runners, which will require live-user evaluations and is a matter for future work. Nevertheless we believe that a retrospective evaluation is an important first-step in understanding the efficacy of the proposed techniques as well as clarifying a opportunities for improvements that may exist.

## 4.1  Setup

The dataset used in this work is a subset of data made available under a data sharing agreement with Strava Inc. and is summarised in Table 1. It consists of over 5,000 unique runners who completed a marathon in Dublin between 2014–2017 with finish-times between 2 and 6 h. For the purpose of producing our case base approximately 20% of these runners experienced a training break ≥14-days in the 6–16 weeks before race-day.

For each task, prediction and explanation, we use a standard cross-fold validation procedure by splitting the data in to test (20%) and training (80%) subsets and average across 5 splits.

**Table 1.** A summary of the dataset used in this study for runners of the Dublin marathon in the period 2014–2017. The table includes personal characteristics and mean and standard deviation for marathon finish time (minutes), number of weekly training activities, and total weekly distance (km).

| Year | Sex | Runners | Age | Race-Time | Active Days/Wk | Distance/Wk |
|------|-----|---------|-----|-----------|----------------|-------------|
| 2014 | F | 85 | 37.2 ± 7.5 | 262.08 ± 35.07 | 2.7 ± 0.87 | 33.4 ± 11.83 |
|      | M | 425 | 36.8 ± 6.6 | 233.23 ± 36.68 | 2.9 ± 1.19 | 37.9 ± 19.24 |
| 2015 | F | 129 | 37.3 ± 7.5 | 260.61 ± 40.97 | 2.8 ± 1.03 | 34.3 ± 14.59 |
|      | M | 642 | 38.2 ± 7.7 | 229.31 ± 37.82 | 3.0 ± 1.23 | 39.4 ± 20.19 |
| 2016 | F | 322 | 38.4 ± 7.6 | 266.67 ± 43.77 | 3.0 ± 1.07 | 35 ± 14.22 |
|      | M | 1249 | 39.1 ± 7.5 | 231 ± 37.85 | 3.0 ± 1.18 | 39.4 ± 19.18 |
| 2017 | F | 562 | 38.4 ± 7.7 | 267.08 ± 39.54 | 3.0 ± 0.96 | 35.8 ± 14.53 |
|      | M | 1865 | 39.5 ± 7.8 | 232.15 ± 38.12 | 3.1 ± 1.19 | 40.3 ± 19.33 |

## 4.2  Evaluating Prediction Accuracy

The prediction task is responsible for (a) predicting whether or not a runner is expected to suffer a future ≥14-day training break and (b) the *risk* of suffering a future ≥14-day training break. To evaluate the former we compare the predicted class to the actual class of the test runners to calculate a prediction error as the fraction of incorrect predictions. Separately, to evaluate the risk score we compute the correlation coefficient between binned the runners based on their risk score, calculated the proportion of true-positive runners (those that did

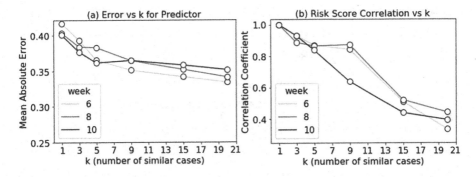

**Fig. 1.** (a) shows the change in mean absolute error of the predictor for different values of k for weeks 6, 8, and 10 weeks from race-day. (b) shows the same but looking at the change in the correlation coefficient of the proportion of runners in the injured class and the risk-score

experience a $\geq$14-day break), and calculated the correlation coefficient between this proportion and the risk-score.

To begin with, Fig. 1 shows the relationship between prediction accuracy and risk-score correlation and $k$, the number of similar cases retrieved during prediction at three different points during training (6, 8, and 10 weeks before race-day). It is noteworthy that prediction accuracy is modest, showing error rates of about 35% in this evaluation, and highlighting the challenging nature of the prediction task. That being said, the prediction error falls for larger values of $k$. However, the risk-score correlation also declines with $k$ meaning that the risk-score is a less reliable indicator of true-injury risk as we retrieve more cases. It is notable, how the risk-score correlation is higher closer to race-time (weeks 6 and 8 compared to week 10) which suggests that it provides a more reliable risk-assessment as runners progress through their training, which is to be expected. In this case, the result suggest that $5 \leq k \leq 10$ offers a good balance between prediction accuracy and risk-score correlation and for the remainder of this evaluation we use $k = 5$.

Figure 2(a) shows a more detailed summary of prediction error and correlation scores ($k = 5$) as training progresses, from 16 weeks before race-day to 6 weeks before race-day. We can see how, closer than 13 weeks from race-day tends to produce relatively stable prediction accuracy and correlation scores. In particular, the correlation score remains about 0.8 throughout this period which means the risk-assessment score serves as a highly correlated indicator of injury risk.

Figure 2(b) separates the prediction results based on the predicted class. The difference between these declines as race-day approaches. Unfortunately, the ability to accurately predict a future training break (*injured*) is compromised during the early stages of training, although it does decline as race-day approaches. In contrast, the ability to predict *healthy* runners is more effective during the early stages of training but the error increases gradually as training proceeds.

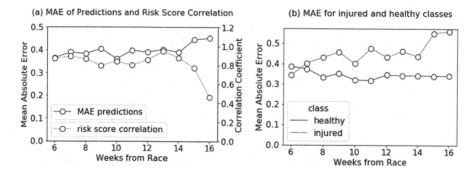

**Fig. 2.** (a) depicts the mean absolute error of the predictor (blue), and the correlation coefficient (orange) for different weeks leading up to race-day. (b) depicts the MAE of the predictor for the healthy (blue) and injured (orange) classes (Color figure online)

At the beginning of this paper we called out the difficulty of this injury-prediction task, which was made all the more difficult by the lack of reliable injury data and the need to use training breaks as a proxy for injuries. It is not surprising therefore to find that the resulting predictions are at best modestly accurate. However, given that today runners benefit from little or no feedback on their injury risk it can be argued that even these modestly accurate predictions can be useful. Moreover, in practice it will likely be more useful to take advantage of the injury risk score as a way to convey this feedback to a runner. More harm than good might be done by predicting a runner will become injured if they do not or vice versa – at the very least it will compromise trust in the system – but the ability to guide the runner with an injury risk-score that is more reliably correlated with injury incidence rates among similar runners will be useful, especially when compared to baseline incident rates. For example, telling a runner that, *"Your injury risk-score is 0.6 meaning that you are twice as likely as the average runner to experience an injury before race-day"* is informative and, when combined with an explanation, may be actionable too.

### 4.3   Evaluating Injury Explanations

In the context of this work, the primary purpose of providing the runner with a prediction explanation is less about justifying the prediction and more about suggesting ways in which the runner might improve their training prospects, by highlighting features that discriminate between supporting and counterfactual cases. In this evaluation we focus on outcomes where the runner is predicted to be at risk of becoming injured, on the grounds that these are outcomes where a runner will be more likely to expect support. The key questions are then:

1. How often can we identify significant feature differences to distinguish between positive and negative cases?
2. What types of features are more likely to be significant when we compare similar injured runners to similar healthy runners?

3. What types of features are more likely to be in the top-3 features recommended to be part of the explanation?

To answer these questions, we use Algorithm 2 to identify the significant features for runners predicted to be at risk of injury and in each case select the top-3 features based on their combined t-value as described previously. Figure 3 shows the distribution of the number of significant features for each injury prediction. We can see how on average there are 2.38 significant differences and how 80% of predictions can be linked to at least 1 significant feature difference. This is a positive result as it highlights that significant differences are commonplace and offers a starting point to provide a runner with an actionable explanation.

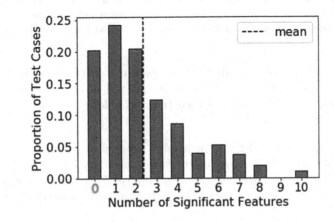

**Fig. 3.** The distribution of the number of features differing significantly among the nearest positive and negative cases

Regarding the types of features that are significant, Fig. 4(a) shows the fraction of injury predictions associated with significant feature differences by feature type. Each bar represents a type of feature and the height of the bar reflects the fraction of times that at least one feature from a given type is significant. In general, the *volume* features (active days and total distance) are significant more often than their *intensity* counterparts (fastest 10 km pace and mean pace). This is an interesting result as it suggests that training volume may be more important than training intensity when it comes to injury, indicating that some runners may protect against injury by reducing volume ahead of intensity.

Similarly, Fig. 4(b) shows the proportions of feature types that occur in the top-3 features, based on the sum of the absolute $t$ statistic values for significant features (see Sect. 3.5). The same ordering of feature types occurs as in Fig. 4(a), with the number of active days and total weekly distance in over 40% of the top-3 recommendations, compared with 20–30% for the intensity features.

Obviously more work on generating explanations is required, and this evaluation represents only the beginning. For example, we have not yet described

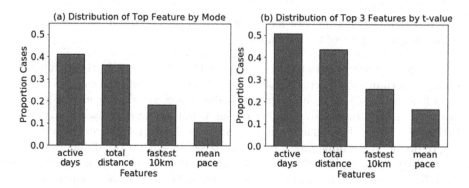

**Fig. 4.** Fraction of test cases (injury predictions) where a given feature was found (a) to be significant when comparing injured and healthy cases and (b) to be recommended as a top-3 feature based on the sum of the $t$ scores for that feature type.

how to translate a set of $k$ significant features into an explanation that the runner can understand and act on. Most likely this will involve communicating to the runner how they might change their training to lower the risk of injury, for example:

> *Your risk of injury is **2x** that of similar runners at this training stage. Runners like you who become injured have a weekly distance that is **10%** **greater** than runners like you who avoid injury. Consider **reducing** your weekly distance from your current level of 60km to **54 km** per week.*

The benefit of this type of explanation is that while helping the runner understand their risk level (2×), relative to similar runners, it also provides information about a possible reason for the higher risk level (total weekly distance) and suggests a course of action (reducing total weekly distance by 10%) to reduce this risk. Evaluating different forms of explanation is planned as future work.

## 5  Conclusions

We attempted to tackle an important problem facing recreational marathoners by (a) trying to predict whether their recent training behaviour is likely to lead to injury and (b) if it is, by explaining the reasons for this in terms of features of their training that they may wish to adjust. We have described and evaluated a case-based reasoning approach using real-world training data. The results are modest but encouraging. Despite a lack of precise injury data, and even with a simple case representation and straightforward similarity assessment, it was nevertheless possible to generate an injury risk-score that reliably correlated with injury incidence rates even if actual injury predictions were less reliable.

Obviously there are some shortcomings in this work. The lack of true injury data is an obvious weakness and area for improvement, although such data is not routinely collected by the current generation of fitness apps. It is also worth highlighting that the nature of the dataset used was such that runners who became injured close to race-day, causing them to drop out, were missing from the dataset, because they never completed the marathon. This means that our existing dataset provides an incomplete account of runners with injury-related training breaks. The features available were also limited to distance and pacing information. In the future, it may become feasible to incorporate more reliable indicators of injury and to harness additional features such as heart-rate to offer greater insights into training and recovery. Finally, it will also be important to evaluate future versions of this work in live user-trials in order to better understand how runners respond to injury feedback such as predictions and explanations. Do they trust in these predictions? Do they change their behaviour? Is their risk of injury reduced as a result? How is overall performance impacted on race-day? We believe that this work has great potential because even modest improvements in managing the injury-risk of runners can have a significant impact on their training and race outcome.

**Acknowledgments.** Supported by Science Foundation Ireland through the Insight Centre for Data Analytics (12/RC/2289_P2) and the SFI Centre for Research Training in Machine Learning (18/CRT/6183).

# References

1. Mann, R.: Biomechanics of running. Running Injuries, pp. 1–20 (1989)
2. Kluitenberg, B., van Middelkoop, M., Diercks, R., van der Worp, H.: What are the differences in injury proportions between different populations of runners? A systematic review and meta-analysis. Sports Med. **45**(8), 1143–1161 (2015). https://doi.org/10.1007/s40279-015-0331-x
3. Napier, C., MacLean, C.L., Maurer, J., Taunton, J.E., Hunt, M.A.: Kinetic risk factors of running-related injuries in female recreational runners. Scand. J. Med. Sci. Sports **28**, 2164–2172 (2018)
4. Vannatta, C.N., Heinert, B.L., Kernozek, T.W.: Biomechanical risk factors for running-related injury differ by sample population: a systematic review and meta-analysis. Clin. Biomech. **75**, 10499 (2020)
5. Nielsen, R.O., Buist, I., Sørensen, H., Lind, M., Rasmussen, S.: Training errors and running related injuries: a systematic review. Int. J. Sports Phys. Ther. **7**, 58–75 (2012)
6. Baltich, J., Emery, C., Whittaker, J., Nigg, B.: Running injuries in novice runners enrolled in different training interventions: a pilot randomized controlled trial. Scand. J. Med. Sci. Sports **27**, 08 (2016)
7. Damsted, C., Parner, E.T., Sørensen, H., Malisoux, L., Nielsen, R.O.: ProjectRun21: do running experience and running pace influence the risk of running injury-A 14-week prospective cohort study. J. Sci. Med. Sport **22**, 281–287 (2019)
8. Kemler, E., Blokland, D., Backx, F., Huisstede, B.: Differences in injury risk and characteristics of injuries between novice and experienced runners over a 4-year period. Phys. Sportsmed. **46**, 485–491 (2018)
9. Agresta, C.E., Peacock, J., Housner, J., Zernicke, R.F., Zendler, J.D.: Experience does not influence injury-related joint kinematics and kinetics in distance runners. Gait Posture **61**, 13–18 (2018)

10. Fokkema, T.: Prognosis and prevention of injuries in recreational runners. Ph.D. thesis, University of Rotterdam (2020)

11. Fokkema, T., et al.: Online multifactorial prevention programme has no effect on the number of running-related injuries: a randomised controlled trial. Br. J. Sports Med. **53**, 1479–1485 (2019)

12. Fokkema, T., Vos, R.-J., Bierma-Zeinstra, S., Middelkoop, M.: Opinions, barriers, and facilitators of injury prevention in recreational runners. J. Orthop. Sports Phys. Ther. **49**, 1–22 (2019)

13. Fields, K.B., Delaney, M., Hinkle, J.S.: A prospective study of type A behavior and running injuries. J. Fam. Pract. **30**, 425–429 (1990)

14. Nielsen, R.O., et al.: Predictors of running-related injuries among 930 novice runners: a 1-year prospective follow-up study. Orthop. J. Sports Med. **1**(1), 2325967113487316 (2013)

15. Thornton, H.R., Delaney, J.A., Duthie, G.M., Dascombe, B.J.: Importance of various training-load measures in injury incidence of professional rugby league athletes. Int. J. Sports Physiol. Perform. **12**, 819–824 (2017)

16. Malisoux, L., Nielsen, R.O., Urhausen, A., Theisen, D.: A step towards understanding the mechanisms of running-related injuries. J. Sci. Med. Sport **18**, 523–528 (2015)

17. Lazarus, B.H., et al.: Proposal of a global training load measure predicting match performance in an elite team sport. Front. Physiol. **8**, 930 (2017)

18. Barros, E.S., et al.: Acute and chronic effects of endurance running on inflammatory markers: a systematic review. Front. Physiol. **8**, 779 (2017)

19. Bowen, L., Gross, A.S., Gimpel, M., Bruce-Low, S., Li, F.-X.: Spikes in acute: chronic workload ratio (ACWR) associated with a 5–7 times greater injury rate in English Premier League football players: a comprehensive 3-year study. Br. J. Sports Med. (2019). https://doi.org/10.1136/bjsports-2018-099422

20. Bornn, L., Ward, P., Norman, D.: Training schedule confounds the relationship between acute: chronic workload ratio and injury, Sloansportsconference Com (2019)

21. Rossi, A., Pappalardo, L., Cintia, P., Iaia, F., Fernández, J., Medina, D.: Effective injury forecasting in soccer with GPS training data and machine learning. PLOS One **13**, e0201264 (2018)

22. Gabbett, T.J.: The training—injury prevention paradox: should athletes be training smarter and harder? Br. J. Sports Med. **50**(5), 273–280 (2016)

23. López-Valenciano, A., et al.: A preventive model for muscle injuries: a novel approach based on learning algorithms. Med. Sci. Sports Exerc. **50**, 915–927 (2018)

24. Claudino, J.G., Capanema, D.O., de Souza, T.V., Serrão, J.C., Machado Pereira, A.C., Nassis, G.P.: Current approaches to the use of artificial intelligence for injury risk assessment and performance prediction in team sports: a systematic review. Sports Med. Open **5**(1), 1–12 (2019). https://doi.org/10.1186/s40798-019-0202-3

25. Carey, D.L., Ong, K.-L., Whiteley, R., Crossley, K.M., Crow, J., Morris, M.E.: Predictive modelling of training loads and injury in Australian football, arXiv preprint arXiv:1706.04336 (2017)

26. Kampakis, S.: Predictive modelling of football injuries, arXiv preprint arXiv:1609.07480 (2016)

27. Rossi, A., Pappalardo, L., Cintia, P., Iaia, F.M., Fernàndez, J., Medina, D.: Effective injury forecasting in soccer with GPS training data and machine learning. PloS One **13**(7), e0201264 (2018)

28. Kampakis, S.: Comparison of machine learning methods for predicting the recovery time of professional football players after an undiagnosed injury. In: MLSA@PKDD/ECML (2013)
29. Rajšp, A., Fister, I.: A systematic literature review of intelligent data analysis methods for smart sport training. Appl. Sci. **10**(9), 3013 (2020)
30. Berndsen, J., Lawlor, A., Smyth, B.: Running with recommendation. In: HealthRecSys@ RecSys, pp. 18–21 (2017)
31. Berndsen, J., Smyth, B., Lawlor, A.: Pace my race: recommendations for marathon running. In: Proceedings of the 13th ACM Conference on Recommender Systems, pp. 246–250. ACM (2019)
32. Smyth, B., Cunningham, P.: Running with cases: a CBR approach to running your best marathon. In: Aha, D.W., Lieber, J. (eds.) ICCBR 2017. LNCS (LNAI), vol. 10339, pp. 360–374. Springer, Cham (2017). https://doi.org/10.1007/978-3-319-61030-6_25
33. Smyth, B., Cunningham, P.: An analysis of case representations for marathon race prediction and planning. In: Cox, M.T., Funk, P., Begum, S. (eds.) ICCBR 2018. LNCS (LNAI), vol. 11156, pp. 369–384. Springer, Cham (2018). https://doi.org/10.1007/978-3-030-01081-2_25
34. Feely, C., Caulfield, B., Lawlor, A., Smyth, B.: Using case-based reasoning to predict marathon performance and recommend tailored training plans. In: Watson, I., Weber, R. (eds.) ICCBR 2020. LNCS (LNAI), vol. 12311, pp. 67–81. Springer, Cham (2020). https://doi.org/10.1007/978-3-030-58342-2_5
35. Feely, C., Caulfield, B., Lawlor, A., Smyth, B.: Providing explainable race-time predictions and training plan recommendations to marathon runners. In: Fourteenth ACM Conference on Recommender Systems, RecSys 2020, New York, NY, USA, pp. 539-544. Association for Computing Machinery (2020)
36. Smyth, B., Lawlor, A., Bernsden, J., Feely, C.: Recommendations for marathon runners, User Modeling and User Adapted Interaction (Unpublished)
37. Keane, M.T., Smyth, B.: Good counterfactuals and where to find them: a case-based technique for generating counterfactuals for explainable AI (XAI). In: Watson, I., Weber, R. (eds.) ICCBR 2020. LNCS (LNAI), vol. 12311, pp. 163–178. Springer, Cham (2020). https://doi.org/10.1007/978-3-030-58342-2_11
38. Adadi, A., Berrada, M.: Peeking inside the black-box: a survey on explainable artificial intelligence (XAI). IEEE Access **6**, 52138–52160 (2018)

# Bayesian Feature Construction for Case-Based Reasoning: Generating Good Checklists

Eirik Lund Flogard[1,2]([⊠]) [iD], Ole Jakob Mengshoel[1] [iD], and Kerstin Bach[1] [iD]

[1] Norwegian University of Science and Technology (NTNU), Sem Sælands Vei 9, Trondheim, Norway
{eirik.l.flogard,ole.j.mengshoel,kerstin.bach}@ntnu.no
[2] Norwegian Labour Inspection Authority, Prinsensgt. 1, Trondheim, Norway
eirik.flogard@arbeidstilsynet.no

**Abstract.** Checklists are used to aid the fulfillment of safety critical activities in a variety of different applications, such as aviation, health care or labour inspections. However, optimizing a checklist for a specific purpose can be challenging. Checklists also need to be trustworthy and user friendly to promote user compliance. With labour inspections as a starting point, we introduce the Checklist Construction Problem. To address the problem, we seek to optimize the content of labour inspection checklists in order to improve the working conditions in every organisation targeted for inspections. To do so, we introduce a hybrid framework called BCBR to construct trustworthy checklists. BCBR is based on case-based reasoning (CBR) and Bayesian inference (BI) and constructs new checklists based on past cases. A key novelty of BCBR is the use of BI for constructing new features in past cases. The augmented past cases are retrieved via CBR to construct new checklists, which ensures justification for the content of the checklists and promotes trust. Experiments suggest that BCBR is more effective than any other baseline we tested, in terms of constructing trustworthy checklists.

**Keywords:** Bayesian CBR · Feature construction · Checklist

## 1 Introduction

**Context.** Every year more than three million workers are victims of serious accidents causing more then 4000 deaths due to poor working conditions in EU alone.[1] World-wide, it has been estimated that there are at least 9.8 million

**Fig. 1.** Conceptual view of NLIA's procedure

---

[1] https://eur-lex.europa.eu/legal-content/EN/TXT/PDF/?uri=CELEX:52014DC 0332.

© Springer Nature Switzerland AG 2021
A. A. Sánchez-Ruiz and M. W. Floyd (Eds.): ICCBR 2021, LNAI 12877, pp. 94–109, 2021.
https://doi.org/10.1007/978-3-030-86957-1_7

people in forced labour (2005) [2]. The most important measure to prevent poor working conditions is regulations. Regulations are usually enforced through labour inspections, which make them a vital part of the strategy employed by many countries to ensure good health, safety, decent work conditions and well-being for workers (see UN's SDGs 3, 8 and 16[2]). Hence it is important to carry out labour inspections efficiently at large scale.

To identify poor working conditions, labour inspection agencies use surveys to check individual organisations for non-compliance [24]. Such procedures vary between different countries and we will use the Norwegian Labour Inspection Authority (NLIA) as an example. NLIA's inspection procedure is shown in Fig. 1. It consists of a checklist which is a set of control points that are answered during the inspection. Every control point is a question that corresponds to a specific regulation. The answer to each question indicates whether the inspected organisation is compliant or not. These answers provide a basis for reactions if non-compliance is found. Checklists for ensuring health and safety are also used in other domains such a surgery or flight procedures to ensure high accuracy of due diligence, and success often relies on correctly applying checklists [5].

**Challenges with Checklists.** Currently, labour inspection agencies operate with a limited, fixed number of static procedures or checklists targeting specific industries that organisations belong to. The inspectors select the checklist they subjectively believe is most relevant to the organisation they are visiting. A drawback with this approach is that the selected checklist can be poorly optimized for its target, while also being limited in terms of scope. This may prevent the inspections from fulfilling their purpose of addressing high risks to the workers' health, environment and safety. Checklists used for other applications such as aviation and health care may have similar problems where poorly optimized checklists can suffer from compatibility issues with users or contexts [5,7]. This can have a negative effect on the users' motivation to use the checklists.

**Contributions.** We introduce the Checklist Construction Problem (CCP): Suppose that we have $N$ unique questions with yes/no answers, where the answer to each question has an unknown probability distribution. Given the questions, construct a checklist for a target entity by selecting $K$ unique questions that maximize the likelihood for obtaining no-answers to every selected question.

This problem could be applied to any domain where checklist optimization is an issue, such as healthcare or aviation. In these domains, the $N$ unique questions may be designed to accomplish a specific task such as surgery or flight check and the target entity may be a patient or an aircraft. Any question with a likely no-answer should then be on the surgery or flight checklist so that yes-answers are obtained instead. However, this work focuses on solving CCP for labour inspections and introduces a new data set as a starting point to do so.

To solve CCP, we introduce BCBR, which is a framework based on Bayesian inference (BI) and case-based reasoning (CBR) for constructing new checklists optimized for a target organisation (entity). BCBR uses CBR to retrieve ques-

---

[2] https://sdgs.un.org/.

tions from checklists which have been used in past cases to survey organisations similar to the target organisation. BI is used to construct features in past cases which ensures that the retrieved questions have high probabilities for non-compliance. The approach starts with a data set of cases containing organisations and questions from previously used checklists. New features are then constructed by means of BI and added to each row in the data set to create augmented cases. The augmented cases are added to a case base which is queried using similarity based retrieval. The query contains a target probability and organisation, which is used to retrieve cases containing the questions for a new checklist (solution).

From a technical perspective, the use of augmented cases is a key novelty of BCBR that can be viewed as a data-driven approach that uses feature construction to embed solution knowledge in cases for case retrieval in CBR [8,15,18]. The use of BI to estimate probability ensures transparency because the estimates are made by counting cases in the data set. The use of similarity based retrieval also promotes trustworthiness and ensures justification of the BI estimates because they are related to past cases. Trustworthiness is important to ensure user compliance with the checklists. The core contributions of this paper are:

- We introduce a formal definition of the Checklist Construction Problem and a new data set of previously used questions (control points) collected from NLIA's labour inspections between 2012 and 2019.
- We present the details for BCBR, which is designed for constructing checklists based on CBR and Bayesian inference.
- We establish an approach for evaluating the checklists constructed by BCBR. The framework is then empirically compared to baselines. The results show that BCBR constructs more efficient checklists than the baselines.

## 2   Related Work

**Hybrid Frameworks Based on CBR and BI.** There are multiple examples of frameworks with combinations of CBR and BI to address uncertainty for applications where some prior belief or information is available. Such frameworks also provide explanations, where CBR has been used to achieve explanation goals [22] (such as transparency and justification) or generate explanations [19]. Nikpour et al.[18] use Bayesian posterior distributions to modify or add features to input case descriptions to increase accuracy of similarity assessments in case retrieval. They also use the same approach to provide explanations for case failures in different domains [17]. This approach is similar to BCBR, but BCBR constructs new features which are also added to the case base-cases rather than modifying input cases. Kenny et al. [12] also use a combination of BI and CBR to exclude outlier cases from case retrieval and to provide explanations by examples. The purpose of the framework is to predict grass growth for sustainable dairy farming. Gogineni et al. [9] combines CBR and BI to retrieve and down-select explanatory cases for underwater mine clearance.

**Similarity Based Retrieval for Trustworthiness.** Lee et al. [13] replaced the output layer of a neural network with $k$-nearest neighbour (kNN) to generate voted predictions and find the nearest neighbour cases to explain the predictions. This also guarantees that every prediction can be justified by a relevant past explanatory case. The justification via explanatory cases increases the reliability of the neural network predictions and promotes trustworthiness. BCBR is also based on the same principle where BI predictions are justified by being embedded in past cases as features.

**Trustworthy Case-Based Recommender Systems.** BCBR aims to select a subset of all possible questions for a new checklist. Similarily, in recommender systems, a user is recommended a subset of items from the space of all possible items. Such systems can be divided into two classes: collaborative and case-based (content or user-based) recommender systems [3], where the latter approach could relate to our work. The case-based approach has been used to predict running-paces for different stages in ultra races, based on cases from similar runners in past cases [16]. CBR has also been used to provide explanatory cases for black-box recommender systems to achieve justification [4,10]. Explanations for such systems can also be created through relations between features (concepts) [11]. However, the quality of explanations for black-box systems in terms of transparency, interpretability and trustworthiness can still be questionable [20]. Some authors also suggest to avoid explainable black box models in cases where they are not needed [21] and to use transparent, interpretable models for high-stakes decision making [20].

# 3   Case and Problem Definition

In this section we introduce the formal case and problem definition used for the rest of the paper.

**Data Set and Cases.** A data set $\mathcal{D}$ for variables $\mathbf{Z}$ is a finite length tuple where a case $\mathbf{d}_j \in \mathcal{D}$ is an instantiation of $\mathbf{Z}$ [6]. A case is a tuple $\mathbf{d} = (e, \mathbf{x}, l)$ where $e$ denotes a question from a checklist, $\mathbf{x}$ is an entity and $l \in \{0,1\}$ denotes the answer of the question. A case in the data set is a past experience where a question $e$ has been applied to $\mathbf{x}$ to obtain the answer $l$. A case description is shown in Table 1.

**Table 1.** Description of a case in the data set

| Name | Description | Type |
|------|-------------|------|
| $x_{isc}$ | Industry subgroup code | Ordinal |
| $x_{igc}$ | Industry group code | Ordinal |
| $x_{ic}$ | Industry code | Ordinal |
| $x_{iac}$ | Industry area code | Ordinal |
| $x_{imac}$ | Industry main area code | Nominal |
| $x_{mnr}$ | Municipality number | Ordinal |
| $x_{fyl}$ | Fylke (county) | Nominal |
| $e$ | Question | Nominal |
| $l$ | Non-compliance | Binary |

**Entity.** Every case $d$ in the data set contains an entity description in the form of an organisation $\mathbf{x}$, defined by its location and industry. The features are organised according to Fig. 2. An organisation can be implicitly defined as $\mathbf{x} = (x_{mnr}, x_{isc})$, since the other features of $\mathbf{x}$ are located higher in the hierarchies.

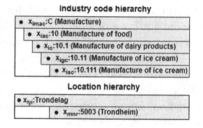

**Fig. 2.** Industry and location hierarchies of an organisation

**Question.** Each case in the data set contains a question (control point) $e$ with a yes/no answer. The question is used to survey the entity $\mathbf{x}$ in the case. A specific question can appear in multiple checklists.

**Checklist.** A checklist $\mathbf{y}$ is defined as a set of yes/no questions constituted by cases in the data set, so that $\mathbf{y} = (e_1 \in \mathbf{d}_1, e_2 \in \mathbf{d}_2 ... e_{nd} \in \mathbf{d}_{nd})$. A question can only appear once per checklist such that $e_i \neq e_j$ for every $e_i \wedge e_j \in \mathbf{y}$.

**Answer.** The label $l$ of a case is the observed answer from applying the question $e$ to the entity $\mathbf{x}$. The answer $l = 1$ means that non-compliance has been found, while $l = 0$ means that $\mathbf{x}$ is compliant.

**The Checklist Construction Problem.** The problem is shown on Fig. 3. Let there be a set of $N$ unique questions and a new target entity $\mathbf{x}^{cnd}$. Each question has an unobserved answer $l$ about $\mathbf{x}^{cnd}$ that belongs to an unknown distribution. Given the $N$ questions, a model $\mathcal{M}$ first needs to correctly estimate the probability for observing $l = 1$ for each question. $\mathcal{M}$ then needs to select $K$ unique questions $(e_1, e_2, ..., e_K)$, with the highest esti-

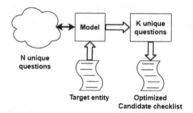

**Fig. 3.** An overview of CCP.

mated probability, for a candidate checklist $\mathbf{y}^{cnd}$. The goal is to observe as many $l = 1$ answers as possible when applying $\mathbf{y}^{cnd}$ to $\mathbf{x}^{cnd}$.

## 4   BCBR Framework

An overview of the BCBR framework is shown in Fig. 4. The motivation for the framework is to solve the CCP problem while also ensuring that every question $e_i \in \mathbf{y}^{cnd}$ can be justified by a relevant past experience (see Sect. 5.3). The framework can be described by the following three steps: (1) A naive Bayesian inference method is used to generate two probability estimates ($\theta^{be}_{x_{isc}}$ and $\theta^{be}_{x_{mnr}}$) for every case $\mathbf{d}_j \in \mathcal{D}$. The estimates are generated by counting the cases in the data set with the same question and entity description as $\mathbf{d}_j$. This is done because many of the cases in the data set contains identical questions and/or identical target entities. Using Bayesian inference also ensures transparency for

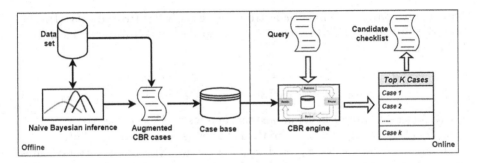

**Fig. 4.** An overview of the BCBR framework. The creation of augmented cases and the case base happens offline. The case base is used for the construction of checklists in the online-part.

the estimates. (2) A case base $\mathcal{CB}$ of augmented CBR cases $c_j$ is created. Each case $c_j \in \mathcal{CB}$ is created by adding both estimates as features to each $d_j \in \mathcal{D}$. (3) A query $q$ is defined, which contains a target entity $\mathbf{x}^{cnd}$ and target values for the probability estimates. The query is used to retrieve a selection of $K$ cases from $\mathcal{CB}$. Each case contains a question $e_i$ for the candidate checklist $\mathbf{y}^{cnd}$.

## 4.1 Bayesian Inference

We use empirical distributions of the data set $\mathcal{D}$ to estimate the probability for observing $l = 1$, to achieve transparency for the BCBR framework. When prior knowledge or belief about $l$ is available, BI can be used instead of the standard maximum likelihood method. An advantage with BI is that it (to some extent) can be used to address inaccurate empirical estimates caused low or zero case counts ("Zero count problem") [6]. The problem may have a negative impact on the quality of the $K$ answers selected by BCBR. To further deal with this problem we use Naive Bayesian inference (NBI) which generates two probability estimates instead of just one. A derivation for this follows below.

**Estimating the Empirical Probability for Non-compliance (1).** By using the definitions from Sect. 3, the empirical distribution of the data set $\mathcal{D}$ can be defined as:

$$\theta_D (\alpha) = \frac{\mathcal{D}\#(\alpha)}{\mathcal{N}} \tag{1}$$

where $\mathcal{D}\#(\alpha)$ is the number of cases in the data set $\mathcal{D}$ which satisfy the event $\alpha$ and $\mathcal{N}$ is the number of cases in $\mathcal{D}$ [6]. We denote the event $L = 1$ as observing the outcome $l = 1$ and $L = 0$ for $l = 0$. From the expression above, the probability for $L = 1$ can then be calculated given $\mathbf{x}$ and $e$:

$$\theta_D (L = 1|\alpha) = \frac{\theta_D (L = 1 \wedge \alpha)}{\theta_D (\alpha)} = \frac{\mathcal{D}\#(L = 1 \wedge \mathbf{X} = \mathbf{x} \wedge E = e)}{\mathcal{D}\#(\mathbf{X} = \mathbf{x} \wedge E = e)} \tag{2}$$

where $\alpha = (\mathbf{X} = \mathbf{x}) \wedge (E = e)$. That is, the event where the entity description is given as $\mathbf{x}$ and the question is given as $e$.

**Naive Bayesian Inference for Estimating Empirical Probability (1).** The posterior probability for an event $L = 1|\alpha$ can be expressed as the mean of a Beta distribution [6]:

$$\theta^{be}(L = 1|\alpha) = \frac{\mathcal{D}\#(L = 1 \wedge \alpha) + \psi_{L=1|a}}{\mathcal{D}\#(L = 1 \wedge \alpha) + \psi_{L=1|a} + \mathcal{D}\#(L = 0 \wedge \alpha) + \psi_{L=0|\alpha}} \tag{3}$$

where $\psi$ is a set of prior belief parameters and where $(\mathcal{D}\#(L = 1 \wedge \alpha) + \psi_{L=1|a})$ and $(\mathcal{D}\#(L = 0 \wedge \alpha) + \psi_{L=0|a})$ are the parameters for a Beta distribution.

From the components $x_{isc}$ and $x_{mnr}$ of $\mathbf{x}$, two NBI probability estimates $\theta^{be}_{x_{isc}}$ and $\theta^{be}_{x_{mnr}}$ can be obtained from Eq. 3 by substituting $\alpha$: $\theta^{be}_{x_{isc}} = \theta^{be}(L = 1|(X_{isc} = x_{isc} \wedge E = e))$ and $\theta^{be}_{x_{mnr}} = \theta^{be}(L = 1|(X_{mnr} = x_{mnr} \wedge E = e))$. Using two probability estimates instead of one is an effective measure against low case counts because $\mathcal{D}\#(X_{isc} = x_{isc} \wedge E = e) \geq \mathcal{D}\#(\mathbf{X} = \mathbf{x} \wedge E = e)$ and $\mathcal{D}\#(X_{mnr} = x_{mnr} \wedge E = e) \geq \mathcal{D}\#(\mathbf{X} = \mathbf{x} \wedge E = e)$. The approach is "naive" since it assumes that $x_{mnr}$ and $x_{isc}$ are independent given $l$ and $e$.

### 4.2   Case Base Creation and CBR Engine

This section defines the details for the augmented CBR cases, case base and similarity based retrieval from Fig. 4.

**Augmented CBR Case and Case Base.** Algorithm 1 shows the creation of a case base $\mathcal{CB}$ with augmented cases c. The algorithm includes two additional features: $\kappa_{x_{mnr}}$ and $\kappa_{x_{isc}}$. The features are included to adjust for the case counts of the probability estimates when retrieving cases. The values for the $\theta^{be}$ and the $\kappa$-features are estimated from $\mathcal{D}$, given $x_{mnr,j}$, $x_{isc,j}$ and $e_j$ from $\mathbf{d}_j \in \mathcal{D}$. The features are added to $\mathbf{d}_j$ to form a case $\mathbf{c}_j$ for $\mathcal{CB}$. An example showing the specific features of the augmented cases can be found in Sect. 4.3.

---

**Algorithm 1.** Creation of a case base $\mathcal{CB}$ with cases $\mathbf{c}_j$

---

**Input:** $\mathcal{D}$;
**Output:** $\mathcal{CB} \leftarrow ()$;
**for each** $\mathbf{d}_j \in \mathcal{D}$ **do**
  $//(x_{isc,j}, x_{mnr,j}, e_j) \in \mathbf{d}_j$
  $\theta^{be}_{x_{isc}} \leftarrow \theta^{be}(L = 1|(x_{isc,j}, e_j))$;
  $\theta^{be}_{x_{mnr}} \leftarrow \theta^{be}(L = 1|(x_{mnr,j}, e_j))$;
  $\kappa_{x_{mnr}} \leftarrow \mathcal{D}\#(L = 1 \wedge X_{mnr} = x_{mnr,j} \wedge E = e_j)$;
  $\kappa_{x_{isc}} \leftarrow \mathcal{D}\#(L = 1 \wedge X_{isc} = x_{isc,j} \wedge E = e_j)$;
  $\mathbf{c}_j \leftarrow Join(\mathbf{d}_j, \theta^{be}_{x_{mnr}}, \theta^{be}_{x_{isc}}, \kappa_{x_{mnr}}, \kappa_{x_{isc}})$;
  $\mathcal{CB} \leftarrow Join(\mathcal{CB}, \mathbf{c}_j)$;
**end for**
**return** $\mathcal{CB}$;

---

**Case Retrieval and Similarity Function.** To retrieve questions $e_i$ for the candidate checklist $\mathbf{y}^{cnd}$, a query case $\mathbf{q}$ and similarity function is used. The query consists of the target entity $\mathbf{x}^{cnd}$ and the desired values for both the probability estimates and the case count features. A similarity function assigns a score $Sim(\cdot, \cdot) \in [0, 1]$ to every pair $(\mathbf{q}, \mathbf{c}_j \in \mathcal{CB})$. A set of unique $e_i$ for $\mathbf{y}^{cnd}$ is then retrieved from the $K$ cases with the highest similarity score. The similarity function is defined according to the equation below:

$$Sim(\mathbf{q}, \mathbf{c}_j) = \frac{1}{\sum w_i} \sum_i w_i \cdot sim_i(\mathbf{q}, \mathbf{c}_j). \tag{4}$$

Where $w_i$ is a weight, $sim_i$ is a local similarity function and $i$ denotes a feature common to the query and the case. Each local similarity function in Eq. (4), yields a score $[0, 1]$ for each feature $(i)$ according to the similarity $sim_i(\mathbf{q}, \mathbf{c}_j)$ between the cases $\mathbf{q}$ and $\mathbf{c}_j$. The local similarity functions and the weights are defined by a domain expert for the purpose of this work (see Sect. 5.1).

### 4.3   Example: NBI Estimates, Case Retrieval and CBR Case

**NBI Estimates.** Let $x_{isc} = 22.230$, $x_{mnr} = 1507$ be features of an entity description $\mathbf{x}$ and $e =$"Did the employer make sure to equip all employees who carry out work at the construction site with a HSE card?" be a question of a case $\mathbf{d} \in \mathcal{D}$. The prior parameters are $\psi_{L=1|\alpha} = 1$ and $\psi_{L=0|\alpha} = 5$ because $l = 1$ is observed in approximately 1 of 6 cases. Given this information, $\theta^{be}_{x_{isc}}$ is estimated by counting cases $\mathbf{d}$ in data set $\mathcal{D}$ which satisfy $X_{isc} = x_{isc}$ and $E = e$. Applying $\alpha = (X_{isc} = x_{isc} \wedge E = e)$ to Eq. 3 yields: $\theta^{be}_{x_{isc}} = \frac{1+1}{1+2+6} \approx 22\%$.

This estimate is more accurate than the empirical probability estimate, which is $\theta_{x_{isc}} = \frac{1}{1+2} \approx 33\%$ (Eq. 2). The difference can be explained by low case count, which affect the quality of both the Bayesian and empirical estimates.

The same procedure is used to calculate: $\theta^{be}_{x_{mnr}} = \frac{89+1}{89+186+6} \approx 32\%$. In this case the Bayesian estimate is approximately the same as the empirical probability estimate, since the case count is high. The estimates are used to create an augmented CBR case $\mathbf{c}$.

**Case Retrieval and Augmented CBR Case.** For this example we assume that a case base of CBR cases has been created and that $K = 1$, for the sake of brevity. The case retrieval starts by defining a query case (Query 1), shown in Table 2. $\theta^{be}_{x_{isc}}$ and $\theta^{be}_{x_{mnr}}$ are set to 100%, which is the target value for the retrieved cases. Both $\kappa_{x_{isc}}$ and $\kappa_{x_{mnr}}$ are set to 70 so that case counts of 70 or higher yield full similarity scores, according to Fig. 5.

**Table 2.** Description of case features, similarity weights, query and retrieved case for the example.

| Feature | w | Query 1 | Case 1 | Query 2 | Case 2 |
|---|---|---|---|---|---|
| $x_{isc}$ | 1 | 22.230 | 22.230 | 22.230 | 22.230 |
| $x_{igc}$ | 2 | 22.23 | 22.23 | 22.23 | 22.23 |
| $x_{ic}$ | 2 | 22.2 | 22.2 | 22.2 | 22.2 |
| $x_{iac}$ | 2 | 22 | 22 | 22 | 22 |
| $x_{imac}$ | 2 | C | C | C | C |
| $x_{mnr}$ | 2 | 1507 | 1507 | 1507 | 1507 |
| $x_{fyl}$ | 2 | MoM | MoM | MoM | MoM |
| $l$ | 0 | – | 0 | – | 0 |
| $e$ | 0 | – | $e_1$ | – | $e_2$ |
| $\theta^{be}_{x_{isc}}$ | 9 | 100% | 22% | – | 7% |
| $\theta^{be}_{x_{mnr}}$ | 4 | 100% | 32% | – | 7% |
| $\kappa_{x_{isc}}$ | 1 | 70 | 1 | – | 0 |
| $\kappa_{x_{mnr}}$ | 1 | 70 | 89 | – | 30 |
| $Sim$ | | – | 0.546 | – | 0.448 |

After applying the similarity function to every pair $(\mathbf{q}, \mathbf{c} \in \mathcal{CB})$, the top $K = 1$ case with highest similarity (Case 1) is retrieved for the candidate checklist $\mathbf{y}^{cnd}$.

For comparison, we also define Query 2 in Table 2 where $\theta^{be}_{x_{isc}}$, $\theta^{be}_{x_{mnr}}$, $\kappa_{x_{isc}}$ and $\kappa_{x_{mnr}}$ are undefined. The $K = 1$ case returned from $\mathcal{CB}$ is Case 2. Case 2 fully matches Query 2 in terms of $\mathbf{x}$, but $\theta^{be}_{x_{isc}}$ and $\theta^{be}_{x_{mnr}}$ sug-

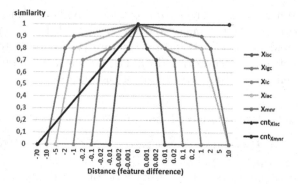

**Fig. 5.** Local similarity functions.

gest that it is unlikely to observe $l = 1$ when $e_2$ is applied to $\mathbf{x}$. This is expected because we removed the part of the query that maximizes the probability for observing $l = 1$.

## 5   Experiments

In this section three experiments are presented. In the first experiment a simple label classification problem is introduced to establish a starting point for comparing ML methods as baselines for the labour inspection CCP. The second experiment aims to measure the justification of checklists constructed by BCBR and the two best-performing baselines from the first experiment. The third experiment aims to measure the performance of BCBR against the baselines from the second experiment.

### 5.1   Experimental Setup

**Measure of Justification.** We introduce Eq. 5 to measure the justification $(J \in [0, 100\%])$ of a checklist $\mathbf{y}$ for a given entity $\mathbf{x}$, according to the proportion of questions $e_i \in \mathbf{y}$ which also exist in past cases $(e_i, \mathbf{x}, \cdot) \in \mathcal{D}$.

$$J(\mathbf{y}, \mathbf{x}, \mathcal{D}) = \frac{|\{e_i \in \mathbf{y} : (e_i, \mathbf{x}, \cdot) \in \mathcal{D}\}|}{|\{e_i \in \mathbf{y}\}|} \tag{5}$$

The expression can be seen as an adaptation of Massie alignment score [14] that measures the percentage of questions $e_i \in \mathbf{y}$ with full alignment to the nearest neighbour case in $\mathcal{D}$.

**BCBR Configuration.** For the experiments, BCBR uses the same configuration as in Sect. 4.3. The only difference is that $K = 15$ is used instead of $K = 1$, so that the constructed checklists consist of 15 questions.

The weights and local similarity functions are set based on domain knowledge and are shown in Table 2 and Fig. 5 respectively. The weights are set according to the importance of each feature, while the similarity functions are defined to model the similarity according to the hierarchical relationship between the ordinal features of the entity $\mathbf{x}$ (see Sect. 3). For the other features not shown in Fig. 5, the default option in the myCBR tool is used to define the local similarity functions.

**Baselines for the Experiments.** The baseline methods used for the experiments are: CBR (CBR-BL), Logistic Regression (LR), Decision tree (DT) and Naive Bayes classifier (NBC), Conditional probability estimates (CP), Bayesian inference (BI), Naive conditional probability (NCP) and NBI.

CBR-BL generates predictions from the label of the closest neighbour case in the training data. CP generates predictions for any pair $(e, \mathbf{x})$ according to Eq. 2. BI uses Eq. 3 with $\psi_{L=1|\alpha} = 1$, $\psi_{L=0|\alpha} = 5$ and $\alpha = (\mathbf{X} = \mathbf{x} \wedge E = e)$. NCP is based on Eq. 2 and is defined as: $\theta\left(L = 1|e, \mathbf{x}\right) = \frac{\theta_{x_{isc}} + \theta_{x_{mnr}}}{2}$. The baseline NBI estimates are calculated using $\psi_{L=1|\alpha} = 1$ and $\psi_{L=0|\alpha} = 5$ according to:

$$\theta\left(L = 1|e, \mathbf{x}\right) = \frac{\theta_{x_{isc}}^{be} + \theta_{x_{mnr}}^{be}}{2}.$$

**Environment.** A Dell XPS 9570 with Intel i9 8950hk, 32 GB RAM and Windows 10 were used for the experiments. Every experiment is conducted in a Python environment using Jupyter Notebook. NBI for BCBR, NBI, BI, CP and NCP are implemented as MSSQL17 queries via PYODBC. The similarity based retrieval for BCBR and CBR-BL are implemented via MyCBR [1]. The rest of the methods are implemented via Scikit-learn 0.24.

**Data Set.** For the experiments we introduce a new data set of questions used in previous inspections conducted by NLIA.[3] The data set is denoted as $\mathcal{D}$ for the rest of this section and consists of 1,111,502 entries from inspections conducted between 01/01/2012 and 01/06/2019. Embedded in these entries are $N = 1,967$ unique questions from checklists used in 59,988 inspections. Each entry (case) in $\mathcal{D}$ is also associated with an $id^4$ which maps to a checklist $\mathbf{y}$ (past solution) used to survey the organisation $\mathbf{x}$ in one of the 59,988 inspections within $\mathcal{D}$.

## 5.2   Experiment 1: Answer Classification Performance (Baselines)

The goal of this experiment is to compare ML methods and select two of the best as baselines for the labour inspection CCP. Because CCP is a complex problem, we here study a new, simple classification problem as a stepping stone.

**The Answer Classification Problem.** Let each $\mathbf{d}_j \in \mathcal{D}$ be a case with a two-class ground truth label $l_j$. A model $\mathcal{M}$ is trained on the cases in $\mathcal{D}$. For any new case $\mathbf{d} = (e, \mathbf{x}, l)$ where $l = 0$ (compliance) or $l = 1$ (non-compliance), the problem goal is for $\mathcal{M}$ to correctly classify the value of $l$ based on $(e, \mathbf{x})$.

**Method.** Each model is validated on the data set $\mathcal{D}$, using 8-fold cross validation with the same partitioning of data for every model. Each model $\mathcal{M}$ outputs a

---

[3] The data set is available at https://dx.doi.org/10.21227/m1t7-hg51.
[4] The $id$ is a "key" for identifying a past checklist/organisation pair (value) in $\mathcal{D}$.

class prediction score for every $(e, \mathbf{x})$. Thus, the classification threshold is set to the median of $\mathcal{M}$'s scores for each validation fold. The results are measured in terms of accuracy, precision and recall which are calculated for per validation fold: $Acc = \frac{TP+TN}{TP+FP+TN+FN}$, $Prec = \frac{TP}{TP+FP}$ and $Rec = \frac{TP}{TP+FN}$.

**Results and Discussion.** The results are shown in Table 3 where the baselines are sorted according to $Avg$, which is the average score of the preceding columns. In terms of the $Avg$-score NBI performs better then standard ML methods such as LR, DT and NBC. NBI also has the best recall and an average runtime of 10.4 s per validation fold, which is significantly less than NBC, DT, LR and CBR-BL. BI has the best performance in terms of accuracy and precision, but it also has poor recall which results in a low aver-

**Table 3.** Results from the experiment. Time is measured in seconds per validation fold.

| Method | Acc | Prec | Rec | Avg | Time |
|--------|------|-------|-------|-------|-------|
| CBR-BL | 0.677 | 0.178 | 0.246 | 0.367 | 60238 |
| Random | 0.500 | 0.161 | 0.500 | 0.387 | – |
| CP | 0.680 | 0.210 | 0.357 | 0.416 | **3.84** |
| BI | **0.760** | **0.270** | 0.288 | 0.439 | 3.89 |
| DT | 0.644 | 0.233 | 0.529 | 0.469 | 122.6 |
| NCP | 0.592 | 0.250 | 0.761 | 0.534 | 9.0 |
| NBC | 0.588 | 0.251 | 0.778 | 0.539 | 67.33 |
| LR | 0.591 | 0.252 | 0.782 | 0.542 | 68.4 |
| NBI | 0.605 | 0.261 | **0.790** | **0.552** | 10.4 |

age score. The worst performing method was CBR-BL where the size of the training data was reduced to 100,000 cases due to long running time.

The results indicate that NBI yields the best average performance, which motivates us to combine NBI with CBR. LR, NBC and NCP also perform well, but we select NBI and LR as baselines for the next experiments. A limitation for this experiment is that it cannot be used to evaluate BCBR, as BCBR is designed for CCP and not ACP.

### 5.3 Experiment 2: Trustworthiness of Constructed Checklists

The goal of this experiment is to measure justification of constructed checklists $\mathbf{y}^{cnd}$ for the CCP. This is done by measuring the average proportion of questions $e_i \in \mathbf{y}^{cnd}$ which are justified by past cases. The experiment is based on Lee et al.'s use of past cases to justify predictions and promote trust [13]. The experiment is conducted on checklists constructed by BCBR and two of the baselines from Sect. 5.2, NBI and LR.

**Method.** Each model $\mathcal{M}$ is trained on the data set $\mathcal{D}$ containing 1,111,502 entries. An evaluation data set $\mathcal{D}_V$ of 59,988 tuples $(\mathbf{x}^{cnd}, \mathbf{y})$ of past entity/checklist pairs is created using every unique $id$ from $\mathcal{D}$. For each $\mathbf{x}^{cnd} \in \mathcal{D}_V$, $\mathcal{M}$ constructs a checklist $\mathbf{y}^{cnd}$ for $\mathbf{x}^{cnd}$ as following depending on the model in question. For $\mathcal{M} = NBI$ or $\mathcal{M} = LR$: $\mathcal{M}$ generates a prediction score for every unique $e_j \in \mathcal{D}$. The $K = 15$ questions with the highest prediction scores are selected as the candidate checklist $\mathbf{y}^{cnd}$ for $\mathbf{x}^{cnd}$. For $\mathcal{M} = BCBR$: a query containing $\mathbf{x}^{cnd}$ is defined to retrieve past cases, containing $K = 15$ unique questions for $\mathbf{y}^{cnd}$.

Each $\mathbf{y}^{cnd}$ constructed by one of the models $\mathcal{M}$ then forms an evaluation pair $(\mathbf{y}^{cnd}, \mathbf{x}^{cnd})$ with each corresponding $\mathbf{x}^{cnd}$ from $\mathcal{D}_V$. Based on Eq. 5, the average justification $(J_\mathcal{M})$ for every pair $(\mathbf{y}^{cnd}, \mathbf{x}^{cnd})$ given $\mathcal{M}$ is:

$$J_{\mathcal{M}}(\mathcal{D}, \mathcal{D}_V) = \frac{\sum_{(\mathbf{y}^{cnd}, \mathbf{x}^{cnd})} J(\mathbf{y}^{cnd}, \mathbf{x}^{cnd}, \mathcal{D})}{|\mathcal{D}_V|} \tag{6}$$

$J_{\mathcal{M}}$ measures the average percentage of questions $e_i \in \mathbf{y}^{cnd}$ where at least one corresponding explanatory case $(e_i, \mathbf{x}^{cnd}, \cdot)$ exists in $\mathcal{D}$. The purpose of the $J_{\mathcal{M}}$ score is to enable a fair comparison between the three models. A higher relative score means higher justification of the checklists constructed by $\mathcal{M}$.

**Results and Discussion.** The results are: $J_{NBI} = 0.6\%$, $J_{LR} = 4.8\%$ and $J_{BCBR} = 64\%$. This suggests that both LR and NBI perform poorly in terms of justification of their constructed checklists. Qualitative assessments of some of the checklists also reveal that many of their questions $(e_i \in \mathbf{y}^{cnd})$ are unrelated to and incompatible with the target entities. Because of the incompatibility issues and that less than 5% of the items on the checklists are justified, LR and NBI are not trustworthy. BCBR scored 64% which is significantly higher. Incompatible questions also seam to appear less frequently in BCBR's checklists.

## 5.4   Experiment 3: Evaluation of Constructed Checklists

The goal of this experiment is to evaluate the performance of the BCBR framework against LR, NBI and the original past checklists from the data set. Since BCBR uses similarity based retrieval, NBI and LR serve as non-similarity based baselines to compare with. Due to the results in Sect. 5.3, a filter is applied to both LR and NBI to ensure that every checklist can be justified by past cases. This is necessary for the evaluation procedure, as it assumes that the questions on the checklists can be justified by past similar cases.

**Method.** The evaluation approach is done on the data set $\mathcal{D}$ which contains 1,111,502 entries. The approach can be summarized as following: The data set $\mathcal{D}$ is partitioned into a training fold $(\mathcal{D}_T)$ and validation fold $(\mathcal{D}_{CB})$, where the training fold is used to calculate probability estimates for the validation cases. The validation fold is used as the case base and for performance evaluation. A model $\mathcal{M}$ is trained on $\mathcal{D}_T$ and the evaluation is done on every checklist $\mathbf{y}^{cnd}$ constructed by $\mathcal{M}$.

A problem with the validation is that since every $\mathbf{y}^{cnd}$ is a new checklist, the ground truths $l$ needed to evaluate $\mathbf{y}^{cnd}$ can be missing. A common solution to this problem is to collect the ground truth empirically [23], but this is not an option for us. To get a meaningful validation result, the performance statistics for the evaluation need to be estimated. To accomplish this, the following assumption is made: Let $\mathbf{d}^{cnd} = (-, \mathbf{x}^{cnd}, -)$ be a case without question component or observed ground truth answer and $\mathbf{d} = (e, \mathbf{x}, l)$ be any validation case with ground truth. If $\mathbf{x}^{cnd}$ and $\mathbf{x}$ are content-wise equal or similar, we assume that the unobserved ground truth answer $l^{cnd}$ from applying $e$ to $\mathbf{x}^{cnd}$ is correctly estimated from an empirical distribution of $l$, conditioned on $\mathbf{x}, e$ and the validation data fold. This is based on the assumption that similar problems have similar solutions [15].

Based on the assumption, we introduce the following procedure to estimate accuracy (Acc), precision (Prec)[5] and recall (Rec) for every model $\mathcal{M}$.

1. Let $\mathcal{D}_T$ be the training fold and $\mathcal{D}_{CB}$ be both the validation fold and case base(for BCBR). Let $\mathcal{D}_V$ be a set of past entity/checklist pairs $(\mathbf{x}^{cnd}, \mathbf{y})$ from $\mathcal{D}_{CB}$, created using every unique $id$ in $\mathcal{D}_{CB}$. A model $\mathcal{M}$ is trained on $\mathcal{D}_T$.

2. For every $\mathbf{x}^{cnd} \in \mathcal{D}_V$, $\mathcal{M}$ selects $K$ unique questions $(e_i)$ for a checklist $\mathbf{y}^{cnd}$ to form a validation pair $(\mathbf{x}^{cnd}, \mathbf{y}^{cnd})$. The questions are selected from $\mathcal{D}_{CB}$.

3. For each pair $(\mathbf{x}^{cnd}, \mathbf{y}^{cnd})$ the number of true positives $(TP)$, false positives $(FP)$, true negatives $(TN)$ and false negatives $(FN)$ are estimated by evaluating each $e_i \in \mathbf{y}^{cnd}$(predicted positives) and $e_j \notin \mathbf{y}^{cnd}$(predicted negatives).

4. For every question $e_i \in \mathbf{y}^{cnd}$, both $TP_{e_i}$ and $FP_{e_i}$ are estimated using the following function: $f(l, \mathbf{x}_0, e_i) = \frac{\mathcal{D}_{CB}\#(L=l \wedge \mathbf{X}=\mathbf{x}_0 \wedge E=e_i)}{\mathcal{D}_{CB}\#(\mathbf{X}=\mathbf{x}_0 \wedge E=e_i)}$, so that $TP_{e_i} = f(1, \mathbf{x}_0, e_i)$ and $FP_{e_i} = f(0, \mathbf{x}_0, e_i)$. If $\mathcal{D}_{CB}\#(\mathbf{X} = \mathbf{x}^{cnd} \wedge E = e_i) > 0$, then $\mathbf{x}_0 = \mathbf{x}^{cnd}$ is applied to $f$. If $\mathcal{D}_{CB}\#(\mathbf{X} = \mathbf{x}^{cnd} \wedge E = e_i) = 0$, then $\mathbf{x}_0 = \mathbf{x}_i$ from the case $(e_i, \mathbf{x}_i, l_i)$, retrieved by BCBR[6] for $\mathbf{y}^{cnd}$, is used because there is no data to evaluate $(e_i, \mathbf{x}^{cnd})$. Each $TP_{e_i}$ and $FP_{e_i}$ is assigned a value $v \in [0, 1]$ via $f$ so that $TP_{e_i} = 1 - FP_{e_i}$.

5. For every unique question $e_j \notin \mathbf{y}^{cnd}$ in $\mathcal{D}_{CB}$, both $TN_{e_j}$ and $FN_{e_j}$ are estimated using the function: $g(l, e_j \notin \mathbf{y}^{cnd}) = \frac{\mathcal{D}_{CB}\#(L=l \wedge \mathbf{X}=\mathbf{x}^{cnd} \wedge E=e_j)}{\mathcal{D}_{CB}\#(\mathbf{X}=\mathbf{x}^{cnd} \wedge E=e_j)}$. The function is used to obtain $TN_{e_j} = g(0, e_j)$ and $FN_{e_j} = g(1, e_j)$, so that each $TN_{e_j}$ and $FN_{e_j}$ receives a value of $v \in [0, 1]$ and that $TN_{e_j} = 1 - FN_{e_j}$.

6. $TP$, $FP$, $FN$ and $TN$ for each candidate checklist $\mathbf{y}^{cnd} \in (\mathbf{x}^{cnd}, \mathbf{y}^{cnd})$ are calculated as following: $TP = \sum_{e_i} TP_{e_i}$, $FP = \sum_{e_i} FP_{e_i}$, $TN = \sum_{e_j} TN_{e_j}$ and $FN = \sum_{e_j} FN_{e_j}$ for every unique $e_i \in \mathbf{y}^{cnd}$ and $e_j \notin \mathbf{y}^{cnd}$ from $\mathcal{D}_{CB}$.

7. Statistics are then calculated for each $\mathbf{y}^{cnd}$: $Acc_{\mathbf{y}^{cnd}} = \frac{TP+TN}{TP+FP+TN+FN}$, $Prec_{\mathbf{y}^{cnd}} = \frac{TP}{TP+FP}$ and $Rec_{\mathbf{y}^{cnd}} = \frac{TP}{TP+FN}$. Repeat from Step 2 until every pair $(\mathbf{x}^{cnd}, \mathbf{y}^{cnd})$ is evaluated.

8. The average Acc, Prec and Rec of every checklist $\mathbf{y}^{cnd}$ constructed by $\mathcal{M}$ is: $Acc = \frac{\sum_{\mathbf{y}^{cnd}} Acc_{\mathbf{y}^{cnd}}}{|\mathcal{D}_V|}$, $Prec = \frac{\sum_{\mathbf{y}^{cnd}} Prec_{\mathbf{y}^{cnd}}}{|\mathcal{D}_V|}$ and $Rec = \frac{\sum_{\mathbf{y}^{cnd}} Rec_{\mathbf{y}^{cnd}}}{|\mathcal{D}_V|}$.

The procedure is used to evaluate BCBR and the other baselines. To evaluate the original checklists, the procedure is applied to the past checklists in the validation fold so that $\mathbf{y}^{cnd} = \mathbf{y}$ for $\mathbf{y} \in \mathcal{D}_V$ in Step 2. Step 2 for NBI and LR is done by generating predictions for every unique question (see Sect. 5.3). Then a filter is applied after prediction and before the selection of the questions for $\mathbf{y}^{cnd}$. The filter excludes any question $(e)$ from selection if $(e, \mathbf{x}^{cnd}, \cdot) \notin \mathcal{D}_{CB}$. This means that every $e_i \in y^{cnd}$ is justified by a past case so that $J_{NBI}$ and $J_{LR}$ is 100% (Eq. 6). The filter is necessary for the evaluation to ensure that NBI and LR construct checklists that satisfy the assumption above. The models use $K = 15$ and are validated using 4,8 and 16-fold cross validation.

---

[5] An additional statistic Prec(gt) is included, which is precision calculated (step 4–8) using only $e_i \in \{\mathbf{y}^{cnd} \cap \mathbf{y}\}$ from cases containing the original ground truth labels.

[6] The condition $\mathcal{D}_{CB}\#(\mathbf{X} = \mathbf{x}^{cnd} \wedge E = e_i) = 0$ only occurs if BCBR is used.

**Results and Discussion.** The results are shown in Table 4. The *Avg* column shows the average of the four preceding columns, where the results suggest that the checklists constructed by NBI, LR and BCBR are more effective than the original checklists. BCBR scores 0.474 which is significantly higher than the original

**Table 4.** 8 fold cross validation results of the constructed vs. the original checklists (Org. CL).

| Method | Acc | Prec (gt) | Prec | Rec | *Avg* |
|--------|-----|-----------|------|-----|-------|
| Org. CL | 0.337 | 0.170 | 0.181 | 0.622 | 0.328 |
| LR | 0.484 | 0.226 | 0.267 | 0.694 | 0.418 |
| NBI | 0.486 | 0.229 | 0.270 | 0.698 | 0.421 |
| BCBR | **0.574** | **0.259** | **0.343** | **0.718** | **0.474** |

checklists and also higher than NBI and LR. Figure 6 shows the results for different numbers of validation folds. The figure suggests that BCBR consistently outperforms NBI and LR in accuracy and precision. Also, both accuracy and precision statistics tend to increase with the size of the validation data sets. We believe this is caused by the fact that $TP$ and $TN$ increases compared to $FP$ and $FN$ as the quality of the retrieved questions increases when more cases are available. Recall also decreases with the size of the validation data sets as the number of predicted positives is fixed ($K = 15$), which entails that $FN$ increases more than $TP$ when the size of the validation set increases. The experiment suggests that BCBR is more effective for constructing checklists than LR or NBI.

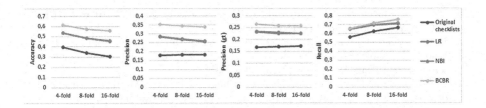

**Fig. 6.** Crossvalidation results for different validation fold sizes

A limitation of this experiment is that the results are based on estimates of *Acc*, *Prec* and *Rec*. For CBR frameworks, the validity of the evaluation results partially depends on high similarity between the **x**-part of the query and retrieved cases. This could be problematic when evaluating and comparing multiple CBR-based frameworks and should be investigated in future work.

## 6  Conclusion

In this paper we studied the problem of constructing checklists for safety critical applications, in particular labour inspections where constructing good high-performance checklists manually is difficult. Thus, we proposed the CCP where we consider the automatic construction of good, justifiable checklists. To address the CCP we introduced BCBR, which uses naive BI to construct features in CBR cases for retrieving questions for the checklists. We conducted three experiments on a data set of past labour inspections, which we introduced for the paper. Because CCP is a fairly complex problem, we conducted our first experiment on a simple answer classification problem. The goal of the experiment was to select two baselines for CCP, which was NBI and LR. In the second experiment we measured the justification of the checklist constructed by BCBR, NBI and LR, where we found that only BCBR constructs checklists which are justified by past cases. Another conclusion from the experiment is that questions selected for the constructed checklists should be justified in terms of prior use in similar entities, because some questions may be closely related to the entities that they originally were designed for. The results from the last experiment also suggest that BCBR is the most effective method for constructing checklists to address poor working conditions in inspected organisations. The checklists constructed by BCBR also perform significantly better than the original checklists.

One of the things that could be addressed in future work is solution adaptation, such as adapting questions after they have been retrieved for a checklist. Another option is to explore data-driven approaches to derive the weights and local functions for BCBR. It could also be interesting to see how BCBR perform in other CCPs such as surgery or preflight checklists.

# References

1. Bach, K., Mathisen, B.M., Jaiswal, A.: Demonstrating the MYCBR rest API. In: ICCBR Workshops, pp. 144–155 (2019)
2. Belser, P.: Forced labour and human trafficking: estimating the profits (2005)
3. Bridge, D., Goker, M.H., McGinty, L., Smyth, B.: Case-based recommender systems. Knowl. Eng. Rev. **20**(3), 315–320 (2005)
4. Caro-Martinez, M., Recio-Garcia, J.A., Jimenez-Diaz, G.: An algorithm independent case-based explanation approach for recommender systems using interaction graphs. In: Bach, K., Marling, C. (eds.) ICCBR 2019. LNCS (LNAI), vol. 11680, pp. 17–32. Springer, Cham (2019). https://doi.org/10.1007/978-3-030-29249-2_2
5. Catchpole, K., Russ, S.: The problem with checklists. BMJ Qual. Saf. **24**(9), 545–549 (2015)
6. Darwiche, A.: Modeling and Reasoning with Bayesian Networks. Cambridge University Press, Cambridge (2009). https://doi.org/10.1017/CBO9780511811357
7. Degani, A., Wiener, E.L.: Human factors of flight-deck checklists: the normal checklist. Ames Research Center (1990)
8. Gabel, T., Godehardt, E.: Top-down induction of similarity measures using similarity clouds. In: Hüllermeier, E., Minor, M. (eds.) ICCBR 2015. LNCS (LNAI), vol. 9343, pp. 149–164. Springer, Cham (2015). https://doi.org/10.1007/978-3-319-24586-7_11
9. Gogineni, V.R., Kondrakunta, S., Brown, D., Molineaux, M., Cox, M.T.: Probabilistic selection of case-based explanations in an underwater mine clearance domain. In: Bach, K., Marling, C. (eds.) ICCBR 2019. LNCS (LNAI), vol. 11680, pp. 110–124. Springer, Cham (2019). https://doi.org/10.1007/978-3-030-29249-2_8
10. Jorro-Aragoneses, J., Caro-Martinez, M., Recio-Garcia, J.A., Diaz-Agudo, B., Jimenez-Diaz, G.: Personalized case-based explanation of matrix factorization recommendations. In: Bach, K., Marling, C. (eds.) ICCBR 2019. LNCS (LNAI), vol. 11680, pp. 140–154. Springer, Cham (2019). https://doi.org/10.1007/978-3-030-29249-2_10

11. Jorro-Aragoneses, J.L., Caro-Martínez, M., Díaz-Agudo, B., Recio-García, J.A.: A user-centric evaluation to generate case-based explanations using formal concept analysis. In: Watson, I., Weber, R. (eds.) ICCBR 2020. LNCS (LNAI), vol. 12311, pp. 195–210. Springer, Cham (2020). https://doi.org/10.1007/978-3-030-58342-2_13

12. Kenny, E.M.: Predicting grass growth for sustainable dairy farming: a CBR system using Bayesian case-exclusion and *Post-Hoc*, personalized explanation-by-example (XAI). In: Bach, K., Marling, C. (eds.) ICCBR 2019. LNCS (LNAI), vol. 11680, pp. 172–187. Springer, Cham (2019). https://doi.org/10.1007/978-3-030-29249-2_12

13. Lee, R., Clarke, J., Agogino, A., Giannakopoulou, D.: Improving trust in deep neural networks with nearest neighbors. In: AIAA Scitech 2020 Forum, p. 2098 (2020)

14. Massie, S., Wiratunga, N., Craw, S., Donati, A., Vicari, E.: From anomaly reports to cases. In: Weber, R.O., Richter, M.M. (eds.) ICCBR 2007. LNCS (LNAI), vol. 4626, pp. 359–373. Springer, Heidelberg (2007). https://doi.org/10.1007/978-3-540-74141-1_25

15. Mathisen, B.M., Aamodt, A., Bach, K., Langseth, H.: Learning similarity measures from data. Progr. Artif. Intell. **9**(2), 129–143 (2019). https://doi.org/10.1007/s13748-019-00201-2

16. McConnell, C., Smyth, B.: Going further with cases: using case-based reasoning to recommend pacing strategies for ultra-marathon runners. In: Bach, K., Marling, C. (eds.) ICCBR 2019. LNCS (LNAI), vol. 11680, pp. 358–372. Springer, Cham (2019). https://doi.org/10.1007/978-3-030-29249-2_24

17. Nikpour, H., Aamodt, A.: Fault diagnosis under uncertain situations within a Bayesian knowledge-intensive CBR system. Progr. Artif. Intell. **10**(3), 245–258 (2021). https://doi.org/10.1007/s13748-020-00227-x

18. Nikpour, H., Aamodt, A., Bach, K.: Bayesian-supported retrieval in BNCreek: a knowledge-intensive case-based reasoning system. In: Cox, M.T., Funk, P., Begum, S. (eds.) ICCBR 2018. LNCS (LNAI), vol. 11156, pp. 323–338. Springer, Cham (2018). https://doi.org/10.1007/978-3-030-01081-2_22

19. Recio-Garcia, J.A., Diaz-Agudo, B., Pino-Castilla, V.: CBR-LIME: a case-based reasoning approach to provide specific local interpretable model-agnostic explanations. In: Watson, I., Weber, R. (eds.) ICCBR 2020. LNCS (LNAI), vol. 12311, pp. 179–194. Springer, Cham (2020). https://doi.org/10.1007/978-3-030-58342-2_12

20. Rudin, C.: Stop explaining black box machine learning models for high stakes decisions and use interpretable models instead. Nat. Mach. Intell. **1**(5), 206–215 (2019). https://doi.org/10.1038/s42256-019-0048-x

21. Rudin, C., Radin, J.: Why are we using black box models in AI when we don't need to? Harv. Data Sci. Rev. **1**(2) (2019)

22. Sørmo, F., Cassens, J., Aamodt, A.: Explanation in case-based reasoning-perspectives and goals. Artif. Intell. Rev. **24**(2), 109–143 (2005). https://doi.org/10.1007/s10462-005-4607-7

23. Wang, C., Agrawal, A., Li, X., Makkad, T., Veljee, E., Mengshoel, O., Jude, A.: Content-based top-n recommendations with perceived similarity. In: 2017 IEEE International Conference on Systems, Man, and Cybernetics (SMC) (2017)

24. Weil, D.: If osha is so bad, why is compliance so good? RAND J. Econ. **27**(3), 620 (1996)

# Revisiting Fast and Slow Thinking in Case-Based Reasoning

Srashti Kaurav[1]($\boxtimes$), Devi Ganesan[1], Deepak P[1,2], and Sutanu Chakraborti[1]

[1] Indian Institute of Technology, Madras, Chennai 600036, TN, India
{cs19s013,gdevi,sutanuc}@cse.iitm.ac.in
[2] Queen's University Belfast, Belfast, UK
deepaksp@acm.org

**Abstract.** A dichotomous Case-Based Reasoning (CBR) model is one in which two kinds of reasoning mechanisms are employed; these may be for realizing fast and slow problem-solving as demanded by the nature of the incoming query. Such dichotomous operation is inspired by Daniel Kahneman's seminal work on the two modes of thinking observed in humans. In this paper, we present the following three directions of refinement for a dichotomous CBR model: selection of attributes for a fast thinking model based on parsimonious CBR, switching from fast to slow thinking based on constraints derived from domain knowledge and arriving at a complexity measure for evaluating dichotomous models. For all the three improvements identified, we discuss the results on real-world data sets and empirically analyse the effectiveness of the same.

**Keywords:** Fast and slow thinking · Dichotomous CBR models · Cognitive CBR

## 1 Introduction

Case-Based Reasoning (CBR) is based on the idea of experiential problem solving, the idea that past problem solving experiences can be reused to solve new problems. CBR has found interesting applications in several real world tasks in domains such as diagnostics, planning, design and configuration [13]. However, it is indeed paradoxical that while CBR was inspired initially by models of human problem solving, many practical CBR realizations have made design choices that considerably compromised on CBR's cognitive appeal. In this paper, we attempt to explore avenues to realize CBR in ways that can mirror the dichotomy between *slow* and *fast* thinking, as elucidated in Daniel Kahneman's seminal work "Thinking, Fast and Slow" [10]. This paper reports follow-up work based on our earlier recent work [12], which introduced the possibility of realizing Kahneman's ideas within the CBR paradigm.

Kahneman's central thesis is that human cognition operates in two modes: (a) fast thinking, which is quick, intuitive and largely involuntary, and (b) slow thinking, which is deliberate, effortful and often involves complex computations.

© Springer Nature Switzerland AG 2021
A. A. Sánchez-Ruiz and M. W. Floyd (Eds.): ICCBR 2021, LNAI 12877, pp. 110–124, 2021.
https://doi.org/10.1007/978-3-030-86957-1_8

Kahneman identifies two systems corresponding to these two modalities, which he calls System 1 and System 2, respectively. These two systems are merely conceptual abstractions and do not imply any physiological or biological separation within the brain. It is tempting to realize these two modalities in the CBR context since we often encounter applications where we would like a CBR system to effectively trade-off effectiveness for time efficiency or vice versa. This can be achieved by a CBR system that uses the fast thinking mode to solve a vast majority of queries (target problems), and switches over to slow thinking, only if it recognizes the target problem as hard. This raises two central questions: (a) On what dimensions are slow and fast thinking mechanisms different? (b) On what basis would we judge a target problem to be solvable by fast thinking, and what are the mechanisms of switching from fast to slow thinking? With respect to the first question, [12] presents two different schemes: the first in which fast thinking uses a subset of features used by slow thinking, and the second in which slow thinking makes use of time-consuming adaptation processes, which fast thinking does away with. With respect to the second question, a couple of switching strategies are presented in [12], details of which are outlined in the next section.

Despite making preliminary attempts to realize slow and fast thinking within CBR, several important questions remained unaddressed in [12]. This paper is aimed at identifying and addressing three of these gaps. Firstly, in a mechanism where fast thinking uses a subset of the attributes used in slow thinking, we need a principled approach to decide on feature selection. This is clearly an optimization problem, and the objective function is decided by the nature of trade-offs required in the domain of interest. We formalize the problem and report empirical findings in a general setting where similarity estimation over different features have different time requirements, and relative importances to effectiveness and time efficiency can be flexibly tuned. Our second contribution is motivated by the observation that a student solving a high school problem in physics based on reusing his experience in solving similar problems in the past often realizes that he made a mistake when he observes that his solution fails to satisfy the constraints as demanded by the laws of physics. Thus domain constraints play an important role in effecting the switching from fast to slow thinking. We present novel CBR realizations that are inspired by this idea. The third contribution of the current work is to present a principled approach to quantify the tradeoff between effectiveness and efficiency in a CBR realization of the fast-slow dichotomy. Such a quantification also allows us to have a-priori insights on how the nature of the domain influences the design choices of a CBR system that trades off speed for accuracy or vice versa. The structure of the paper is as follows.

We present a brief background and an overview of related works in this area in Sect. 2. Sections 3, 4 and 5 present our central contributions, as outlined above, along with a critical analysis of empirical findings. Section 6 presents potential extensions of the work and summarizes its key contributions.

## 2    Background

Looking at the picture of an angry face, we instinctively and involuntarily seem to relate to the emotion expressed in the picture. In contrast, the problem of multiplying two three-digit numbers needs slow and deliberate effort. Kahneman's work discriminates between these two approaches of problem solving, calling them *fast* and *slow* thinking, respectively. It often happens that both modes of operation are at work in the same task. Given a chessboard setting, fast thinking may instinctively suggest a move, which in hindsight appears to be a wrong one, and hence slow thinking based on more elaborate cost-benefit analysis is called into action. Similarly, while solving a numerical physics problem, one may be tempted to reuse the steps in a set of problems solved previously, which on-the-surface appear to be similar, though more careful inspection reveals subtle differences that require one to revisit first principles and work from scratch. Thus, one central role of slow thinking is often to correct errors made in fast thinking. Kahneman demonstrates this tension between two aspects of our cognition: the availability heuristic, which is a cognitive bias that refers to our propensity to solve a complex problem by quickly reusing the solution of what appears to be a simple, more familiar problem, and metacognition, the ability to reason about our thinking processes and thus correct failures.

To the best of our knowledge, Craw and Aamodt [4] make the first effort to build a bridge between the dichotomous cognitive mechanisms in Kahneman's work and CBR. The central observation in [4] is that fast thinking can be realized in settings where the knowledge of similarity is straightforward and similar problems are likely to have similar solutions. In contrast, slow thinking may be necessitated when retrieved neighbours have conflicting solutions, and hence more complicated retrieval mechanisms or adaptation knowledge may be involved. Kannengiesser and Gero [11] have used the dual System theory proposed by Kahneman in case-based design task. Case-based design has the following phases: problem anticipation, search, match, retrieve, select, modify and repair, where each phase involves either System 1 or System 2 operations. Plaza and Aamodt [1] discussed about using Kahneman's dual system model in an integrated view of CBR encompassing both data-driven and knowledge-intensive processes. They have suggested that System 1 which is associative can involve data-driven approaches like deep learning.

We have seen in the earlier section that realizing the fast-slow dichotomy within CBR involves identifying the mechanisms that separate the two modalities, as also the criteria for switching from *fast* to *slow*, given a target problem. Our earlier work [12] presents three distinct models, referred to as Models 1, 2 and 3, which differ with respect to each other in terms of the mechanisms to realize fast and slow thinking. Two distinct approaches for switching from fast to slow thinking were also presented. We summarize the key ideas below.

Let us consider a classification problem. Model 1 uses a standard model-based Machine Learner like Support Vector Machine (SVM) to realize the fast thinking module. Classifiers like SVM output a score corresponding to the confidence of the classification. If the confidence is low, the system switches to slow thinking where a slower instance-based learner like a CBR system solves the target problem. Unlike Model 1, Model 2 uses CBR for both slow and fast thinking, except that in the fast thinking mode, the CBR system uses only a subset of the entire set of features used in the slow thinking mode, for similarity estimation purposes. Model 3 too uses CBR for realizing both fast and slow thinking, but here slow thinking makes use of adaptation, while fast thinking publishes its result without adaptation. Lets us now turn our attention to the switching mechanisms presented in [12]. The first mechanism, called tag-based switching, involves examining each case in the case base and tagging it as simple/$FT$ or complex/$ST$, based on the nature of its neighbourhood. If the class label (solution) of a case is largely in agreement with those of its neighbours, it is tagged simple; otherwise, it is labelled complex. Given a target case, a decision is made to switch from fast thinking to slow thinking if many of the neighbours of the target case are tagged as complex. This is justified, since in such cases, the solutions of the neighbours are likely to be poorly indicative of the true target case solution, and hence additional evidence is needed to arrive at a confident prediction. A second switching mechanism called oracle-based switching is also presented in [12], which is based on partial feedback from an expert on the solution generated by the fast thinking module. In particular, an expert can declare whether the solution is correct or not without giving the true solution; if the solution is incorrect, the system switches to slow thinking.

In the current work, we confine our attention to realizations where both the fast and slow thinking modules are implemented using CBR. In particular, we choose Model 2 as the basis of our research. The current work is based on the observation that several questions needed to be answered in the earlier work [12], in order to make it more rigorous. For one, the mathematical premise for choosing a subset of features from the entire pool for fast thinking in Model 2 remained unclear. We show how this can be cast as an optimization problem that can be solved under diverse criteria pertaining to diverse similarity estimation time requirements across different features, and also the nature of effectiveness v/s time efficiency trade-offs required in the domain of interest. Secondly, we address the need of a principled study based on quantifying the gains achieved by splitting a CBR system between fast and slow thinking modalities. In some domains, the gains are expected to be higher than in some others. We generalize the basic idea of footprint-based competence models [16], and empirically demonstrate the advantages of this conceptual extension in facilitating a systematic study in this context. In addition to these contributions, we focus purely on the tag-based switching mechanism, while also presenting a novel mechanism called *constraint-based switching* inspired by Kahneman's original work. The broad idea is to trigger a shift from fast to slow thinking when the system realizes that the solution generated by the fast thinking module fails to comply with certain underlying domain constraints.

Though the work presented in this paper has very little overlap with past work in terms of its specific goals and contributions, it may be worth highlighting a few other efforts to strike a trade-off between effectiveness of retrieval and time efficiency in CBR, and in the wider Machine Learning context. Dileep and Chakraborti [5] presented an approach where text classification, instead of treating each test document uniformly, resorted to fast classification in case the title or some noteworthy features were reckoned to be adequately indicative of the category label, and fell back to more elaborate slow thinking mechanisms only on demand. [7] is one of the earliest works in the CBR community, where retrieval failures trigger introspective reasoning processes, which in turn are used to refine indices in a CBR planner. Another interesting work is that on anytime algorithms [17], which are designed with the goal of giving increasingly better results when allowed more computation time.

## 3    Feature Selection

We now consider a feature selection problem which is targeted towards improving the fast response process, within the *'fast and slow thinking'* dichotomy.

### 3.1    Feature Selection for Parsimonious Search

Looking back at [12], it may be seen that the two systems (System 1 and System 2) in Model 2 differ based on the sufficiency and insufficiency of a small subset of attributes in solving the task. In other words, if a small subset of features are sufficient to solve a problem, it gets assigned the "*FT* tag", which indicates that the problem can be solved using fast thinking. This leads us to a natural question: *how do we choose a small subset of features to instantiate the parsimonious search in Model 2?* Towards addressing this question, we may observe two key aspects relating to the choice of features:

- Similarity computation over the chosen subset of features should be fast.
- The similarity space over the chosen subset of features should be meaningful for problem solving.

*First*, one may observe that different features differ in the computation that would be incurred for estimating similarities. As an example, numeric attributes lead to fast similarity computation since numeric comparisons could be realized using swift bit-wise operations. On the other hand, similarities between text attributes and set-valued attributes call for costly operations such as subsequence finding and set intersection computation. Thus, posed with two features that are equally relevant for problem solving, we may intuitively choose the feature whose similarity computation is relatively lighter computationally. *Second*, features may differ in the amount of utility towards solving a problem. Consider the scenario of estimating the chances of an individual's risk of fatal outcome if she were to contract COVID-19. Features relating to the past history of respiratory diseases would be considered far more relevant than, say, an attribute such as gender. Thus, even when comparing respiratory disease histories may be more onerous computationally, we may choose to include it in preference to the gender attribute. In simple terms, people who are similar in their respiratory disease histories may have similar COVID-19 risks as compared to people with similar gender. To summarize, the set of features to be chosen for the parsimonious search depends on two criteria; that of computational load in estimating similarities, and problem solving utility.

### 3.2  Modelling Desirability of Feature Sets

In the feature selection approach we will describe, we will use *case base alignment* as a measure of problem solving utility. Given a set of chosen features, case base alignment measures the extent to which the maxim of *similar problems lead to similar solutions* holds within the similarity space defined by the chosen features. In our model, we use a popular case base alignment measure i.e. alignCorr [14] which considers the correlation between the problem similarities, and the solution similarities of cases, as a measure of alignment.

Our desirability objective for a set of features $F$, is modelled as the following:

$$D(F) = \frac{\left(Alignment(X, F)\right)^{\alpha}}{\left(k \times \sum_{f \in F} t(f)\right)^{1-\alpha}} \tag{1}$$

$$F^* = \arg\max_{F' \subseteq \mathcal{F}} D(F') \tag{2}$$

where, $\mathcal{F}$ is the set containing all the features.

In other words, the task is to choose a subset of features, $F^*$ from $\mathcal{F}$, balancing between the need for higher alignment and the need for fast computation. $Alignment(X, F)$ denotes the case base alignment of the case base $X$ as measures over the feature set $F$. Each feature, $f$, also has a computation time, $t(f)$. The cumulative computation time appears in the denominator of Eq. 1 and is weighted by a weighting parameter $k$. The parameter $\alpha$ controls the relative importance of the alignment and computation time considerations. For higher values of $\alpha$, the discovered subsets would be better in terms of alignment, and vice versa for smaller values of $\alpha$. To put it another way, for scenarios where real-time responses are desired, $\alpha$ may be set to a very low value. $F^*$, as may be obvious by now, denotes the ideal choice of features based on our modelling of the objective.

### 3.3  Brute Force Approach

The simplest possibility of identifying the optimal desirability feature subset would be to inspect all possible non-empty feature subsets of $\mathcal{F}$, where $\mathcal{F}$ is the set containing all the features. This very simple but computationally expensive approach can be summarized in the procedural steps below:

- $F = \phi, V = 0$
- $\forall F' \in PowerSet(\mathcal{F})$

- $Val(F') = D(F')$
- $if(Val(F') > V)$
    * $F = F'$
    * $V = Val(F')$
- return $F$ as $F^*$

$PowerSet(\mathcal{F})$ contains all possible non-empty feature subsets of $\mathcal{F}$. The above procedure leads to an expensive computation since the number of elements in $PowerSet(\mathcal{F})$ would be in $\mathcal{O}(2^{|\mathcal{F}|})$. However, this process, by design, would be able to always identify the optimal desirability feature subset since all subsets are being evaluated.

### 3.4 Greedy Approach

The greedy approach reins in the computational load by channelizing the search towards certain candidates based on an estimated likelihood of them leading to desirable feature subsets. The procedural steps are outlined below:

- $F = \mathcal{F}, V = D(\mathcal{F})$
- $while\ (F \neq \phi)$
    - $F' = \arg\max_{f \in F} D(F - \{f\})$
    - $if(D(F') > V)$
        * $F = F'$
        * $V = D(F')$
- return $F$ as $F^*$

This greedy approach starts with all the features as among the chosen features. This is followed by progressively dropping the feature that leads to the smallest drop in desirability, until all features are exhausted. This search excluding one feature at a time could miss the optimal $F^*$ if it doesn't fall within that search path, but the heuristic search ensures that such misses are not very likely. In return for such approximation, the greedy search is able to achieve a quadratic response time, i.e., in $\mathcal{O}(|\mathcal{F}|^2)$, a massive improvement from the brute force method which was in $\mathcal{O}(2^{|\mathcal{F}|})$.

### 3.5 Experimental Analysis

**Experimental Setup:** In Model 2 mentioned in Sect. 2 we need to select features for System 1, such that it has high global alignment and the response time for a given query is less. We have experimented to solve the optimization problem given in Eq. 1 using Greedy approach with two datasets i.e. *AutoMPG Dataset* (AMPG) [6] and *Pima Indians Diabetes Dataset* (PID) [15] with 399 and 768 instances respectively and 8 features each. We have observed that the time taken for similarity computation by each attribute is almost the same for both datasets. Therefore, we have also experimented by externally modifying the time vector $t$ to study how the results differ when the most important attributes take more time than other attributes. We expect that for smaller $\alpha$ values where the time factor is given more importance, the time consuming features must be neglected in the System 1 feature set. Whereas these features can be included for high values of $\alpha$ since the weightage for alignment will be more.

**Experimental Results:** Table 1 and Table 2 show the results for the AMPG and PID datasets respectively where the selected (and not selected) features are reported in the form of binary vectors. For example, the binary vector $[0, 0, 0, 1, 0, 0, 0]$ indicates that only the fourth feature was chosen, while all other features were discarded. The timing for similarity computation is also represented using a similar vector of length 8; for example, $[1, 1, 1, 10, 1, 1, 1, 1]$ denotes that $t(f)$ for the fourth feature is 10 whereas for all other features, it is 1. This suggests that computing similarities for the fourth feature would be ten times as expensive as computing it over the other features. Tables 1 and 2 show the results for varying values of $\alpha$. For lower $\alpha$ (i.e., closer to 0 given that $\alpha \in [0, 1]$), we would expect that the computation time of feature are given primacy; on the other hand, for $\alpha$ closer to 1.0, even computationally costly features may be selected if they help improve case base alignment. In Table 1, moving from left to right, we are increasing the timing for features progressively; for example, the second column has $t(f_4) = 10$, whereas for the third and fourth columns, $t(f_2)$ and $t(f_1)$ are additionally set to 10 respectively. From the first column, where the favorite choice seems to be the choice of $f_4$. When $t(f_4)$ is set to 10 for the second column, the choice is seen to shift to $f_2$. For $\alpha = 0.93$, the choice with uniform timing vector is $\{f_4\}$; however, when $f_4$ is made a costly feature in the second column, observe that the method is forced to choose multiple features to heed to the high impetus on case alignment (since $\alpha = 0.93$) in order to compensate for the inability to choose the best feature. This shows a meaningful movement of choices based on how the computational expense can be traded off for problem solving utility as measured using case base alignment. Similar trends are observed in Table 2 for the PID dataset also, where the choice of desirable features shift away from computationally expensive features with higher settings of $\alpha$. Figure 1 shows the tradeoff between the $\alpha$ and global alignment value [14] of System 1 with respect to the selected features.

## 4    Constraint-Based Switching

In our earlier work [12], we proposed various switching techniques to switch from System 1 to System 2. In the context of Model 2, having parsimonious CBR as fast thinking, the switching strategies are briefly described as follows: (a) Oracle-based switching, which involves a partial feedback given by an expert or an oracle. The partial feedback is such that the reasoner is informed whether the System 1 solution is *correct* or *incorrect*. (b) Tag-based switching is mainly proposed for the classification setting. It involves tagging all the cases in the case-base as *FT* , *ST* or *N*. Given a case base, each case $C$ in the case base is solved using a leave one out simulation. If $C$ can be successfully solved using fast thinking, it is tagged as *FT*. Otherwise, it is tagged as *ST*, indicating that

**Table 1.** Results for AutoMPG Dataset for various timing vectors

| $\alpha$ value | Timing vectors, $t$ | | | |
|---|---|---|---|---|
| | $[1,1,1,1,1,1,1,1]$ | $[1,1,1,10,1,1,1,1]$ | $[1,10,1,10,1,1,1,1]$ | $[10,10,1,10,1,1,1,1]$ |
| 0.1 | $[0,0,0,1,0,0,0,0]$ | $[0,1,0,0,0,0,0,0]$ | $[1,0,0,0,0,0,0,0]$ | $[0,0,1,0,0,0,0,0]$ |
| 0.5 | $[0,0,0,1,0,0,0,0]$ | $[0,1,0,0,0,0,0,0]$ | $[1,0,0,0,0,0,0,0]$ | $[0,0,1,0,0,0,0,0]$ |
| 0.8 | $[0,0,0,1,0,0,0,0]$ | $[0,1,0,0,0,0,0,0]$ | $[1,0,0,0,0,1,0,0]$ | $[0,0,1,0,0,0,0,0]$ |
| 0.9 | $[0,0,0,1,0,0,0,0]$ | $[0,1,0,0,0,1,0,0]$ | $[1,0,1,0,0,1,0,0]$ | $[0,0,1,1,0,1,0,0]$ |
| 0.93 | $[0,0,0,1,0,0,0,0]$ | $[1,1,1,0,0,1,1,0]$ | $[1,0,1,0,0,1,0,0]$ | $[0,0,1,1,0,1,0,0]$ |
| 0.94 | $[0,0,0,1,0,0,0,0]$ | $[1,1,1,0,0,1,1,0]$ | $[1,0,1,0,0,1,1,0]$ | $[0,0,1,1,0,1,0,0]$ |
| 0.95 | $[0,0,1,1,0,1,0,0]$ | $[1,1,1,0,0,1,1,0]$ | $[1,0,1,1,0,1,1,0]$ | $[0,0,1,1,0,1,0,0]$ |
| 0.99 | $[1,1,1,1,1,1,1,0]$ | $[1,1,1,1,1,1,1,0]$ | $[1,1,1,1,1,1,1,0]$ | $[1,1,1,1,1,1,1,0]$ |

**Table 2.** Results for Pima Diabetes Dataset for various timing vectors

| $\alpha$ value | Timing vectors, $t$ | | | |
|---|---|---|---|---|
| | $[1,1,1,1,1,1,1,1]$ | $[1,100,1,1,1,1,1,1]$ | $[100,100,1,1,1,1,1,1]$ | $[100,100,1,1,1,1,1,100]$ |
| 0.1 | $[0,1,0,0,0,0,0,0]$ | $[1,0,0,0,0,0,0,0]$ | $[0,0,0,0,0,0,0,1]$ | $[0,0,0,0,1,0,0,0]$ |
| 0.5 | $[0,1,0,0,0,0,0,0]$ | $[1,0,0,0,0,0,0,0]$ | $[0,0,0,0,0,0,0,1]$ | $[0,0,0,0,1,0,0,0]$ |
| 0.7 | $[0,1,0,0,0,0,0,0]$ | $[1,0,0,0,0,0,0,0]$ | $[0,0,0,0,0,0,0,1]$ | $[0,0,0,0,1,1,0,0]$ |
| 0.85 | $[0,1,0,0,0,0,0,0]$ | $[1,0,0,0,1,1,1,1]$ | $[0,1,0,0,0,1,0,0]$ | $[0,1,0,0,0,1,0,0]$ |
| 0.9 | $[1,1,0,0,0,0,0,0]$ | $[1,1,0,0,0,1,1,0]$ | $[0,1,0,0,0,1,0,0]$ | $[0,1,0,0,0,1,0,0]$ |
| 0.95 | $[1,1,0,0,0,0,0,0]$ | $[1,1,0,0,0,1,1,0]$ | $[1,1,0,0,0,1,1,0]$ | $[1,1,0,0,0,1,1,0]$ |
| 0.99 | $[1,1,0,0,0,1,1,0]$ | $[1,1,0,0,0,1,1,0]$ | $[1,1,0,0,0,1,1,0]$ | $[1,1,0,0,0,1,1,0]$ |

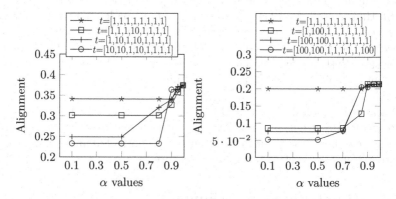

**Fig. 1.** Alignment represents the global alignment i.e. alignCorr value of the case base with respect to the feature subset. The left and right graph shows Alignment v/s $\alpha$ values for AMPG and PID Dataset respectively.

it can be solved using slow thinking. A case $C$ is tagged as $N$, i.e. none tag when, it can neither be solved using System 1 nor System 2. When a query $q$ is encountered, if the majority of the $k$ nearest neighbours to the query are tagged as $FT$, then the System 1 solution is predicted for the query, else it is solved using System 2. (c) Hybrid switching considers both tag-based and oracle-based switching. When majority of the nearest neighbours to the query are tagged as

*FT* the System 1 solution is predicted. Otherwise, the feedback value is checked, and switching is done accordingly. In our previous work [12] the three switching strategies for Model 2, i.e. Oracle-based, Tag-based and Hybrid Switching, lead to Model 2A, 2B and 2C respectively.

In this section, we propose a novel switching technique involving the use of domain constraints. In continuation with the previous work, we will call the model involving constraint based-switching as Model 2D. In his book, "Thinking, Fast and Slow"[9], Kahneman mentions an example where different subjects were asked, "How many murders occur in the state of Michigan in one year?" Most people responded with lower values than what they had answered when the same question was asked for Detroit, a high-crime city in the state of Michigan. Kahneman explains with this example that the fast, intuitive thinking failed to recognize that the number of crimes in Michigan will be inclusive of the number of crimes in Detroit. What if the subjects have verified their intuitive response against the domain constraints that Detroit is a high crime city in Michigan state, therefore Michigan will have at least as many crimes as Detroit? We believe they will be able to understand the fault in their intuitive answer. Similarly, when a person trying to solve a physics problem observes that his solution fails to satisfy the constraints of the laws of physics. He switches to a slower form of thinking to solve that particular problem. This motivated us to use such constraints to decide whether to invoke System 2 or not. The next subsection will describe constraint-based switching in detail.

## 4.1   Model 2D

**Constraint:** The constraint in the Detroit example mentioned above was that the no of crimes in any state would be inclusive of the number of crimes in any city of that state. In the second example, the laws of physics serve as constraints while solving a physics problem. While these are examples of constraints that could be derived from domain-specific knowledge, it is also possible to use domain-independent constraints. In the CBR paradigm, a structural constraint is generally imposed on the case base that *similar problems have similar solutions*. This structural constraint can be exploited in the context of a dichotomous CBR model where System 1 is realized as a *parsimonious CBR system*. That is, System 1 fetches cases by calculating similarity based on only a *subset of attributes*. In order to validate if the retrieval is going to be useful or not, we can apply a constraint that similarity of those features that were neglected by System 1 should be above a minimum threshold. If the above constraint is violated, then perhaps, a retrieval that uses only the parsimonious features may not be able to fetch cases that are sufficiently similar to solve the given query. Hence, the reasoner can switch to slow thinking where the relevant cases are identified by computing similarity based on all the attributes.

Experimentally, we have verified the utility of such constraints on fast thinking by realizing a dichotomous model (Model 2D) where System 1 is Parsimonious CBR and System 2 is Full CBR. In this Model, along with the tag-based switching, we introduce an additional level of constraint checking explained above. In other words, it is a hybrid of tag-based switching and constraint-based switching. Given a query q, System 1 solution is published if either all the kNN to the query are tagged as *FT* or, the constraint is satisfied i.e. the similarity between the query and kNN to the query is more than some threshold. The similarity for constraint checking is based on the attributes that are not used in System 1.

### 4.2 Experimental Analysis

**Experimental Setup:** We have experimented the Model 2D i.e. the hybrid model of Tag-based switching and Constraint-based switching with two classification datasets. *Red Wine Quality Dataset* (RWQ) [3] and *Fetal Health Classification Dataset* (FHC) [2] have 2 and 3 classes respectively. The RWQ has 1599 and FHQ dataset 2126 data instances. This experiment aims to study the improvement in the performance i.e., accuracy and per query response time, of the proposed model by adding the constraint checking layer. We compare performance of Model 2D with basic CBR and the Model 2A, which was using only tag-based switching. The features for System 1 are computed using *Greedy Optimization Approach* mentioned in Sect. 3. The similarity threshold for constraint checking is computed empirically.

**Experimental Results:** The results for both the datasets i.e., RWQ and FHC are shown in Table 3. It can be observed that more cases are being solved using System 1 when the switching is done based on the constraints in addition to the kNN tags. Therefore, in Model 2D, the average per query response time is reduced as compared to Model 2A where switching was based only on the kNN tags. The accuracy of Model 2D is also improved as compared to 2A, therefore we can say that the Model 2D is performing better in terms of both effectiveness and efficiency.

**Table 3.** Results for Switching Strategies. "ST%" is % query cases passed on to System 2 and "Time" represents average per query response time

| Model | Switching strategy | RWQ | | | FHC | | |
|---|---|---|---|---|---|---|---|
| | | Accuracy | ST% | Time (ms) | Accuracy | ST% | Time (ms) |
| Model 2A | Tag-based | 86.56 | 13.25 | 430.86 | 86.59 | 12.37 | 732.87 |
| Model 2D | Tag or constraint based | 87.87 | 6.68 | 381.72 | 86.79 | 10.32 | 653.24 |
| Full CBR | | 84.56 | | 1054.94 | 85.36 | | 2759.59 |

## 5    Complexity Measure for Dichotomous Models

So far, we have analysed the proposed dichotomous models in terms of prediction time and accuracy. In this section, we attempt to quantitatively assess the benefit of a dichotomous model over a singular model. For this, we propose a measure that is derived from the idea of footprint size reduction discussed in [8,9]. In the following paragraphs, we introduce the concept of footprint, its application to quantification of case base knowledge and the proposed measure.

## 5.1  Footprint Size

The idea of footprint set was proposed by Smyth and McKenna [16]. By definition, a footprint set is a minimal subset of the case base having the same problem-solving ability as that of the entire case base. To compute the footprint set of a given case base we need to first identify the *competence group* of every case. A *competence group* is a set of cases that makes an independent contribution to the competence of the whole case base and a footprint set contains the representative case(s) from every *competence group*. Construction of a competence group involves the use of the following sets of cases: *Coverage Set*, *Reachability Set* and *Related Set*. The set of all the target problems that can be solved by a case $c$ is called the *coverage set* of $c$. *Reachability set* for a given target problem $t$ is the set of the cases in the case base that can be used to solve the target problem $t$. A case $c_1$ is said to solve a case $c_2$ if and only if $c_1$ can be retrieved for $c_2$ and the solution of $c_1$ can be adapted to solve $c_2$. The *Related set* of case $c$ is the union of its *Coverage* and *Reachability* sets. Two cases $c_1$ and $c_2$ are said to exhibit a *shared coverage* if their related sets are overlapping. Any two cases having a non-empty *shared coverage* set belong to the same *competence group*. After calculating the *competence groups*, cases from each *competence group* are added to the footprint set, discarding the cases whose *coverage set* is incorporated by the other cases.

**Footprint Size to Quantify Knowledge:** In [8], the authors have proposed the use of footprint size to quantify the knowledge contained in a case base. By definition, a footprint set is constructed by identifying a minimal subset of non-redundant cases that has the same competence as the entire case base. Due to interplay among containers, the authors have observed that addition of any useful domain knowledge to the vocabulary, similarity or adaptation containers leads to a reduction in the number of non-redundant or footprint cases. Further, footprint size can be used as a complexity measure to assess the generalization ability of the reasoner. The lower the footprint size, the greater will the generalization ability of the reasoner be [9]. In the next section, we discuss how the definition of *solves* relation in the footprint construction algorithm can be modified to accommodate a time-constrained competence measurement that may be needed in the domains of interest for dichotomous CBR models.

## 5.2  Footprint with Time

An unique advantage of a dichotomous CBR model is that a reasoner could resort to fast thinking to satisfy any constraint imposed on its response time by an end user. This motivated us to emphasize the need for factoring response time into measurement of system competence for dichotomous models. For this, we have modified the definition of *solves* relation. In the original footprint construction algorithm, a case $c$ is said to solve some target problem $t$ if and only if $c$ can be retrieved and adapted to solve $t$. In the modified definition of the *solves* relation, a case $c$ is said to solve some target problem $t$ if and only if $c$ can be retrieved and adapted to solve $t$ *within a prescribed time limit $T$*. The time limit $T$ in the modified *solves* definition is to be provided by the CBR end user.

We use the modified *solves* relation in the following two contexts: firstly, to compare the reduction in footprint size brought about by a dichotomous model over a singular model and also by different types of dichotomous models; secondly, to compare the response time per query when the modified *solves* relation is used for the selection of features. The time restrictions included in the *solves* relation will favour that dichotomous model in which System 1 is having computationally less expensive features.

## 5.3  Experimental Analysis

**Experimental Setup:** In this experiment we will measure the footprint sizes of a singular Full CBR model and the dichotomous models 2A, 2B and 2D, which are briefly explained in Sect. 4. We have experimented with two different binary classification datasets i.e. *Pima Indians Diabetes Dataset* (PID) [15] and *Red Wine Quality Dataset* (RWQ) [3] with 768 and 1599 instances respectively. For models 2A, 2B and 2D, we have used the principled *Greedy Approach* mentioned in Sect. 3 to select the features for System 1. The footprint size is computed using the modified *solves* function where the $T$ value can be provided as per the users requirements.

**Experimental Results:** Tables 4 and 5, report the results for RWQ and PID datasets respectively. We observe that the size of footprint set computed for the dichotomous model 2B is smaller than that of the other dichotomous models 2A and 2D as well as the singular Full CBR system. Since model 2B has access to an external feedback value, the effectiveness of its switching mechanism is high as indicated by its higher accuracy. Further, when the CBR system is operating in a time-constrained setting (low $T$ value) where quicker response times are needed, the Full CBR model has a large footprint size, indicating that such models may not be a good choice when a low response time is needed. This evaluation measure quantifies the generalization ability of the reasoner in a time-restricted setting and gives an idea of the suitability of the model for the given domain.

**Table 4.** Results for evaluation using RWQ having 1279 cases in the case base

| Model | Accuracy (%) | Average per query response time (ms) | Footprint size for $T = 1500$ ms | Footprint size for $T = 500$ ms |
|---|---|---|---|---|
| Full CBR | 84.56 | 1054.94 | 480.2 | 1279 |
| Model 2A | 86.56 | 430.86 | 467 | 467 |
| Model 2B | 92.13 | 451.34 | 425.8 | 425.8 |
| Model 2D | 87.87 | 381.72 | 448 | 448 |

**Table 5.** Results for evaluation using PID Dataset having 614 cases in the case base

| Model | Accuracy (%) | Average per query response time (ms) | Footprint size for $T = 200$ ms | Footprint size for $T = 160$ ms |
|---|---|---|---|---|
| Full CBR | 73.9 | 191.49 | 289.4 | 614 |
| Model 2A | 75.97 | 156.58 | 281.2 | 281.2 |
| Model 2B | 84.8 | 155.82 | 238.2 | 238.2 |
| Model 2D | 77.6 | 152.4 | 255 | 255 |

# 6    Conclusion and Future Work

We have presented three different directions that can contribute towards realizing the dichotomy of fast and slow thinking within CBR, and in particular, help in building CBR systems that flexibly trade-off time efficiency and effectiveness based on specific domain needs. *Firstly*, we have proposed an approach that can help in a principled selection of a subset of features to be used for fast thinking. We have formulated this as an optimization problem with an objective function that can cater to domain specific requirements, such as prioritizing effectiveness over time efficiency or vice versa. Empirical results suggest that the proposed approach successfully exploits the large differences in the similarity estimations of diverse attributes. *Secondly*, we have proposed a novel switching mechanism that exploits domain-specific constraints to decide when to transfer control over from fast to slow thinking and demonstrated the effectiveness of this idea empirically. *Thirdly*, we have proposed a novel extension of the footprint approach that accounts for time efficiency while estimating the ability of a case to solve another; this leads to a fresh conceptual basis that allows us to quantify the gains that slow-fast dichotomy may bring about, in any given domain.

With respect to the first direction, future work can involve investigation into non-greedy approaches such as genetic algorithms or simulated annealing that incorporate randomness to explore the solution space more exhaustively. In the context of the second contribution, more work needs to go into large-scale real-world experiments that can demonstrate the effectiveness of incorporating domain constraints in the form of rules or models that encapsulate what experts already know from first principles. This can be tried out in diagnostic or planning tasks, where specific expert knowledge can be encapsulated as constraints. We envisage that the extended footprint approach presented in Sect. 5 may have wider applications in the context of CBR maintenance.

# References

1. Aamodt, A., Plaza, E.: Case-based reasoning and the upswing of AI (2017)
2. Ayres-de Campos, D., Bernardes, J., Garrido, A., Marques-de Sa, J., Pereira-Leite, L.: Sisporto 2.0: a program for automated analysis of cardiotocograms. J. Matern. Fetal Neonatal. Med. **9**(5), 311–318 (2000)
3. Cortez, P., Cerdeira, A., Almeida, F., Matos, T., Reis, J.: Modeling wine preferences by data mining from physicochemical properties. Decis. Support Syst. **47**(4), 547–553 (2009)
4. Craw, S., Aamodt, A.: Case based reasoning as a model for cognitive artificial intelligence. In: Cox, M.T., Funk, P., Begum, S. (eds.) ICCBR 2018. LNCS (LNAI), vol. 11156, pp. 62–77. Springer, Cham (2018). https://doi.org/10.1007/978-3-030-01081-2_5
5. Dileep, K.V.S., Chakraborti, S.: Eager to be lazy: towards a complexity-guided textual case-based reasoning system. In: Goel, A., Díaz-Agudo, M.B., Roth-Berghofer, T. (eds.) ICCBR 2016. LNCS (LNAI), vol. 9969, pp. 77–92. Springer, Cham (2016). https://doi.org/10.1007/978-3-319-47096-2_6
6. Dua, D., Graff, C.: UCI machine learning repository (2017). http://archive.ics.uci.edu/ml
7. Fox, S., Leake, D.: Using introspective reasoning to re ne indexing. In: IJCAI, vol. 391397 (1995)
8. Ganesan, D., Chakraborti, S.: An empirical study of knowledge tradeoffs in case-based reasoning. In: IJCAI, pp. 1817–1823 (2018)
9. Ganesan, D., Chakraborti, S.: A reachability-based complexity measure for case-based reasoners. In: The Thirty-Second International Flairs Conference (2019)
10. Kahneman, D.: Thinking, Fast and Slow. Macmillan (2011)
11. Kannengiesser, U., Gero, J.S.: Design thinking, fast and slow: a framework for Kahneman's dual-system theory in design. Design Sci. **5**, 1–21 (2019)
12. Kaurav, S., Ganesan, D., Padmanabhan, D., Chakraborti, S.: Thinking fast and slow: a CBR perspective. In: 34th Florida Artificial Intelligence Research Society Conference (2021)
13. Kolodner, J.: Case-Based Reasoning. Morgan Kaufmann, Burlington (2014)
14. Raghunandan, M.A., Chakraborti, S., Khemani, D.: Robust measures of complexity in TCBR. In: McGinty, L., Wilson, D.C. (eds.) ICCBR 2009. LNCS (LNAI), vol. 5650, pp. 270–284. Springer, Heidelberg (2009). https://doi.org/10.1007/978-3-642-02998-1_20
15. Smith, J.W., Everhart, J., Dickson, W., Knowler, W., Johannes, R.: Using the ADAP learning algorithm to forecast the onset of diabetes mellitus. In: Proceedings of the ASCA in Medical Care, p. 261 (1988)
16. Smyt, B., McKenna, E.: Footprint-based retrieval. In: Althoff, K.-D., Bergmann, R., Branting, L.K. (eds.) ICCBR 1999. LNCS, vol. 1650, pp. 343–357. Springer, Heidelberg (1999). https://doi.org/10.1007/3-540-48508-2_25
17. Zilberstein, S.: Using anytime algorithms in intelligent systems. AI Mag. **17**(3), 73 (1996)

# Harmonizing Case Retrieval and Adaptation with Alternating Optimization

David Leake$^{(\boxtimes)}$ and Xiaomeng Ye

Luddy School of Informatics, Computing, and Engineering, Indiana University, Bloomington, IN 47408, USA
leake@indiana.edu, xiaye@iu.edu

**Abstract.** Case-based reasoning (CBR) research has developed numerous methods for learning to improve case retrieval and adaptation knowledge. Learning for each type of knowledge is usually pursued independently. However, it is well known that the knowledge containers of CBR are tightly coupled, in that changes in one can affect requirements for another, which suggests potential benefit for coupling learning across knowledge containers. This paper proposes applying alternative optimization to learn retrieval and adaptation knowledge together, in order to harmonize their behaviors. For a testbed system using neural network based similarity and adaptation, this study compares alternative optimization, independent learning, and learning by prioritizing adaptation for adaptation-guided retrieval. Results support that alternative optimization can help to balance both components and achieve good performance.

**Keywords:** Adaptation-guided retrieval · Alternating optimization · Case adaptation learning · Case retrieval learning · Neural-network-based adaptation · Siamese network · Similarity

## 1 Introduction

CBR systems solve a new problem by retrieving a similar prior case and applying its solution to the new situation, potentially adapting the solution to address differences between the old and new problem situations (e.g., [11]). Much CBR research has focused on learning to improve case retrieval, often through refinement of case indices and similarity criteria. Another current of research has focused on learning to improve case adaptation. Recently, considerable interest has emerged in the use of network models to learn case retrieval and adaptation.

It is natural to expect that strengthening any component of a CBR system will strengthen the system as a whole. This assumption implicitly underlies research that focuses on learning for a single knowledge container such as similarity knowledge or adaptation knowledge. However, it is well known that similarity/retrieval and case adaptation knowledge containers are intimately connected:

© Springer Nature Switzerland AG 2021
A. A. Sánchez-Ruiz and M. W. Floyd (Eds.): ICCBR 2021, LNAI 12877, pp. 125–139, 2021.
https://doi.org/10.1007/978-3-030-86957-1_9

Smyth and Keane [22] showed that an adaptation-guided retrieval (AGR) approach to case-based planning, which bases retrieval on adaptability, can increase efficiency of plan generation. Richter [20] observed that strengths in one form of knowledge can compensate for weaknesses in another; a consequence is that it may be valuable to focus learning in one form of knowledge to address gaps in another (e.g., responding to gaps in the case base by learning the adaptation knowledge most useful to alleviate the gaps). Uncoordinated learning may even be harmful: Leake, Kinley, and Wilson [7] present a study in which uncoordinated case and adaptation learning degrades system efficiency, but overall efficiency is improved when case and adaptation knowledge are coordinated by learning adaptation-based similarity.

Even if equivalent efficiency or solution quality can be achieved by learning either retrieval or adaptation knowledge, the type of knowledge learned can affect explainability. A system with very strong adaptation might be capable of successfully adapting very distant cases. However, if retrieved cases are to be used to explain system conclusions to a user (e.g., [16]), explanations based on the distant cases might be less compelling than those based on more intuitively relevant cases. Therefore it may be preferable to learn to retrieve closer cases rather than learning stronger adaptation, even if both provide equivalent solution quality. On the other hand, if the user's conception of similarity is not important, it may be appropriate to retrieve the most adaptable cases, no matter how dissimilar they appear to the user, to increase system efficiency. If retrieval and adaptation are trained independently, or in a fixed sequence of one before the other, the balance between the two cannot be optimized for such considerations.

As an alternative to learning retrieval and adaptation knowledge independently or successively in a fixed sequence, this paper proposes applying alternating optimization (AO) [1]. Use of AO coordinates learning of retrieval and adaptation knowledge to harmonize their behaviors according to criteria for a desired balance. This method is applicable to any CBR system in which gradient descent is used to train retrieval and adaptation, as for network-based retrieval and adaptation approaches.

The paper presents an application of the AO approach to a regression task, using network-based learning methods for both retrieval and adaptation. In the testbed system, retrieval is based on a Siamese network for similarity measure and adaptation is based on a neural network based case difference heuristic (NN-CDH) approach for adaptation learning [9]. Experiments compare prediction error for three training schemes. The first is independent training of the two stages. The second is training adaptation first and then training retrieval. As this training gives precedence to adaptation, it provides fully adaptation-guided retrieval (AGR), and will be referred to as AGR training. The third is alternating optimization. The schemes are tested for five data sets. Experimental results show that under independent training, the two components may be poorly balanced; for example, retrieval may be strong while adaptation may provide little benefit or sometimes may even worsen the initial solution provided by retrieval. Results also show that under AGR training, retrieval may learn to retrieve cases that are adaptable but that have distant solutions, decreasing explainability. AO generally decreases the incidence of undesirable behaviors. Because AO harmonizes the retrieval stage and adaptation stage, the CBR system retrieves cases similar to queries and requiring limited adaptation, resulting in good accuracy.

The paper first presents background on retrieval and adaptation learning and introduces alternating optimization. It then presents loss functions for retrieval and adaptation and an algorithm for training them together in the AO framework. It closes with an evaluation of a testbed system, guidelines for application of AO for CBR components, and future directions.

## 2   Background

**Learning for Retrieval and Adaptation:** An extensive body of CBR research has studied machine learning methods to improve retrieval and case adaptation. Retrieval learning is generally focused on adjusting feature weights to improve similarity assessment (see Wettschereck et al. [23] for a survey). Many such methods can be seen as adjusting parameters to optimize accuracy. Recently, there has been much interest in network-based similarity learning by optimizing a loss function. For example, Martin et al. [12] use a Siamese network to learn a similarity measure for retrieval; Mathisen et al. [13] propose using an extended Siamese network, which is the basis of the retrieval method in the testbed system presented in this paper.

Likewise, substantial research has been done on learning to improve case adaptation (*e.g.*, [2,4–9,14,15,21]). Again, network methods have prompted strong interest. Policastro, Carvalho, and Delbem [18] use an adaptive resonance theory neural network for retrieval and multi-layer perceptron for adaptation. Leake, Ye, and Crandall [9], Liao, Liu, and Chao [10], and Policastro, Carvalho, and Delbem [19] use neural networks to learn adaptation knowledge from pairs of cases, using methods based on the case difference heuristic approach [5].

In existing work learning of retrieval and/or adaptation knowledge is coordinated in either of two ways. First, training of retrieval and adaptation may be *independent*, where one is trained without using knowledge of the other [18,24]. Second, they may be trained *in-order*, with one trained (or pre-defined) before the other and the latter trained with the knowledge of the former [17].

**Alternating Optimization:** As defined by Bezdek and Hathaway [1], alternating optimization (AO) is an iterative procedure to minimize a scalar field $f : R^s \rightarrow R$, under certain assumptions, in the nonlinearly constrained optimization problem:

$$\min_{x \in R^s} \{f(x)\}, \text{ subject to constraint functions}$$

$$c_i(x) = 0, i = 1, ..., k; \text{ and}$$

$$c_i(x) \geq 0, i = k + 1, ..., m.$$

AO partitions the parameters $x = (x_1, ..., x_s)$ into $t$ non-overlapping subsets $(X_1, ..., X_t)$, with $X_i \in R^{p_i}$ for $i = 1, ..., t$ and $\sum_{i=1}^{t} p_i = s$. $p_i$ is an integer that controls the number of parameters in subset $X_i$. AO minimizes $f(x)$ iteratively. Each iteration includes $t$ time steps. At each time step $i$ ($i = 1, ..., t$), instead of attempting to optimize all parameters $x$ of a model, AO minimizes $f$ by optimizing $X_i$, while keeping all other subsets $X_j (j \neq i)$ fixed. After one iteration, all subsets $X_i$ for $i = 1, ..., t$ are optimized exactly once. AO can carry out multiple iterations, until $x$ converges and cannot be further optimized, or until a fixed limit of iterations is reached. Examples of alternating optimization algorithms are k-means clustering and the expectation-maximization algorithm.

## 3   Alternating Optimization of Retrieval and Adaptation

Case-based reasoning has four processing stages: retrieval, reuse (adaptation), revision and retention. If each stage is considered as operating based on a collection of parameters $x_i$, then in principle, all stages could be refined in an AO process over all parameters $x = (x_1, x_2, x_3, x_4)$. This paper focuses on optimizing the retrieval function $R$ and adaptation function $A$. Let $CB$ be the case base, $q$ be a query from a query set $Q$, $x_r$ be the parameters of the retrieval function $R$, $x_a$ be the parameters for the adaptation function $A$, and $f$ be a loss function for the CBR system as a whole (e.g., error of the system solution for $q$). The retrieval function $R$ retrieves case(s) $r$ from the case base $CB$ according to parameterization $x_r$ (for example, $x_r$ could determine a similarity measure).

$$r = R(q, CB, x_r), r \in CB$$

The adaptation function $A$ produces a final solution $a$ by adapting the solution of retrieved case $r$ to solve the query $q$ according to $x_a$ (for example, for network-based adaptation, parameters could determine adaptation network weights).

$$a = A(q, r, x_a)$$

Then the problem of optimizing the CBR system is equivalent to finding $x_r$ and $x_a$ to minimize the loss $f$ over $Q$. This can be formulated as

$$\underset{x_a, x_r}{\arg\min} \sum_{q \in Q} \{f(A(q, r, x_a), q)\}, \tag{1}$$

The proposed AO process trains the parameters for retrieval and adaptation in a back-and-forth manner. A training iteration involves two steps, first seeking $x_r$ according to:

$$\underset{x_r}{\arg\min} \sum_{q \in Q} \{f(A(q, R(q, CB, x_r), x_a), q)\}, \text{ while } x_a \text{ is fixed}, \tag{2}$$

and then setting $x_a$ according to:

$$\underset{x_a}{\arg\min} \sum_{q \in Q} \{f(A(q, r, x_a), q)\}, \text{ while } x_r \text{ is fixed}. \tag{3}$$

The optimization runs multiple iterations, alternating between steps applying (2) and (3), until $x_r$ and $x_a$ both converge (*e.g.*, there is no significant update within a preset number of iterations) or an iteration limit is reached.

Goals for performance of CBR systems can be finer-grained than overall criteria such as system accuracy. For example, a system designer could be concerned about the usefulness of cases provided by the retrieval function as explanations, about efficiency of retrieval, about the efficiency or explainability of adaptation given a retrieval, or about some combination. To reflect various goals, we refine (2) for optimizing retrieval as:

$$\arg\min_{x_r} \sum_{q \in Q} \{\alpha g(r, q) + (1-\alpha)f(A(q, r, x_a), q)\}, r = R(q, CB, x_r) \text{ while } x_a \text{ is fixed,} \quad (4)$$

where $f$ is the adaptation loss (*e.g.*, measuring the adaptability of the retrieved case), while $g$ is the retrieval loss (*e.g.*, reflecting whether the case is a convincing explanation). In our testbed implementation, $f$ calculates the error of the adapted solution for $q$, and $g$ calculates the error of the retrieved solution for $q$.

The metaparameter $\alpha$ controls the emphasis between the two losses during the training of $R$. If $\alpha = 0$, then formula (4) is reduced to formula (2), where the retrieval loss is ignored and retrieval $R$ is trained to retrieve the case that minimizes the adaptation loss. If $\alpha = 1$, then adaptation loss is ignored and retrieval $R$ is trained to retrieve the case that minimizes the retrieval loss.

As shown in the evaluation, setting $\alpha = 0$ or $\alpha = 1$ may be problematic, leading to a form of codependence of the two stages. If the retrieval step converges rapidly and can retrieve very similar cases, little learning may be required from adaptation, potentially resulting in adaptation that struggles to handle novel cases. If the adaptation step converges rapidly and can adapt a wide range of differences, little learning may be required from retrieval, because even retrievals distant from the query result in accurate solutions. In that case, the retrievals may be less compelling as explanations (in the extreme, a system could retrieve the same case for all problems, for adaptation to generate the solution from scratch). To mitigate the codependence effect, we choose an $\alpha$ value so that formula (4) considers both retrieval loss and adaptation loss. Alternation between training retrieval and training adaptation also helps mitigate codependence.

## 4   Testbed System Design

To study the impact of alternating optimization on harmonizing retrieval and adaptation, we designed a testbed system for case-based regression, with optimization based on solution accuracy. We note that other loss functions would be possible. For example, previous research on adaptation-guided retrieval focused on adaptation efficiency [22].

### 4.1    Loss Function

As loss function $f$, we use squared error of the final solutions (post-adaptation) compared to the actual solutions.

$$f(a, q) = (a - sol(q))^2. \tag{5}$$

Formula (3) becomes

$$\arg\min_{x_a} \sum_{q \in Q} \{(A(q, r, x_a) - sol(q))^2\}, \text{ while } x_r \text{ is fixed,} \tag{6}$$

As loss function $g$, we use squared error between the retrieved solutions and the actual solutions. This could be considered as a proxy for suitability as an explanation: If the error is already small, the retrieved case is a good candidate to explain the solution for the query. We note that in general the suitability as an explanation also depends on the user's criteria for problem similarity (which may differ from that of the system).

Retrieved cases within a set error threshold $E$ are considered correct, with loss set to zero.

$$g(r, q) = max(0, (sol(r) - sol(q))^2 - E^2)$$

Formula (4) is thus expanded into

$$\arg\min_{x_r} \sum_{q \in Q} \{\alpha \cdot max(0, (sol(r) - sol(q))^2 - E^2) + (1 - \alpha)(A(q, r, x_a) - sol(q))^2\}. \tag{7}$$

### 4.2    Testbed Retrieval and Adaptation

In the testbed system, the retrieval function $R$ performs 1-nearest neighbor retrieval. The similarity measure is learned using a Siamese network following the example of eSNN [13]. This network involves a base network (which extracts features from input cases) and an element-wise distance layer (which subtracts the two features), followed by a fully connected layer for final output. It is trained by backpropagation using formula (7) as the loss function, where the parameter $x_r$ represents the weights and biases of the network.

The adaptation function $A$ is an NN-CDH, as discussed in Sect. 2. It involves a Siamese base network (not related to the Siamese network used in retrieval) and an adaptation network. The Siamese base network is trained to extract feature vectors from the input query $q$ and retrieved case $r$, and the adaptation network takes as input both feature vectors and outputs a predicted difference $d$ of the solutions of the two cases. The predicted difference is added to the retrieved solution to produce the final solution. The NN-CDH adaptation network is trained through back propagation using formula (6) as the loss function, where the parameter $x_a$ contains the weights and biases of the NN-CDH. The networks used in retrieval and adaptation are illustrated in Fig. 1.

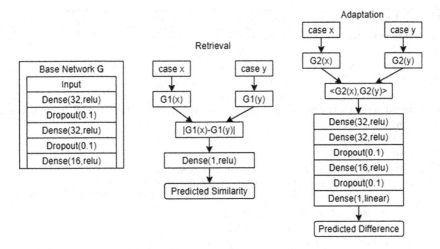

**Fig. 1.** Network Architecture of the Testbed System. Left: Base network used to extract features. Middle: Extended Siamese network for similarity. Right: Extended Siamese network for solution difference prediction

The retrieval function and adaptation functions were chosen based on several factors: They can be independent of or dependent on each other, allowing comparison between the paradigms; they can be trained using formula (3) and formula (4); and they are both powerful enough to solve majority of the queries in the experiment datasets.

### 4.3    Testbed Training and Testing Procedures

A general training process for retrieval and adaptation, applied in the testbed system, is described in Algorithm 1. Given a query $q$, the retrieval stage collects a batch of case pairs by retrieving neighboring cases from $CB$. Retrieval is based on the similarity measure determined by $x_r$. The batch is then used to train $x_a$ and $x_r$. Afterwards, retrieval collects a new batch based on the updated $x_r$ and the batch is used for the next iteration of training. This process repeats until either a predefined iteration limit is reached or the parameters $x_a$ and $x_r$ converge. The steps to update $x_a$ and $x_r$, lines 17 and 18, are only described at a high level because these steps vary across specific adaptation and retrieval components. In our testbed system this training is done by backpropagation to minimize loss functions (6) and (7). The system alternates between training retrieval and training adaptation across batches, with respective loss functions enabling consideration beyond overall performance, for example, explainability or adaptation efficiency.

Parameters in the algorithm enable fine tuning. The parameter $num\_neighbor$ controls the number of neighboring cases to retrieve for a query. The retrieval and adaptation stages train from pairs of a query and its neighbor(s). In principle, pairs of random cases can be used to provide more compre-

---

**Algorithm 1.** Training for Retrieval and Adaptation

---

 1: **procedure** R($q, CB, num\_neighbor, x_r$)
 2:     $r \leftarrow \{\}$
 3:     sort $CB$ by similarity to $q$, using similarity measure determined by $x_r$
 4:     $r$.append($num\_neighbor$ top neighbors of $q$ from $CB$)
 5:     **return** $r$
 6: **procedure** AO_Train($x_r, x_a, batch\_size, CB, R\_epochs, A\_epochs, R\_loss, A\_loss$)
 7:     $pairs \leftarrow \{\}$
 8:     $batch\_counter \leftarrow 0$
 9:     **for** each $q_i$ in $CB$ **do**
10:         $remaining\_cases \leftarrow CB$ not including $q_i$
11:         $r \leftarrow$ R($q_i, remaining\_cases, num\_neighbor, x_r$)
12:         **for** each $r_j$ in $r$ **do**
13:             $pairs$.append($< q_i, r_j >$)
14:         $batch\_counter \leftarrow batch\_counter + 1$
15:         **if** $batch\_counter = batch\_size$ **then**
16:             $batch\_counter \leftarrow 0$
17:             Train $x_a$ with $pairs$ using $A\_loss$ for $A\_epochs$
18:             Train $x_r$ with $pairs$ using $R\_loss$ for $R\_epochs$

---

hensive training samples [6,8], but this is not used here. Lines 17 and 18 could be exchanged depending on the choice of whether to train retrieval or adaptation first.

AO training is a general version of the traditional training scheme used to learn retrieval or adaptation in a CBR system. If $batch\_size$ is set to the size of $CB$ in line 6, then retrieval and adaptation are only trained once. We tested two training schemes for comparison with AO: First, when $\alpha = 1$ and $batch\_size = size(CB)$, the retrieval and adaptation stages are trained independently of each other (**independent training**). Second, when $\alpha = 0$ and $batch\_size = size(CB)$, adaptation is trained first and then the retrieval is trained solely based on adaptation loss (**adaptation-guided retrieval (AGR) training**).

For independent training or AGR training, $x_r$ and $x_a$ are trained until convergence in lines 17 and 18. However, AO uses parameter $R\_epochs$ and $A\_epochs$ to respectively train $x_r$ and $x_a$ only for a set number of epochs before alternating. Forcing more frequent alternation is intended to prevent one parameter from converging too fast, reducing risk of a local minimum.

In this study, we do not explicitly implement a scheme for retrieval-guided adaptation (the counterpart of AGR). In Algorithm 1, a query is paired with a neighboring case found through the *Retrieval* procedure, and the case pair is in turn used for training both retrieval and adaptation. However, this is not strictly retrieval-guided adaptation as pairs are retrieved before $x_r$ is trained, so adaptation is trained with pairs assembled from an untrained retrieval.

Algorithm 2 illustrates the testing mode for the system, which applies trained retrieval followed by trained adaptation.

**Algorithm 2.** Using the CBR System in Testing Mode

1: **procedure** Solve($q, CB, x_r, x_a$)
2:    $r \leftarrow$ R($q, CB, 1, x_r$)
3:    **Return** $A(q, r, x_a)$

## 5    Evaluation

This section evaluates the performance of AO in comparison to independent training and AGR training in terms of quality of retrieved solution, quality of adapted solution, and incremental benefit from adaptation.

### 5.1    Experimental Settings

This study carries out experiments on five regression data sets. Four of these data sets, Energy Efficiency (EE), Yacht Hydrodynamics (YH), Student Performance in Math (SP), and Airfoil Self-noise (AS) are from UCI machine learning repository [3]. EE, YH, and AS are datasets of physics phenomena, and SP is a dataset about predicting students' math grades with social and economical attributes. The fifth data set is an artificial data set (Ar). Ar has 4 attributes $\{x_1, x_2, x_3, x_4\}$, each in $[0, 5]$, and a target value $y$ set by:

$$y = \sum_{i=1}^{4} 3 - |x_i - 3|$$

This is a simple data set for which neighboring cases have similar solutions and adaptation can be done with two simple adaptation rules:

$$\Delta y = \begin{cases} \Delta x_i & \text{if } x_i \leq 3 \\ -\Delta x_i, & \text{otherwise} \end{cases}$$

The testbed system is trained by Algorithm 1 and tested by Algorithm 2. For replicability, the source code and parameter settings are available online.[1]

For experimental runs, each data set is split into a training set (90%) and a test set (10%). The testbed system is trained under three different training schemes, and then tested on the test set. Under independent and AGR training, 10% of the pairs collected are used for validation. Under AO, the training happens per batch and 10% of the pairs in a batch is often too small for validation. For this reason, under AO, 10% of the cases in the training set are separated and used as validation cases. Each validation case *val_case* forms a validation pair with the case base by using the function Retrieve(*val_case*, $CB, 1$). Note that with this experimental design, the three schemes use same number of case

---

[1]  https://github.com/Heuzi/AOtrainingCBR.

pairs in training and in validation. They are also tested on the same number of cases during testing.

In all experiments, we used $num\_neighbor = 1$ in all the retrievals. We train the test-bed system using three training schemes: Independent training, AGR, and AO. Depending on the performance of independent and AGR training, we fine tune $E$ and $\alpha$ to balance the emphasis between retrieval and adaptation in AO. These detailed parameter settings are omitted because they vary across the data sets but can be found in the source code. Instead, Sect. 6 provides general guidelines on how to set the parameters. We evaluate the testbed system performance by mean square prediction error.

We note that the training procedures may not be optimal for each individual training scheme, but this approach ensures fairness between training schemes, making it suitable for comparing their effects.

## 5.2   Experimental Results

All experimental results are shown in Table 1. Each row represents the results over one data set. The histograms provide comparative information about accuracy with retrieval alone, with retrieval followed by trained adaptation, and the accuracy difference between the two, under the three training schemes.

The X axis of each histogram reflects a level of error. Between rows, different scales are used for different data sets, and outliers are included as the left- and right-most buckets. In the first two columns, bars in each histogram reflect the number of trials of a given method for which the error was at a given level. Consequently, higher bars to the left of the histograms reflect better performance. In the third column, bars represent the number of times the difference between error after adaptation and error of the solution of the retrieved case was at a particular level, i.e., how adaptation changes error. The bars for independent training are black, for AGR are light gray, and for AO are dark gray.

Ideally, retrieval will retrieve a similar case and adaptation will improve the solution. This is shown for all runs of AO for EE and AS, for the overwhelming majority of runs for YH, and for the majority of runs for SP and Ar. However, independent training and AGR training show less ideal effects as follows:

A **Independent training scheme (black bars)**: Here adaptation does not necessarily improve the results of retrieval. The retrieval stage may retrieve a case with a solution close to the real solution that the adaptation stage changes for the worse. This can be observed for the majority of runs with EE, YH, Ar, and SP.

B **AGR training scheme (light gray bars)**: Here the retrieval stage generally provides poor initial solutions while the adaptation stage makes substantial corrections (*i.e.*, results in a large negative change to the error). This can be observed for all data sets.

Even when adaptation can generate a successful solution, if moderate adaptation from the retrieved solution is important for explanation, the large adaptations in this condition are undesirable. There are also extreme scenarios in which

**Table 1.** Black: independent training; light gray: AGR training; dark gray: AO

| Data | Retrieval Only | After Adaptation | Δ by Adaptation |
|------|----------------|------------------|-----------------|
| EE | | | |
| YH | | | |
| SP | | | |
| AS | | | |
| Ar | | | |

**Table 2.** Average error by systems under different training schemes

|             | EE    | YH    | SP    | AS   | Ar    |
| ----------- | ----- | ----- | ----- | ---- | ----- |
| Independent | 0.291 | 0.172 | 0.123 | 1.58 | 0.693 |
| AGR         | 0.149 | 0.226 | 0.306 | 1.74 | 0.323 |
| AO          | 0.106 | 0.164 | 0.158 | 1.61 | 0.209 |

retrieval is so poor that even a strong adaptation stage generates bad solutions, as seen in experiments with YH, SP, AS, and Ar.

In comparison, under AO, the retrieval stage generally provides a good initial case, and adaptation further modifies the solution to be closer to the correct solution. In all experiments except with SP, AO shows this reliable behavior while the other two training schemes suffer codependence effects to some degree. As shown in Table 2, the average errors under AO are often better than those of other schemes.

AO decreases the codependence effects between retrieval and adaptation according to parameter settings in Algorithm 1. To alleviate effect (A), training is done on batches of training samples, instead of all the samples; Lines 17 and 18 train for a set number of epochs/steps instead of until convergence, so that neither retrieval nor adaptation fully assumes the accuracy burden before the other is trained. To alleviate effect (B), formula (2) is expanded into formula (7), and fine-tuning $\alpha$ and $E$ enables more control over how each stage is trained.

## 6    Guidelines for Applying AO to Train CBR Components

The experiments illustrate the benefit of AO in the testbed system. This section provides general questions and guidelines for applying AO to other CBR systems.

- **When can AO be applied?**
  AO is applicable if both retrieval and adaptation have training procedures which iteratively minimize loss functions. If loss functions can be formulated for formula (3) and formula (4), then AO can be used to harmonize training.
- **When may AO be useful?**
  AO may be useful in circumstances such as (1) when the effective scope of adaptations is limited, so that the effectiveness of adaptation depends strongly on the cases from which adaptation starts (in contrast, if adaptation can succeed from a wide range of starting points, AO would be expected to be less useful), (2) when successive training might lead to a local minimum in overall loss (*e.g.*, error or adaptation cost), but not a global minimum, (3) when it is desirable to control the balance between contributions of retrieval and adaptation, *e.g.*, to increase explainability by retrieving cases requiring only small adaptations.
- **How can a CBR system be trained with AO?**
  Applying Algorithm 1 requires three steps:
  1. Design the loss functions $f$ and $g$ for formula (3) and formula (4). The adaptation loss $f$ reflects quality of the adapted solution and/or

adaptation process (*e.g.*, solution error, adaptation cost, or a combination), and the retrieval loss $g$ reflects quality of the retrieved case (*e.g.*, accuracy of the retrieved case solution without adaptation, or explainability factors such as similarity between the problems of the query and retrieved case and closeness of the retrieved solution to the system solution). The loss functions should be compatible with training processes for the retrieval and adaptation stages, such that gradient descent can guide training.

2. Choose $\alpha$ and $E$. This can be done by training retrieval and adaptation independently and noting losses induced by $g$ and $f$. The choice of $\alpha$ may emphasize training the process with more room for improvement or be based on other criteria.

3. Determine *batch_size* and the training steps (lines 17 and 18). These design choices influence how fast $x_r$ and $x_a$ converge. For a goal of balanced training, these should be tuned so that retrieval and adaptation converge at a similar rate.

– **Does AO guarantee an optimal solution?**
  No. As detailed in Bezdek and Hathaway [1], the convergence of AO relies on the assumption that each subproblem will converge. In our study, this means that AO will converge if both training of retrieval and adaptation can converge. Additionally, AO may converge to a local minimum or saddle points instead of a global minimum. However, these issues are beyond the scope of this study.

# 7 Future Work

**AO and Explainability:** In this study we used the error of the retrieved solution as a proxy for the explainability of the system solution. Other factors could affect explainability, such as problem similarity according to user criteria (which might differ from system criteria) or the adaptation distance. Such criteria and the benefit of AO for their optimization would be interesting to investigate.

**Alternative Tasks and Loss Functions:** It would be interesting to examine the comparative value of AO for tasks beyond regression and for other loss functions (*e.g.*, efficiency, solution execution cost). A challenge for a broader class of tasks is the development of appropriate loss functions and training procedures.

**Extending AO to the Full CBR Cycle:** The work in this paper explores AO for retrieval and adaptation. However, whether to retain a new case in the case base depends strongly on, and also influences, both retrieval and adaptation capabilities. Consequently, an interesting direction is an extended version of AO to harmonize retention as well.

For example, retention training could optimize to retain cases for which changing adaptation capabilities are most accurate (or most efficient), or could adjust indexing as the similarity criteria for retrieval change. If a suitable loss function and training process could be defined, this could be done within the AO framework; if not, adjustments of retention could still be interleaved with AO for retrieval and adaptation. This raises questions of processing cost and how to determine when the additional processing would be worthwhile.

## 8   Conclusion

Extensive research has examined learning methods for improving particular CBR knowledge containers such as retrieval and adaptation knowledge. However, it is well known that the knowledge containers of CBR are closely connected and overlapping. More knowledge in one sometimes compensates for less in another [20] and coordination between containers can improve overall performance [7,22]. The interaction between retrieval and adaptation has primarily been addressed by treating one component as fixed and adjusting the other. In such scenarios, it is possible that the overall system will reach an undesirable balance between retrieval and adaptation, or will reach a local performance maximum instead of a global one. By iteratively harmonizing both, alternating optimization can achieve a combination of good retrieval and good adaptation. In our experiments, independent training and AGR training suffer codependence effects in all five data sets while AO works well in four of the five. This work also suggests the potential to explore using AO to harmonize other processes of CBR.

**Acknowledgments.** We acknowledge support from the Department of the Navy, Office of Naval Research (Award N00014-19-1-2655), and the US Department of Defense (Contract W52P1J2093009).

## References

1. Bezdek, J.C., Hathaway, R.J.: Some notes on alternating optimization. In: Pal, N.R., Sugeno, M. (eds.) AFSS 2002. LNCS (LNAI), vol. 2275, pp. 288–300. Springer, Heidelberg (2002). https://doi.org/10.1007/3-540-45631-7_39
2. Craw, S., Jarmulak, J., Rowe, R.: Learning and applying case-based adaptation knowledge. In: Aha, D.W., Watson, I. (eds.) ICCBR 2001. LNCS (LNAI), vol. 2080, pp. 131–145. Springer, Heidelberg (2001). https://doi.org/10.1007/3-540-44593-5_10
3. Dua, D., Graff, C.: UCI machine learning repository (2017). http://archive.ics.uci.edu/ml
4. D'Aquin, M., Badra, F., Lafrogne, S., Lieber, J., Napoli, A., Szathmary, L.: Case base mining for adaptation knowledge acquisition. In: Proceedings of the Twentieth International Joint Conference on Artificial Intelligence (IJCAI-07), pp. 750–755. Morgan Kaufmann, San Mateo (2007)
5. Hanney, K., Keane, M.T.: Learning adaptation rules from a case-base. In: Smith, I., Faltings, B. (eds.) EWCBR 1996. LNCS, vol. 1168, pp. 179–192. Springer, Heidelberg (1996). https://doi.org/10.1007/BFb0020610
6. Jalali, V., Leake, D.: Enhancing case-based regression with automatically-generated ensembles of adaptations. J. Inf. Syst. **46**(2), 237–258 (2015). https://doi.org/10.1007/s10844-015-0377-0
7. Leake, D., Kinley, A., Wilson, D.: Learning to integrate multiple knowledge sources for case-based reasoning. In: Proceedings of the Fourteenth International Joint Conference on Artificial Intelligence, pp. 246–251. Morgan Kaufmann (1997)

8. Leake, D., Ye, X.: On combining case adaptation rules. In: Bach, K., Marling, C. (eds.) ICCBR 2019. LNCS (LNAI), vol. 11680, pp. 204–218. Springer, Cham (2019). https://doi.org/10.1007/978-3-030-29249-2_14

9. Leake, D., Ye, X., Crandall, D.: Supporting case-based reasoning with neural networks: an illustration for case adaptation. In: Proceedings of AAAI Spring Symposium AAAI-MAKE 2021: Combining Machine Learning and Knowledge Engineering (2021). https://www.aaai-make.info/program

10. Liao, C., Liu, A., Chao, Y.: A machine learning approach to case adaptation. In: 2018 IEEE First International Conference on Artificial Intelligence and Knowledge Engineering (AIKE), pp. 106–109 (2018)

11. López de Mántaras, R., et al.: Retrieval, reuse, revision, and retention in CBR. Knowl. Eng. Rev. **20**(3) (2005)

12. Martin, K., Wiratunga, N., Sani, S., Massie, S., Clos, J.: A convolutional Siamese network for developing similarity knowledge in the selfback dataset. In: ICCBR (2017)

13. Mathisen, B.M., Aamodt, A., Bach, K., Langseth, H.: Learning similarity measures from data. Progr. Artif. Intell. **9**, 129–143 (2019). https://doi.org/10.1007/s13748-019-00201-2

14. McDonnell, N., Cunningham, P.: A knowledge-light approach to regression using case-based reasoning. In: Roth-Berghofer, T.R., Göker, M.H., Güvenir, H.A. (eds.) ECCBR 2006. LNCS (LNAI), vol. 4106, pp. 91–105. Springer, Heidelberg (2006). https://doi.org/10.1007/11805816_9

15. Minor, M., Bergmann, R., Gorg, S.: Case-based adaptation of workflows. Inf. Syst. **40**, 142–152 (2014)

16. Nugent, C., Cunningham, P.: A case-based recommender for black-box systems. Artif. Intell. Rev. **24**(2), 163–178 (2005)

17. Petrovic, S., Khussainova, G., Jagannathan, R.: Knowledge-light adaptation approaches in case-based reasoning for radiotherapy treatment planning. Artif. Intell. Med. **68**, 17–28 (2016)

18. Policastro, C.A., Carvalho, A.C.P.L.F., Delbem, A.C.B.: Hybrid approaches for case retrieval and adaptation. In: Günter, A., Kruse, R., Neumann, B. (eds.) KI 2003. LNCS (LNAI), vol. 2821, pp. 297–311. Springer, Heidelberg (2003). https://doi.org/10.1007/978-3-540-39451-8_22

19. Policastro, C.A., Carvalho, A.C., Delbem, A.C.: Automatic knowledge learning and case adaptation with a hybrid committee approach. J. Appl. Log. **4**(1), 26–38 (2006)

20. Richter, M.M.: Introduction. In: Lenz, M., Burkhard, H.-D., Bartsch-Spörl, B., Wess, S. (eds.) Case-Based Reasoning Technology. LNCS (LNAI), vol. 1400, pp. 1–15. Springer, Heidelberg (1998). https://doi.org/10.1007/3-540-69351-3_1

21. Shiu, S., Yeung, D., Sun, C., Wang, X.: Transferring case knowledge to adaptation knowledge: an approach for case-base maintenance. Comput. Intell. **17**(2), 295–314 (2001). https://doi.org/10.1111/0824-7935.00146

22. Smyth, B., Keane, M.: Adaptation-guided retrieval: Questioning the similarity assumption in reasoning. Artif. Intell. **102**(2), 249–293 (1998)

23. Wettschereck, D., Aha, D., Mohri, T.: A review and empirical evaluation of feature-weighting methods for a class of lazy learning algorithms. Artif. Intell. Rev. **11**(1–5), 273–314 (1997). https://doi.org/10.1023/A:1006593614256

24. Wiratunga, N., Craw, S., Rowe, R.: Learning adaptation knowledge to improve case-based reasoning. Artif. Intell. **170**, 1175–1192 (2006)

# Adaptation Knowledge Discovery Using Positive and Negative Cases

Jean Lieber and Emmanuel Nauer[✉]

CNRS, Inria, LORIA, Université de Lorraine, 54000 Nancy, France
Emmanuel.Nauer@loria.fr

**Abstract.** Case-based reasoning usually exploits positive source cases, each of them consisting in a problem and a correct solution to this problem. Now, the general issue of exploiting also negative cases— i.e., problem-solution pairs where the solution answers incorrectly the problem—can be raised. Indeed, such cases are "naturally" generated by a CBR system as long as it sometimes proposes incorrect solutions. This paper aims at addressing this issue for adaptation knowledge (AK) discovery: how positive and negative cases can be used for this purpose. The idea is that positive cases are used to propose adaptation rules and that negative cases are used to filter out some of these rules. In a preliminary work, this kind of AK discovery has been applied using frequent closed itemset (FCI) extraction on variations within the case base and tested on a toy Boolean use case, with promising first results. This paper resumes this study and evaluates it on 4 benchmarks, which confirms the benefit of exploiting negative cases for AK discovery. This involves some adjustments in the data preparation and in adaptation rule filtering, in particular because FCI extraction works only with Boolean features, hence some methodology lessons learned for AK discovery with positive and negative cases.

**Keywords:** Adaptation knowledge discovery · Negative cases · Closed itemset extraction · Case-based reasoning

## 1 Introduction

Case-based reasoning (CBR) [13] aims at solving a new problem—the *target problem*—thanks to a set of cases (the *case base*), where a case is a pair consisting of a problem and a solution to this problem. A *source case* is a case from the case base, consisting of a *source problem* and one of its solutions. The classical approach to CBR consists in selecting source cases *similar* to the target problem and adapting them to solve it. The adaptation step may use different approaches, one of them is the use of adaptation knowledge (AK) which can be acquired automatically using AK discovery processes. For this purpose, CBR usually exploits positive source cases [1–3,5]. However, CBR systems sometimes produce also incorrect solutions, generating negative cases, i.e. problem-solution

© Springer Nature Switzerland AG 2021
A. A. Sánchez-Ruiz and M. W. Floyd (Eds.): ICCBR 2021, LNAI 12877, pp. 140–155, 2021.
https://doi.org/10.1007/978-3-030-86957-1_10

pairs where the solution answers incorrectly the problem. These negative cases may play an important role, especially in the knowledge discovery process, by improving the quality of the AK being extracted and, so, by improving the results of the CBR system itself [8].

The objective of this paper is to study how the exploitation of negative cases (in addition to positive cases), which was only evaluated on a toy use case, can be used on different applications. Several existing machine learning datasets have been used for this purpose. The initial process of [8], which requires the representation of the problem and of the solution with Boolean attributes requires two major adaptations. First, the cases which are represented by other types of attributes in particular by nominal values, has to be transformed to fit the Boolean representation. Second, adaptation rules extracted by the AK process have to be filtered: this is an application-dependent task. The idea, beside this filtering, is to use the most efficient adaptation rules to solve a new problem.

The paper is organized as follows. Section 2 introduces the motivations and the preliminaries for this work. A reminder about frequent closed itemset (FCI) extraction and CBR are presented. Section 3 presents the specific approach for exploiting positive and negative cases in an AK discovery process. Section 4 introduces a dataset selection coming from a machine learning test database, and presents the evaluation of the AK discovery approach based on positive and negative cases on 4 selected benchmarks. Section 5 concludes and highlights the methodology lessons learned for AK discovery and then points out lines for future research.

## 2 Motivations and Preliminaries

Many types of CBR applications are concerned by the use of AK and especially adaptation rules to solve new problems. For example, it has been shown that using AK can benefit to TAAABLE, a cooking system [5] which adapts cooking recipes. In [8], a preliminary study on automatically generated datasets has shown that exploiting negative cases in addition to positive ones for AK discovery improves the quality of the adaptation rules being extracted, which, in turn, improves the result of the CBR system. The approach is based on a Boolean representation of the cases. The AK discovery process uses variations between pairs of cases and FCI extraction to compute the adaptation rules.

This Sect. 1 introduces some assumptions and notations about CBR. The Boolean setting which is at the basis of this work is then explained, and in particular, it is explained how the cases are represented in this Boolean framework, how the variations between cases are represented and computed, and, finally, how FCI extraction can be applied to compute adaptation rules that can be used to solve a new problem.

### 2.1 Assumptions and Notations About CBR

Let $\mathcal{P}$ and $\mathcal{S}$ be two sets. A *problem* (resp., a *solution*) is an element of $\mathcal{P}$ (resp., of $\mathcal{S}$). The existence of a binary relation with the semantics "has for solution" is assumed, though it is not completely known to the CBR system. Moreover, it is assumed that this relation is functional (every problem has exactly one correct solution). Let $f$ be the function from $\mathcal{P}$ to $\mathcal{S}$ such that $y = f(x)$ if $y$ is the solution of $x$. A *case* is a pair $(x, y) \in \mathcal{P} \times \mathcal{S}$ where $y = f(x)$. The fact that "has for solution" is functional is useful for making the evaluation process automatic: $y$ is a *correct* solution to $x$ iff $y = f(x)$, hence if $(x, y)$ belongs to the test set and $\hat{y}$ is a solution returned by the CBR system, this solution is correct iff $\hat{y} = y$.

A CBR system on $(\mathcal{P}, \mathcal{S}, f)$ is built with a knowledge base $\mathrm{KB} = (\mathrm{CB}, \mathrm{DK}, \mathrm{RK}, \mathrm{AK})$ where $\mathrm{CB}$, the case base, is a finite set of cases, $\mathrm{DK}$ is the domain knowledge, $\mathrm{RK}$ is the retrieval knowledge (in this work, $\mathrm{RK} = \mathtt{dist}$, a distance function on $\mathcal{P}$), and $\mathrm{AK}$ is the adaptation knowledge that takes the form of adaptation rules.

A CBR system on $(\mathcal{P}, \mathcal{S}, f)$ aims at associating to a query problem $x^{\mathrm{tgt}}$ a $y^{\mathrm{tgt}} \in \mathcal{S}$, denoted by $y^{\mathrm{tgt}} = f_{\mathrm{CBR}}(x^{\mathrm{tgt}})$. The function $f_{\mathrm{CBR}}$ is intended to be an approximation of $f$. It is built thanks to the following functions:

- the retrieval function, with the profile $\mathtt{retrieval} : x^{\mathrm{tgt}} \mapsto (x^s, y^s) \in \mathrm{CB}$;
- the adaptation function, with the profile $\mathtt{adaptation} : ((x^s, y^s), x^{\mathrm{tgt}}) \mapsto y^{\mathrm{tgt}} \in \mathcal{S}$; it is usually based on $\mathrm{DK}$ and $\mathrm{AK}$. $((x^s, y^s), x^{\mathrm{tgt}})$ is an *adaptation problem*.

Thus $f_{\mathrm{CBR}}(x^{\mathrm{tgt}}) = \mathtt{adaptation}(\mathtt{retrieval}(x^{\mathrm{tgt}}), x^{\mathrm{tgt}})$.

With no domain knowledge and no adaptation knowledge ($\mathrm{DK} = \emptyset$ and $\mathrm{AK} = \emptyset$), the adaptation consists usually of a mere copy of the solution. This process is called *null adaptation*:

$$\mathtt{null\_adaptation} : ((x^s, y^s), x^{\mathrm{tgt}}) \mapsto y^s$$

**Adaptation principle using adaptation rules.** Generally speaking, an adaptation rule $\mathtt{ar}$ is a function mapping an adaptation problem $((x^s, y^s), x^{\mathrm{tgt}}) \in \mathrm{CB} \times \mathcal{P}$ to $y^{\mathrm{tgt}} \in \mathcal{S} \cup \{\mathtt{failure}\}$. A cases of failure ($y^{\mathrm{tgt}} = \mathtt{failure}$) occurs when no adaptation rule $\mathtt{ar}$ is applicable on this adaptation problem. Else, $y^{\mathrm{tgt}}$ is a proposed solution to $x^{\mathrm{tgt}}$, by adaptation of $(x^s, y^s)$ according to $\mathtt{ar}$.

A score $\mathtt{supp}(\mathtt{ar}) \geq 0$, called the support of the rule $\mathtt{ar}$, is associated with $\mathtt{ar}$; the higher is $\mathtt{supp}(\mathtt{ar})$, the more $\mathtt{ar}$ is preferred.

The adaptation consists in selecting the subset $\mathrm{AAR}$ of $\mathrm{AK}$ of applicable adaptation rules with maximum support: $\mathtt{ar} \in \mathrm{AAR}$ iff $\mathtt{ar}((x^s, y^s), x^{\mathrm{tgt}}) \neq \mathtt{failure}$ and there exists no $\mathtt{ar}' \in \mathrm{AAR}$ such that $\mathtt{supp}(\mathtt{ar}') > \mathtt{supp}(\mathtt{ar})$.

### 2.2 Boolean Setting Illustrated with a Boolean Function Example

Initially, the Boolean setting has been chosen in [8] to validate the AK discovery approach by experiments using automatically generated Boolean functions as $f$. Let $\mathbb{B} = \{0, 1\}$ be the set of Boolean values. The Boolean operators are denoted by the connector symbols of propositional logic: for $a, b \in \mathbb{B}$, $\neg a = 1 - a$, $a \wedge b = \min(a, b)$, $a \vee b = \max(a, b)$. Let $p \geq 1$. In the examples, an element of $\mathbb{B}^p$ is noted without parentheses and commas: $(0, 0, 1, 1, 0, 0)$ is simply noted by $001100$. The

**Table 1.** An example of 4 problems of $\mathbb{B}^6$ with their solution on $\mathbb{B}$ for $\mathbf{f}(x) = x_1 \vee (x_2 \wedge x_3) \vee \neg x_4$.

|       | $x_1$ | $x_2$ | $x_3$ | $x_4$ | $x_5$ | $x_6$ | $\mathbf{f}$ |
|-------|-------|-------|-------|-------|-------|-------|--------------|
| $c^1$ | 0     | 0     | 1     | 1     | 0     | 0     | 0            |
| $c^2$ | 0     | 1     | 0     | 0     | 1     | 1     | 1            |
| $c^3$ | 0     | 1     | 1     | 0     | 0     | 0     | 1            |
| $c^4$ | 1     | 0     | 1     | 1     | 1     | 0     | 1            |

Hamming distance $\mathtt{dist}$ on $\mathbb{B}^p$ is defined by $\mathtt{dist}(a, b) = \sum_{i=1}^{p} |b_i - a_i|$. For example, with $p = 6$, $\mathtt{dist}(001100, 101110) = 2$.

Let $m, n \in \mathbb{N}^*$, $\mathcal{P} = \mathbb{B}^m$, $\mathcal{S} = \mathbb{B}^n$ and $\mathbf{f} : \mathcal{P} \to \mathcal{S}$, be a Boolean function to be approximated. A CBR system is considered on $(\mathcal{P}, \mathcal{S}, \mathbf{f})$ with $\mathtt{DK} = \emptyset$, $\mathtt{RK} = \mathtt{dist}$, the Hamming distance on $\mathcal{P}$, and AK a set of adaptation rules.

Table 1 introduces 4 cases on $\mathcal{P} = \mathbb{B}^6$, $\mathcal{S} = \mathbb{B}$, with a given $\mathbf{f}$.

**Adaptation rule language.** The adaptation rule language used in this work is based on the notion of variations between Booleans, as described hereafter. Given $\ell, r \in \mathbb{B}$ ($\ell$ stands for *left*, $r$ for *right*), the variation from $\ell$ to $r$ is represented by *variation symbols*. Each of the 4 ordered pairs $(\ell, r)$ is represented by a variation symbol $v$:

- $(\ell, r) = (1, 0)$ is represented by $v = -$;
- $(\ell, r) = (0, 1)$ is represented by $v = +$;
- $(\ell, r) = (0, 0)$ is represented by $v = = 0$;
- $(\ell, r) = (1, 1)$ is represented by $v = = 1$.

Given two cases $c^1 = (\mathbf{x}^1, \mathbf{y}^1)$ and $c^2 = (\mathbf{x}^2, \mathbf{y}^2)$, the variation $V^{12}$ from $c^1$ to $c^2$ is encoded by the set of the expressions $\mathbf{x}_i^v$ and $\mathbf{y}_j^w$ such that $v$ (resp., $w$) is a variation symbol from $\mathbf{x}_i^1$ to $\mathbf{x}_i^2$ (resp., from $\mathbf{y}_j^1$ to $\mathbf{y}_j^2$). For example, the variations from $c^2 = ((010011), 1)$ to $c^3 = ((011000), 1)$ are $V^{23} = \{\mathbf{x}_1^{=0}, \mathbf{x}_2^{=1}, \mathbf{x}_3^+, \mathbf{x}_4^{=0}, \mathbf{x}_5^-, \mathbf{x}_6^-, \mathbf{y}_1^{=1}\}$.

An *adaptation rule* $\mathtt{ar}$ is a set of expressions $\mathbf{x}_i^v$ and $\mathbf{y}_j^w$. It is applicable on an adaptation problem $((\mathbf{x}^s, \mathbf{y}^s), \mathbf{x}^{\mathtt{tgt}})$ if there exists $\mathbf{y}^{\mathtt{tgt}} \in \mathbb{B}^n$ such that $V^{st} \supseteq \mathtt{ar}$ (where $V^{st}$ represents the variation from $(\mathbf{x}^s, \mathbf{y}^s)$ to $(\mathbf{x}^{\mathtt{tgt}}, \mathbf{y}^{\mathtt{tgt}})$). If it is applicable, then its application consists in choosing such a $\mathbf{y}^{\mathtt{tgt}}$. If several $\mathbf{y}^{\mathtt{tgt}}$'s exist, the chosen one is the closest to $\mathbf{y}^s$ according to the Hamming distance on $\mathcal{S} = \mathbb{B}^n$, meaning that if $\mathtt{ar}$ gives no constraint on some $\mathbf{y}_j^{\mathtt{tgt}}$ then $\mathbf{y}_j^{\mathtt{tgt}} = \mathbf{y}_j^s$. For example:

if $\mathtt{ar} = \{\mathbf{x}_1^{=0}, \mathbf{x}_2^-, \mathbf{x}_3^+, \mathbf{y}_1^+\}$, $(\mathbf{x}^s, \mathbf{y}^s) = (010, 00)$ and $\mathbf{x}^{\mathtt{tgt}} = 001$

then $\mathtt{ar}$ is applicable on $((\mathbf{x}^s, \mathbf{y}^s), \mathbf{x}^{\mathtt{tgt}})$ and $\mathtt{ar}((\mathbf{x}^s, \mathbf{y}^s), \mathbf{x}^{\mathtt{tgt}}) = \mathbf{y}^{\mathtt{tgt}} = 10$

### 2.3  Itemset Extraction

Itemset extraction is a collection of data-mining methods for extracting regularities into data, by aggregating object items appearing together. Like formal concept analysis [7], itemset extraction algorithms start from a *formal context*

**Table 2.** An example of formal context representing 12 case variations described by 28 properties: 24 properties describing the variations on the problem and 4 properties describing the variations on the solution.

| | $x_1{=}1$ | $x_1{=}0$ | $x_1-$ | $x_1+$ | $x_2{=}1$ | $x_2{=}0$ | $x_2-$ | $x_2+$ | $x_3{=}1$ | $x_3{=}0$ | $x_3-$ | $x_3+$ | $x_4{=}1$ | $x_4{=}0$ | $x_4-$ | $x_4+$ | $x_5{=}1$ | $x_5{=}0$ | $x_5-$ | $x_5+$ | $x_6{=}1$ | $x_6{=}0$ | $x_6-$ | $x_6+$ | $y_1{=}1$ | $y_1{=}0$ | $y_1-$ | $y_1+$ |
|---|---|---|---|---|---|---|---|---|---|---|---|---|---|---|---|---|---|---|---|---|---|---|---|---|---|---|---|---|
| $V^{12}$ | × | | | | | | × | | × | | | | × | | | | × | | | | | × | | | | | | × |
| $V^{13}$ | × | | | | × | | | | × | | | | × | | | | × | | | | | × | | | | | | × |
| $V^{14}$ | | | × | | | | × | | × | | | | | | × | | × | | | | | × | | | | | | × |
| $V^{21}$ | × | | | | × | | | | × | | | | × | | | | × | | | | | × | | | | | × | |
| $V^{23}$ | × | | | | | | | × | × | | | | × | | | | × | | | | | × | | | | × | | |
| $V^{24}$ | | | × | | × | | | | × | | | | | | | × | × | | | | | × | | | | × | | |
| $V^{31}$ | × | | | | × | | | | × | | | | × | | | | × | | | | | × | | | | | × | |
| $V^{32}$ | × | | | | | | | × | | | × | | × | | | | × | | | | | | × | | | × | | |
| $V^{34}$ | | | × | | × | | | | × | | | | × | | | | × | | | | | × | | | | × | | |
| $V^{41}$ | | | | × | | | × | | × | | | | | | × | | × | | | | | × | | | | | × | |
| $V^{42}$ | × | | | | × | | | | × | | | | × | | | | | | | × | | × | | | | × | | |
| $V^{43}$ | × | | | | × | | | | × | | | | × | | | | × | | | | | × | | | | × | | |

$K$, defined by $K = (G, M, I)$, where $G$ is a set of objects, $M$ is a set of items, and $I$ is the relation on $G \times M$ stating that an object is described by an item [7]. Table 2 shows an example of context, in which the 4 cases of Table 1 have been used to generate 12 case variations (variation combinations between all the possible pairs of distinct cases $c^1$, $c^2$, $c^3$, $c^4$) are described by 28 properties ($x_i^v$ and $y_j^w$ for $i \in [1, 6]$, $j \in \{1\}$ and $v, w \in \{-, +, = 0, = 1\}$). $G$ is a set of 12 objects ($V^{12}$, $V^{13}$, ..., $V^{43}$), $M$ is a set of 28 variation items.

An *itemset* $I$ is a set of items, and the *support* of $I$, $\mathrm{supp}(I)$, is the number of objects of the formal context having every items of $I$. $I$ is frequent, with respect to a threshold $\tau_{\mathrm{supp}}$, whenever $\mathrm{supp}(I) \geq \tau_{\mathrm{supp}}$. $I$ is closed if it has no proper superset $J$ ($I \subsetneq J$) with the same support. For example, $\{x_2^{=1}\}$ is an itemset and $\mathrm{supp}(\{x_2^{=1}\}) = 2$ because exactly 2 cases have the property $x_2^{=1}$. However, $\{x_2^{=1}\}$ is not a closed itemset, because $\{x_2^{=1}, y_1^{=1}\}$ has the same support. For $\tau_{\mathrm{supp}} = 6$, the frequent closed itemsets (FCIs) of this context are $\{x_1^{=0}\}$, $\{y_1^{=1}\}$, and $\{x_3^{=1}, x_6^{=0}\}$.

The experiments use CORON, a software platform which implements efficient algorithms for symbolic data mining and especially FCI computation [14].

## 3   Exploiting Case Variations for Adaptation Knowledge Discovery with Positive and Negative Cases

Exploiting case variations is not a new idea. [9] introduces this approach of AK learning based on pairwise comparisons of cases. This approach, also called *Case Difference Heuristic* in [10], has been applied in various domains such as medicine [3] or cooking [1,5].

So, for ordered pairs of cases $(c^\ell, c^r)$ associated to their variations $V^{\ell r}$ forming a formal context, as the one presented in Table 2, an AK discovery process based on FCI can be run. Each extracted FCI produces an adaptation rule $\mathtt{ar}$ with a support $\mathtt{supp(ar)}$, i.e. the number of $V^{\ell r}$ containing $\mathtt{ar}$. For example, for $\tau_{\mathtt{supp}} = 2$, $\{x_1^{=0}, x_2^{+}, x_4^{-}, y_1^{+}\}$ is an FCI which produces an adaptation rule. For a source case $(00\cdot1\cdot\cdot, 0) \in \mathbb{B}^6 \times \mathbb{B}$, where each $\cdot$ represents 0 or 1, This adaptation rule can be used for solving any target problem $01\cdot0\cdot\cdot$. The proposed solution is 1, obtained by applying a $+$ variation on the source solution 0.

The adaptation rule set extracted using this FCI approach depends on how the formal context is built, and in particular whether it is built only using positive source cases, or using also negative source cases. For a case $(\mathbf{x}, \mathbf{y})^s = (\mathbf{x}^s, \mathbf{y}^s) \in \mathtt{CB}$, the case $(\mathbf{x}, \mathbf{y})^s$ is said *positive* if $\mathbf{y}^s$ is a correct solution for $\mathbf{x}^s$ ($\mathbf{y}^s = \mathtt{f}(\mathbf{x}^s)$) and *negative* otherwise. $\mathtt{CB}^+$ (resp. $\mathtt{CB}^-$) denotes the set of positive (resp. negative) cases of $\mathtt{CB}$, with $\mathtt{CB} = \mathtt{CB}^+ \cup \mathtt{CB}^-$ and $\mathtt{CB}^+ \cap \mathtt{CB}^- = \emptyset$.

Starting from the two sets of cases $\mathtt{CB}^+$ and $\mathtt{CB}^-$, ordered pairs of cases $(\mathbf{c}^1, \mathbf{c}^2)$ are formed, with $\mathbf{c}^1 = (\mathbf{x}^1, \mathbf{y}^1) \in \mathtt{CB}^+$, $\mathbf{c}^2 = (\mathbf{x}^2, \mathbf{y}^2) \in \mathtt{CB}$ and $\mathbf{x}^1 \neq \mathbf{x}^2$. Each such pair is encoded by a set $V^{12}$ of the variations from $\mathbf{x}_i^1$ to $\mathbf{x}_i^2$ and from $\mathbf{y}_j^1$ to $\mathbf{y}_j^2$, as presented before. When $\mathbf{c}^2 \in \mathtt{CB}^+$, the variation from $\mathbf{c}^1$ to $\mathbf{c}^2$ can be considered as a positive example of adaptation rule (i.e. the application of the rule produces a correct answer). When $\mathbf{c}^2 \in \mathtt{CB}^-$, the variation from $\mathbf{c}^1$ to $\mathbf{c}^2$ can be considered as a negative example of adaptation rule (i.e. the application of the produces an incorrect answer).

The AK learning process based on FCI extraction takes as input a set of $V^{\ell r}$ which is used to build the formal context. In [8], two approaches have been used to build the formal context. The first one consists in using each $V^{\ell r}$ as an object with variations between pairs of cases of $\mathtt{CB}^+$ as properties. This approach is denoted by $AK^+$ in the following. A limit of $AK^+$ is that it may produce too general adaptation rules (this issue has been discussed in [4]). For example, the formal context presented in Table 2 produces 15 rules containing a variation on $y_1$ for $\tau_{\mathtt{supp}} = 2$, e.g. $\{x_5^{=1}, y_1^{=1}\}$. This rule expresses the fact that only by knowing the variation $= 1$ on $x_5$, the outcome is $\mathbf{y}^1 = 1$.

As the application of such a general rule is likely to give an incorrect answer, a second approach exploiting negative cases to filter too general adaptation rules is also considered. This idea is inspired from machine learning approaches based on version spaces [12] or its link with formal concept analysis [6,11]: generating adaptation rules covering positive examples without covering negative ones. This approach exploiting both positive and negative cases is denoted by $AK^{\pm}$ in the following. Suppose that $\mathbf{c}^5 = (001111, 1) \in \mathtt{CB}^-$ (i.e. 1 is an incorrect solution for 001111 w.r.t. $\mathtt{f}$). Using this negative case in the AK process eliminates some rules, and in particular the rule $\{x_5^{=1}, y_1^{=1}\}$, because $\mathbf{c}^5$ is a counterexample of the application of the rule. Applying this rule on 001111, the problem part of $\mathbf{c}^5$, produces 1, the solution of $\mathbf{c}^5$, which is an incorrect result.

To illustrate the benefit of $AK^{\pm}$ w.r.t. $AK^+$ but also w.r.t. the classical nearest neighbor approach, consider the cases of Table 1 and let $\mathbf{x}^{\mathtt{tgt}} = 000111$. For the nearest neighbor approach, according to $\mathtt{dist}$, if considering the closest case of the case base w.r.t 000111 which is $(010011, 1)$ with $\mathtt{dist}(000111, 010011) = 2$, the result will be 1 (by $\mathtt{null\_adaptation}$). For the $AK^+$ solving approach, considering the same closest case than for the nearest neighbor approach and the rules computed on variations only between positive cases, the higher support rule $\{x_5^{=1}, y_1^{=1}\}$ can be applied with $c^2 = (010011, 1)$ and its application gives 1 as result. Considering now the same closest case than before but using rules computed on variations between positive cases but taking into account the negative examples. The previous rule $\{x_5^{=1}, y_1^{=1}\}$ is removed because of $c^5 = (001111, 0)$ which is a negative example for this rule. Another rule $\{x_1^{=0}, x_2^-, x_4^+, y_1^-\}$ can then be applied because of the variations $(x_1^{=0}, x_2^-, x_3^{=0}, x_4^+, x_5^{=1}, x_6^{=1}\}$ between $x^s = 010011$ and $x^{tgt} = 000111$. The variation $y_1^-$ can be applied on $y^s = 1$, producing 0 as (correct) result.

## 4   Experiments on Benchmarks

To test our approach on real applications, we have chosen to use benchmarks describing problems with their solutions. For this, the UCI Machine Learning Repository[1] and the Open ML website[2] have been used. These resources contain a great number of benchmarks. However, our Boolean representation formalism constrains to choose datasets where the problem and its solution are described by Boolean features or by features which can be transformed easily into Boolean ones (e.g. if an attribute is nominal, it can be transformed in several Boolean attributes, the value being encoded by only one true attribute among all these Boolean attributes).

For the experiments presented in this paper, we focused on datasets:

- where the solution is Boolean or nominal, typically benchmarks addressing a classification problem;
- containing enough data (i.e. at least 100 cases) because the experiments require enough data to build a learning dataset to extract adaptation rules and to have a testing dataset in order to evaluate the different problem solving approaches (nearest neighbors, $AK^+$ and $AK^\pm$).

The following sections not only present the execution on different benchmarks but also the methodological work of data preparation.

### 4.1   Experiment Setting and Evaluation Methodology

The objective of the evaluation is to study, on various benchmarks, how the Boolean approach exploiting negative cases in addition to positive ones improves the results of the CBR system.

For each dataset, the raw data describing the problem and its solution are transformed into a Boolean encoding. The result of this encoding is the case base CB. The size of CB, denoted by |CB|, depends on the dataset. For each run, CB is split in 2 subsets of the same size by a random selection of cases: $CB = CB_L \cup CB_T$, with $CB_L$, a set of cases used for learning AK, and $CB_T$, a set of cases to test the different approaches. $CB_L$ is then split in 2 subsets of the same size, also

---

[1]   https://archive.ics.uci.edu/ml/index.php.
[2]   https://www.openml.org/search?q=&type=data.

by a random selection of cases: $CB_L = CB^+ \cup CB^-$. $CB^+$ is the set of positive cases. To build $CB^-$, the set of negative cases, each negative source case $(x, y^-)$ is generated from a positive source case $(x, y)$ by a modification of the solution part. For $S = \mathbb{B}^n$, with $n = 1$, the modification of the solution part consists in taking the negation of y: $y^- = \neg y$. This is the case for the applications described in Sects. 4.2, 4.3 and 4.4. When $n \geq 2$, the modification of the solution part depends on the application. Such modification is detailed further, in the context of the application described in Sect. 4.5.

Three adaptation approaches are tested: $AK^+$, $AK^\pm$, and $NN$, the classical nearest neighbor approach with null_adaptation for adaptation function. For $NN$, retrieval consists in selecting the 3 most similar source cases to the target problem (according to dist) and adaptation consists in making a vote among their solution parts.

For the two approaches based on AK, all source cases for which adaptation rules can be applied participate to the problem solving. A vote on the results computed from the retrieved cases is used to associate a unique answer. Moreover, a vote is also used when using adaptation rules: 3 adaptation rules with the higher supports are used to adapt each of the source cases and the most frequent result wins.

The two AK-based approaches depends on two parameters: the number of rules being extracted by the knowledge discovery process and the specificity of the rules (i.e. the minimal size of the problem part of the rule). The impact of these two parameters on the results is illustrated and analyzed.

All the approaches are evaluated according to two measures: the precision prec and the correct answer rate car. Let ntp be the number of target problems posed to the system (ntp = $|CB_T|$), na be the number of (correct or incorrect) answers (ntp − na is the number of target problems for which the system fails to propose a solution), and nca be the number of correct answers. So, the precision prec is defined as the average of the ratios $\frac{nca}{na}$, and the correct answer rate car is defined as the average of the ratios $\frac{nca}{ntp}$. The average is computed on 100 runs for each evaluation, for different number of rules used for adaptation and for different sizes of rules. The way the rules are filtered is now detailed.

**Filtering Rules.** As mentioned above, using too general adaptation rules may produce incorrect results. General rules contain only variations on few variables, as more specific rules contain variations on more variables. In the TAAABLE project, some studies using adaptation rules on cooking recipes have shown that using more specific rules gives better results than using too general ones [5]. According to the idea that, all other things being equal, more a rule contains variables, less it is risky to use it, a given number of variables that must appear in the rules can be set (in particular for the variables $x_i$ linked to the problem description).

4 benchmarks have been chosen, each of them raising different issues. For each benchmark, a description of the dataset and its purpose are first presented, followed by the results of the experiments and the analysis of these results.

**Table 3.** `prec` (first 3 lines) and `car` (last 3 lines) of the three approaches for different numbers of rules and different minimal problem sizes for the Congressional Voting benchmark.

| nAR | 100 | | | | | 200 | | | | | 300 | | | | | 400 | | | | |
|---|---|---|---|---|---|---|---|---|---|---|---|---|---|---|---|---|---|---|---|---|
| minPS | 1 | 2 | 3 | 4 | 5 | 1 | 2 | 3 | 4 | 5 | 1 | 2 | 3 | 4 | 5 | 1 | 2 | 3 | 4 | 5 |
| prec | | | | | | | | | | | | | | | | | | | | |
| $NN$ | .87 | .88 | .88 | .88 | .87 | .88 | .88 | .87 | .87 | .88 | .87 | .88 | .88 | .88 | .87 | .88 | .87 | .87 | .87 | .88 |
| $AK^+$ | .77 | .84 | .91 | .93 | .95 | .78 | .82 | .91 | .92 | .94 | .77 | .78 | .91 | .93 | .93 | .82 | .83 | .91 | .92 | .93 |
| $AK^\pm$ | **.94** | **.94** | **.95** | **.96** | **.96** | **.94** | **.94** | **.94** | **.95** | **.95** | **.94** | **.94** | **.94** | **.95** | **.95** | **.95** | **.94** | **.94** | **.94** | **.95** |
| car | | | | | | | | | | | | | | | | | | | | |
| $NN$ | .87 | .88 | .88 | **.88** | **.87** | .88 | .88 | .87 | .87 | **.88** | .87 | .88 | .88 | .88 | **.87** | .88 | .87 | .87 | .87 | .88 |
| $AK^+$ | .77 | .84 | .91 | .88 | .82 | .78 | .82 | .91 | .90 | .86 | .77 | .78 | .91 | .92 | .87 | .82 | .83 | .91 | .91 | .89 |
| $AK^\pm$ | **.91** | **.91** | **.91** | .86 | .80 | **.94** | **.93** | **.93** | **.91** | .87 | **.93** | **.93** | **.94** | **.93** | .86 | **.94** | **.94** | **.94** | **.93** | **.90** |

**Table 4.** `prec` and `car` for $AK^+$ approach on the Congressional Voting Records benchmark with the use of the same number of positive cases for $AK^+$ adaptation rule extraction as for $AK^\pm$.

| nAR | 100 | | | | | 200 | | | | | 300 | | | | | 400 | | | | |
|---|---|---|---|---|---|---|---|---|---|---|---|---|---|---|---|---|---|---|---|---|
| minPS | 1 | 2 | 3 | 4 | 5 | 1 | 2 | 3 | 4 | 5 | 1 | 2 | 3 | 4 | 5 | 1 | 2 | 3 | 4 | 5 |
| prec | .76 | .83 | .91 | .92 | .94 | .82 | .87 | .92 | .91 | .92 | .80 | .87 | .93 | .91 | .91 | .79 | .87 | .93 | .91 | .92 |
| car | .76 | .83 | .91 | .88 | .84 | .82 | .87 | .92 | .90 | .87 | .80 | .87 | .93 | .90 | .88 | .79 | .87 | .93 | .91 | .88 |

## 4.2 Congressional Voting Records

This dataset[3] contains votes for each of the U.S. House of Representatives Congressmen on 16 key votes (e.g. education budget or duty free export). The value, for each key vote, can be 'yes', 'no', or 'no vote'. These 16 attributes, describing the votes of a given congressman (and which can be considered as the problem description), are linked to his/her political party, i.e. republican or democrat (which is the solution). So, the aim, for this benchmark, is to determine the political party of the voter from his/her votes. In this experiment, $\mathcal{P} = \mathbb{B}^{16}$ and $\mathcal{S} = \mathbb{B}$. The dataset contains 435 records: 232 records with a complete information about the problem ('yes' or 'no' for the 16 votes) and 203 with at least one 'no vote' value. For this first experiment, it has been chosen to remove the 203 records which do not have the complete information, and thus to keep only the 232 complete records. These 232 records produce 152 cases as there are identical records (exactly the same vote on the 16 questions). The distribution for the solution of these 152 cases is 95 democrats and 57 republicans.

Let `nAR` be the number of adaptation rules being used in the AK-based approaches, and `minPS`, the minimal problem size, i.e. the minimal number of variables linked to the problem that has to appear in the rules. Table 3 presents the results, for `nAR` $\in \{100, 200, 300, 400\}$ and $1 \leq$ `minPS` $\leq 5$. The first three lines give the `prec` score, the last three lines give the `car` score for 100 runs.

Some important points can be noticed:

---

[3] https://archive.ics.uci.edu/ml/datasets/Congressional+Voting+Records.

1. $AK^{\pm}$ is *systematically* better in prec than $NN$ and $AK^{+}$ with also a better car, independently of the values of the 2 parameters nAR and minPS.
2. For $AK^{+}$, if too less rules (100 or 200) and too general rules (minPS $\leq$ 2) are used, $AK^{+}$ gives not so good results: the prec and car scores are lower than for $NN$. However, with more specific rules, $AK^{+}$ overcomes $NN$ for prec and car.
3. The choice of rules of good quality (i.e. specific rules) has an important impact on the results for $AK^{+}$: for a given number of rules, the prec increases when minPS increases. Indeed, this parameter has a real impact in filtering too general rules. For $AK^{\pm}$, the impact of this parameter is lower because bad adaptation rules are already filtered thanks to the negative cases.
4. The number of rules parameter has a minimal impact on prec compared to the specificity of the rules. For a given minPS, the prec is quite the same, independently of the number of rules. However, the number of rules parameter impacts the car because using more rules improves the car.

To summarize, the experiments on this dataset show that, except for a low number of rules or with an acceptance of too general rules, $AK^{+}$ gives better results than $NN$. But, the most important fact is that $AK^{\pm}$ overcomes $NN$ and $AK^{+}$ in any situation. Moreover, $AK^{\pm}$ is very *stable* for the prec, and is not really impacted by the number of rules nor the specificity of the rules parameters. This is because the adaptation rules extracted for $AK^{\pm}$ are, in all situations, of better quality than for $AK^{+}$.

**Is Using both Positive and Negative Cases Better than Using only Positive Cases?** In all our experiments (the previous one but also the following ones), all approaches use the same case base CB$^{+}$ as source cases. CB$^{+}$ is also used to compute the adaptation rules in both AK-based approaches, but $AK^{\pm}$ uses in addition other cases: the negative cases of CB$^{-}$. So, there is a kind of disparity between $AK^{+}$ and $AK^{\pm}$ because $AK^{\pm}$ uses the double number of cases to acquire adaptation rules. This is justified by the fact that, for a CBR application, CB$^{+}$ is used for both AK approaches, while CB$^{-}$ can only be used by $AK^{\pm}$, knowing that negative cases can be acquired with a little effort during the use of the system. However, in order to compare $AK^{+}$ and $AK^{\pm}$ more "fairly", the following question has been raised: what kind of result could $AK^{+}$ give when using the same number of cases than $AK^{\pm}$ to build its rule set? To examine this issue, an experiment has been run focusing on the exploitation of the same number of cases for $AK^{+}$ and for $AK^{\pm}$. Instead of using only CB$^{+}$ for $AK^{+}$, the complete CB$_{L}$ is now used (without modification of cases to build CB$^{-}$).

The results, presented in Table 4, show that $AK^{\pm}$ outperforms $AK^{+}$ when using the same number of cases: prec and car of $AK^{+}$ in Table 4 are always lower than the prec and car of $AK^{\pm}$ in Table 3. This strenghens the idea that using negative cases in addition to positive cases improves the AK process.

**Table 5.** `prec` and `car` of the three approaches for different numbers of rules and different minimal problem sizes for the Tic Tac Toe benchmark.

| nAR | 100 | | | | | 200 | | | | | 300 | | | | | 400 | | | | |
|---|---|---|---|---|---|---|---|---|---|---|---|---|---|---|---|---|---|---|---|---|
| minPS | 1 | 2 | 3 | 4 | 5 | 1 | 2 | 3 | 4 | 5 | 1 | 2 | 3 | 4 | 5 | 1 | 2 | 3 | 4 | 5 |
| prec | | | | | | | | | | | | | | | | | | | | |
| $NN$ | .60 | .61 | .60 | .61 | .61 | .61 | .60 | .61 | .60 | .60 | .60 | .60 | .61 | .61 | .60 | .60 | .61 | .60 | .60 | .61 |
| $AK^+$ | .58 | .60 | .58 | .60 | .64 | .57 | .60 | .59 | .60 | .63 | .57 | .60 | .59 | .61 | .63 | .59 | .60 | .59 | .60 | .63 |
| $AK^\pm$ | **.70** | **.72** | **.71** | **.69** | **.70** | **.71** | **.71** | **.70** | **.68** | **.69** | **.71** | **.71** | **.70** | **.69** | **.70** | **.71** | **.70** | **.70** | **.70** | **.69** |
| car | | | | | | | | | | | | | | | | | | | | |
| $NN$ | .60 | .61 | .60 | **.61** | **.61** | .61 | .60 | .61 | .60 | **.60** | .60 | .60 | .61 | .61 | **.60** | .60 | .61 | .60 | .60 | .61 |
| $AK^+$ | .58 | .60 | .56 | .51 | .41 | .57 | .60 | .59 | .57 | .49 | .57 | .60 | .59 | .60 | .55 | .59 | .60 | .59 | .60 | .58 |
| $AK^\pm$ | **.61** | **.61** | **.61** | .54 | .42 | **.66** | **.66** | **.65** | **.62** | .53 | **.68** | **.68** | **.67** | **.65** | .58 | **.69** | **.68** | **.69** | **.67** | **.62** |

## 4.3 Tic Tac Toe Endgame

This dataset[4] contains the possible situations of the Tic Tac Toe[5] where the player with the 'x' mark starts. There are 9 attributes describing the problem. Each attribute represents one position of the board, its value can be 'x' (resp, 'o') if an 'x' mark (resp. an 'o' mark) has been played on this position. A third value 'b' indicates that the position is empty (i.e. neither 'x', neither 'o' appears). The solution is 1 if the player with the 'x' mark wins, and 0 otherwise. So, the aim of this dataset is, knowing which marks appears on some positions, to infer whether the player with the 'x' mark wins or not.

**Data simplification:** for this dataset, we have chosen to simplify the initial representation of 3 possible values for each variable, to a representation in which a variable has only 2 possible Boolean values: 1 if an 'x' mark is on the square and 0 otherwise (i.e. the square is marked by 'o' or 'b' in the raw data). In this experiment, $\mathcal{P} = \mathbb{B}^9$ and $\mathcal{S} = \mathbb{B}$. With this transformation, the dataset, containing initially 958 records, produces 272 cases, with a rather balanced solution distribution of 118 'x wins' and 154 'x does not win'.

Table 5 presents the results of the three adaptation approaches. This experiment gives similar results than for the first experiment. The hypothesis about $AK^\pm$ to be the better approach is confirmed, demonstrating once again the benefit of exploiting negative cases.

## 4.4 Cardiac Diagnosis

This dataset[6] describes diagnosing of cardiac Single Proton Emission Computed Tomography (SPECT) images. Each patient of the dataset is described by 22 binary features and is classified into two categories: normal (0) and abnormal (1). In this experiment, $\mathcal{P} = \mathbb{B}^{22}$ and $\mathcal{S} = \mathbb{B}$. The dataset, containing initially

---

[4]  https://archive.ics.uci.edu/ml/datasets/Tic-Tac-Toe+Endgame.

[5]  Tic Tac Toe is a game where 2 players, one playing with an 'x' mark and one playing with an 'o' mark have to align 3 of their marks on a 3 × 3 board.

[6]  https://archive.ics.uci.edu/ml/datasets/SPECT+Heart.

**Table 6.** prec and car of the three approaches for different numbers of rules and different minimal problem sizes for the Cardiac diagnosis benchmark.

| nAR | 100 | | | | | 200 | | | | | 300 | | | | | 400 | | | | |
|---|---|---|---|---|---|---|---|---|---|---|---|---|---|---|---|---|---|---|---|---|
| minPS | 1 | 2 | 3 | 4 | 5 | 1 | 2 | 3 | 4 | 5 | 1 | 2 | 3 | 4 | 5 | 1 | 2 | 3 | 4 | 5 |
| prec | | | | | | | | | | | | | | | | | | | | |
| $NN$ | .80 | .80 | .80 | .79 | .80 | .79 | .79 | .80 | .79 | .80 | .79 | .79 | .79 | .79 | .79 | .80 | .80 | .79 | .80 | .80 |
| $AK^+$ | .85 | .85 | .85 | .84 | .83 | .85 | .85 | .85 | .84 | .83 | .85 | .86 | .85 | .84 | .83 | .85 | .85 | .84 | .84 | .83 |
| $AK^\pm$ | **.88** | **.87** | **.88** | **.88** | **.86** | **.87** | **.87** | **.87** | **.86** | **.86** | **.86** | **.87** | **.88** | **.86** | **.86** | **.86** | **.86** | **.86** | **.86** | **.86** |
| car | | | | | | | | | | | | | | | | | | | | |
| $NN$ | .80 | .80 | .80 | **.79** | **.80** | .79 | .79 | .80 | .79 | **.80** | .79 | .79 | .79 | .79 | **.79** | .80 | .80 | .79 | .80 | **.80** |
| $AK^+$ | **.85** | **.85** | **.83** | .78 | .72 | **.85** | **.85** | **.84** | **.80** | .75 | **.85** | **.86** | **.85** | **.81** | .76 | **.85** | **.85** | **.84** | **.82** | .78 |
| $AK^\pm$ | .81 | .80 | .76 | .71 | .64 | .82 | .82 | .81 | .77 | .70 | .83 | .84 | .82 | .79 | .73 | .84 | .84 | .82 | .80 | .75 |

267 records, produces 219 unique cases, with a distribution of 186 abnormal and 33 normal diagnoses. This biased distribution of the solutions may have an impact on the results, but as the 3 approaches which are compared are all concerned by this bias, it has been decided, for this dataset, not to balance this distribution and to keep all the cases to run the experiments. The results, presented in Table 6, show again the best behavior of the $AK^\pm$ approach, even with this bias in the dataset. For prec, the improvement of $AK^\pm$ is, in average, of +8% comparing to $NN$.

### 4.5 Car Evaluation

This section addresses Car evaluation.[7] Each model of car is described by 6 features:

- the buying price which values can be *vhigh* (very high), *high*, *med* or *low*;
- the maintenance price: *vhigh*, *high*, *med* or *low*;
- the number of doors: 2, 3, 4 or *5more*;
- the capacity in persons to carry: 2, 4 or *more*;
- the luggage boot size: *small*, *med* or *big*;
- the estimated safety: *low*, *med* or *high*.

According to these features, the car models are classified into 4 categories, from the worst one to the best one: *unacc* (unacceptable), *acc* (acceptable), *good*, *vgood* (very good). There are 1728 car descriptions with the following biased distribution for the solution in the initial dataset: 1210 *unacc*, 384 *acc*, 69 *good*, and 65 *vgood*. To test now the approaches on an unbiased distribution of the solutions, only the maximum possible number of cases with the same number of cases in each category has been kept, so 65 cases per category, forming a dataset of 260 cases.

In this dataset, none of the features is Boolean. Each feature has exactly 1 value belonging to a set of more than 2 possible values. Moreover, the solution takes also its value in a set of more than 2 values. A specific transformation is

---

[7]  https://archive.ics.uci.edu/ml/datasets/Car+Evaluation.

required to encode these non Boolean values into Boolean ones, and a specific rule filtering must be applied to ensure that the result is one category value. These two adjustments for applying the Boolean approach are now detailed.

**Encoding Non Boolean Attributes into Boolean Ones Using One-Hot Encoding.** Let $a$ be an attribute and let $v_a$ be a value for $a$. In addition to attributes which are already described by Boolean values ($v_a \in \{0,1\}$), nominal values can also be considered. For an attribute $a$ with nominal values, i.e. $v_a \in \{a_1, a_2, \ldots, a_n\}$, $n$ Boolean attributes can be used to encode each of these $n$ possible values. The value $v_a = a_i$ is encoded with 1 on the $i^{th}$ attribute, all the other of the $n$ attributes being set to 0. For example, for the buying price, the possible values are *vhigh*, *high*, *med* or *low*. So, 4 Boolean variables are used to encode the initial value, with only one of these variables set to 1. For example, a *low* buying price is encoded by 1000, *med* by 0100, *high* by 0010, and *vhigh* by 0001. This encoding process is used on the 6 features describing the problem, as well as for encoding the solution. The illustration below shows an example of the complete problem/solution encoding for the initial record/case $((vhigh, med, 5more, 4, big, high), vgood)$.

$$\underbrace{0\ 0\ 0\ 1}_{1}\ \underbrace{0\ 0\ 1\ 0}_{2}\ \underbrace{0\ 0\ 0\ 1}_{3}\ \underbrace{0\ 1\ 0}_{4}\ \underbrace{0\ 0\ 1}_{5}\ \underbrace{0\ 0\ 1}_{6} \rightarrow \underbrace{0\ 0\ 0\ 1}_{7}$$

with 1, 2, 3, 4, 5 and 6 respectively encoding the buying price ($vhigh$), the maintenance price ($med$), the number of doors ($5more$), the number of persons to carry (4), the luggage boot size ($big$), the safety ($high$), and 7 encodes the *solution* category of the car ($vgood$).

**Generating a negative case when $n \geq 2$ for $S = \mathbb{B}^n$.** For a correct solution $y \in \mathbb{B}$, $y$ is transformed in a incorrect solution $y'$ simply by $y' = \neg y$. However, when a solution is encoded on more than one variable, (e.g. for this dataset, on 4 variables), it requires a specific approach to transform a positive case into a negative one. For the experiment with $n \geq 2$, but with the assumption that only 1 variable of the solution is set to 1, an incorrect solution is randomly selected among the solutions encoding syntactically correct solutions, i.e. a solution with only 1 variable set to 1 and which is not equal to the correct solution. For example, if the correct solution is 0001, the possible incorrect solutions are 0010, 0100 and 1000. An incorrect solution is randomly chosen among these 3 possibilities.

**Filtering Out Rules that Incompletely Describe the Solution Variation.** When a solution is encoded on more than one variables, (e.g. for this dataset, on 4 variables), obtaining a result which correspond to 1 of the possible values as solution requires rules containing information on all the solution variables. For example, a rule like $\{x_1^{=1}, x_2^+, y_1^+\}$, containing only 1 solution variable ($y_1$), applied to a source case solution 0010 produces 1010 which does not correspond to a possible correct answer. So, for this dataset, only rules containing the 4 variables of the solution are kept.

**Table 7.** prec and car of the three approaches for different numbers of rules and different minimal problem sizes for the Car benchmark.

| nAR | 100 | | | | | 200 | | | | | 300 | | | | | 400 | | | | |
|---|---|---|---|---|---|---|---|---|---|---|---|---|---|---|---|---|---|---|---|---|
| minPS | 1 | 2 | 3 | 4 | 5 | 1 | 2 | 3 | 4 | 5 | 1 | 2 | 3 | 4 | 5 | 1 | 2 | 3 | 4 | 5 |
| prec | | | | | | | | | | | | | | | | | | | | |
| $NN$ | .67 | .67 | .68 | .68 | .67 | .67 | .67 | .67 | .67 | .68 | .67 | .66 | .67 | .67 | .68 | .67 | .67 | .67 | .67 | .67 |
| $AK^+$ | .71 | .71 | .71 | .72 | .72 | .70 | .72 | .71 | .73 | .72 | .72 | .72 | .72 | .72 | .71 | .73 | .72 | .72 | .72 | .72 |
| $AK^\pm$ | **.81** | **.81** | **.80** | **.80** | **.81** | **.81** | **.80** | **.80** | **.80** | **.81** | **.79** | **.80** | **.80** | **.80** | **.80** | **.80** | **.80** | **.80** | **.80** | **.81** |
| car | | | | | | | | | | | | | | | | | | | | |
| $NN$ | .67 | .67 | .68 | .68 | **.67** | .67 | .67 | .67 | .67 | .68 | .67 | .66 | .67 | .67 | .68 | .67 | .67 | .67 | .67 | .67 |
| $AK^+$ | .66 | .66 | .67 | .67 | .59 | .68 | .69 | .68 | .69 | .69 | .68 | .69 | .70 | .69 | .68 | .70 | .70 | .69 | .70 | .69 |
| $AK^\pm$ | **.73** | **.73** | **.73** | **.73** | .64 | **.77** | **.75** | **.75** | **.74** | **.77** | **.75** | **.76** | **.77** | **.76** | **.77** | **.77** | **.78** | **.76** | **.76** | **.77** |

Table 7 presents the results on the Car benchmark with $|CB^+| = |CB^-| = 26$, i.e. $\frac{|CB|}{10}$ instead of $\frac{|CB|}{4}$ for the other experiments. The idea is to examine, in addition, the behavior of the approach on a smallest learning case base. One more time, $AK^\pm$ improves the results w.r.t. $NN$, by an average of +11% in prec. Moreover, with this benchmark, $AK^\pm$ improves also well the car by an average of +10%.

## 4.6   Results and Discussion

The different experiments, on different benchmarks, with various dimensions of the problem space (from 9 for the Tic Tac Toe benchmark to 22 for the cardiac diagnosis benchmark) has shown the efficiency of using negative examples for acquiring adaptation rules. The results are similar for the 4 benchmarks and, in any situation, $AK^\pm$ is better than $AK^+$ and $NN$ for precision, and most of the time for the car measure.

So, exploiting negative cases improves the CBR system results. For the prec measure, approaches based on AK built only on positive examples gives most of the time better results than the $NN$ baseline approach, when the number and the length of rules are sufficient to avoid the use of too general rules. However, when introducing the exploitation of negative cases, the prec measure really increases, with various improvement: around +7% for the Congressional Voting, Tic Tac Toe and Cardiac diagnosis benchmarks, to even around +14% for the Car Evaluation benchmark. But the most important fact is that the results shows that from the precision point of view, $AK^\pm$ has given *always* better results than $AK^+$ and $NN$. For the car measure, $AK^+$ and $AK^\pm$ give results sometimes under the ones of the $NN$ approach. However, these car scores have to be considered in regards to the prec score because increasing the precision has most of the time a negative impact on the number of answers (i.e. the system answers better but less often), and so, on the car measure. For the Congressional Votes, $AK^\pm$ car measure is almost always better than for $NN$ and, in that case, also with a significant improvement, especially when minPS and nAR are high. Combining with a significant improvement of the prec measure as well, the results can be considered as excellent. For the Cardiac diagnosis, the car measure of $AK^{\pm}$ and $NN$ are quite similar and for the Car evaluation, the car measure overcomes all the time, with an improvement around $+11\%$, the car of $NN$. In conclusion, the results of the approach based on negative cases must be highlighted, because of the increasing of the precision without having an impact on the decreasing of the car, which is most of the time higher than for $NN$.

## 5    Conclusion

This paper shows the benefits of exploiting negative cases in addition to positive ones to extract, using frequent closed itemsets, adaptation rules of higher quality which improves the results of a CBR system. A methodology has also been presented to transform a case description originally not encoded by Boolean attributes into a Boolean encoding, and to filter adaptation rules. The rule filtering can be based on the support of the rule (i.e. how many times this rule is found in the learning dataset), on the length of the rule (i.e. the more variables the rule uses, the best it is). Applying this methodology on 4 benchmarks with different characteristics shows that similar results are obtained for these applications, which argue for the interest of exploiting negative cases.

The quality of the results depends on numerous parameters. In this work, the impact of the number of rules as well as the size of the rules has been examined. However, it could be interesting to study the impact of other parameters, like, for example, the size of $CB_L$, $CB^+$ and $CB^-$.

Finally, another interesting future work is to use more finely the negative cases. In this work, the negative cases are used to filter out adaptation rules, usually with a great support value but which are sometimes too general. However, like in the version space model, instead of simply removing a rule which could be useful in some problem solving situations, we could imagine to specialize this rule, to avoid its use only in a given context. For example, an approach based of FCI on variations between positive and negative cases could bring out elements which could be exploited to refine adaptation rules built only on variations between positive cases.

## References

1. Badra, F., Cordier, A., Lieber, J.: Opportunistic adaptation knowledge discovery. In: McGinty, L., Wilson, D.C. (eds.) ICCBR 2009. LNCS (LNAI), vol. 5650, pp. 60–74. Springer, Heidelberg (2009). https://doi.org/10.1007/978-3-642-02998-1_6
2. Berasaluce, S., Laurenço, C., Napoli, A., Niel, G.: An experiment on mining chemical reaction databases. In: Le Thi, H.A., Dinh, T.P. (eds.) Modelling, Computation and Optimization in Information Systems and Management Sciences - MCO'04, pp. 535–542. France, Hermes Science Publishing, London, Metz (2004)
3. d'Aquin, M., Badra, F., Lafrogne, S., Lieber, J., Napoli, A., Szathmary, L.: Case base mining for adaptation knowledge acquisition. In: Veloso, M.M., (ed.) Proceedings of the 20th International Joint Conference on on Artificial Intelligence (IJCAI 2007), Morgan Kaufmann, Inc., pp. 750–755 (2007)

4. Dufour-Lussier, V., Lieber, J., Nauer, E., Toussaint, Y.: Improving case retrieval by enrichment of the domain ontology. In: Ram, A., Wiratunga, N. (eds.) ICCBR 2011. LNCS (LNAI), vol. 6880, pp. 62–76. Springer, Heidelberg (2011). https://doi.org/10.1007/978-3-642-23291-6_7
5. Gaillard, E., Lieber, J., Nauer, E.: Adaptation knowledge discovery for cooking using closed itemset extraction. In: The Eighth International Conference on Concept Lattices and their Applications - CLA 2011, Nancy, France (2011)
6. Ganter, B., Kuznetsov, S.O.: Hypotheses and version spaces. In: Ganter, B., de Moor, A., Lex, W. (eds.) Conceptual Structures for Knowledge Creation and Communication, pp. 83–95. Heidelberg, Springer, Berlin Heidelberg, Berlin (2003)
7. Ganter, B., Wille, R.: Formal Concept Analysis: Mathematical Foundations, 284 p. Springer, Berlin, Heidelberg (1999). https://doi.org/10.1007/978-3-642-59830-2
8. Gillard, T., Lieber, J., Nauer, E.: Improving adaptation knowledge discovery by exploiting negative cases: first experiment in a boolean setting. In: ICCBR 2018–26th International Conference on Case-Based Reasoning, Stockholm, Sweden (2018)
9. Hanney, K., Keane, M.T.: Learning adaptation rules from a case-base. In: Smith, I., Faltings, B. (eds.) Advances in Case-Based Reasoning - Third European Workshop, EWCBR 1996. LNAI 1168, Springer, Verlag, Berlin, pp. 179–192 (1996)
10. Jalali, V., Leake, D.B.: CBR meets big data: a case study of large-scale adaptation rule generation. In: ICCBR (2015)
11. Kuznetsov, S.O.: Complexity of learning in concept lattices from positive and negative examples. Discret. Appl. Math. **142**(1), 111–125 (2004)
12. Mitchell, T.: Version Space: An Approach to Concept Learning, Ph.D. Thesis, Stanford University (1978)
13. Richter, M.M., Weber, R.O.: Case representations. In: Case-Based Reasoning, pp. 87–111. Springer, Heidelberg (2013). https://doi.org/10.1007/978-3-642-40167-1_5
14. Szathmary, L., Napoli, A.: CORON: a framework for levelwise itemset mining algorithms. In: Supplementary Proceedings of The Third International Conference on Formal Concept Analysis (ICFCA 2005), Lens, France, pp. 110–113 (2005)

# When Revision-Based Case Adaptation Meets Analogical Extrapolation

Jean Lieber[1]([⊠]), Emmanuel Nauer[1], and Henri Prade[2]

[1] Université de Lorraine, CNRS, Inria, LORIA, 54000 Nancy, France
Jean.Lieber@loria.fr
[2] IRIT, CNRS, Université Paul Sabatier, Toulouse, France

**Abstract.** Case-based reasoning, where cases are described in terms of problem-solution pairs $\texttt{case} = (\texttt{x}, \texttt{y})$, amounts to propose a solution to a new problem on the basis of past experience made of stored cases. On the one hand, the building of the solution to a new problem may be viewed as a form of belief revision of the solution of a retrieved case (whose problem part is similar to the new problem) constrained by domain knowledge. On the other hand, an extrapolation mechanism based on analogical proportions has been proposed. It exploits triplets of cases $(\texttt{case}^a, \texttt{case}^b, \texttt{case}^c)$ whose descriptions of problem parts $\texttt{x}^a$, $\texttt{x}^b$, $\texttt{x}^c$ form an analogical proportion with the new problem $\texttt{x}^{\text{tgt}}$, in such a way that "$\texttt{x}^a$ is to $\texttt{x}^b$ as $\texttt{x}^c$ is to $\texttt{x}^{\text{tgt}}$". Then, the analogical inference amounts to compute a solution $\texttt{y}^{\text{tgt}}$ of $\texttt{x}^{\text{tgt}}$ by solving (when possible) an equation expressing that "$\texttt{y}^a$ is to $\texttt{y}^b$ as $\texttt{y}^c$ is to $\texttt{y}^{\text{tgt}}$" (where $\texttt{y}^a$, $\texttt{y}^b$ and $\texttt{y}^c$ are respectively the solution parts of $\texttt{case}^a$, $\texttt{case}^b$ and $\texttt{case}^c$). The paper investigates how the belief revision view and analogical extrapolation relate. Besides that it constitutes an unexpected bridge between areas which ignore each other, it casts some light on the adaptation mechanism in case-based reasoning. The paper is illustrated by a running example.

**Keywords:** Analogical inference · Analogical proportion · Belief revision · Case-based reasoning · Extrapolation

## 1 Introduction

Belief revision [2] and case-based reasoning [1] (CBR) are two areas of artificial intelligence that are usually thought of as quite distant and unrelated since the former takes place mainly in the setting of logic, while the latter deals with data and is similarity-based. Moreover, belief revision aims at reestablishing consistency after receiving a new piece of information that conflicts with the current state of belief. CBR has a quite different agenda since it is rather a matter of coping with missing information by taking advantage of similarity for completing a new problem with a plausible solution. However, note that the conclusions derived by belief revision or by CBR are only plausible in both approaches.

© Springer Nature Switzerland AG 2021
A. A. Sánchez-Ruiz and M. W. Floyd (Eds.): ICCBR 2021, LNAI 12877, pp. 156–170, 2021.
https://doi.org/10.1007/978-3-030-86957-1_11

In spite of this apparent state of fact, there exists a belief revision-based view of CBR [4,14]. Indeed, completing the new problem with a plausible solution may be viewed as adapting the solution of a retrieved case in order to be consistent with the specificities of the new case, which is a kind of revision. What is revised is clearly not the case base, but it is a copy of a case dealing with a similar problem. In other words, what is really revised is the allegation that the retrieved case can be *applied* (directly) to solve the target problem.

An extrapolation mechanism based on analogical proportions (which are statements of the form "$a$ is to $b$ as $c$ is to $d$") has been proposed for exploiting cases [16]. On the basis of 3 cases whose problem description parts are in analogical proportion with the new problem, one infers a solution for it from the solutions of the 3 cases. This means that an adaptation process takes place inside the inference. This is made possible by the fact that analogical proportions are a matter of both similarity and dissimilarity. Indeed when the problem parts of cases $a$, $b$, $c$, $d$ are described by vectors of features, the analogical proportion holds between the vectors if $a$ differs from $b$ in the same way $c$ differs from $d$ (and vice-versa) [17,21]. In that respect, a pair of cases may be viewed as encoding a kind of rule of adaptation in the sense of CBR [6].

The above points altogether suggest that there is a bridge between the belief revision-based view of CBR and analogical proportion-based extrapolation. This investigation is the topic of this paper. It is organized as follows. In Sect. 2, after offering a short refresher on propositional logic and introducing notations, two backgrounds are provided respectively on the analogical extrapolation of cases and on the belief revision view of CBR. This necessary setting of the problem addressed in this paper makes Sect. 2 rather long all the more as the running example is also introduced there. In order to bridge extrapolation and revision-based adaptation, Sect. 3 first reformulates adaptation by extrapolation as a single case adaptation. Then, a revision operator based on competence of case pairs is defined, which enables to establish that the adaptation by extrapolation can be equivalently obtained by a revision constrained by competent pairs expressing adaptation knowledge. Section 4 presents some related work and some concluding remarks, including a presentation of future work.

## 2    Setting of the Problem and Running Example

This section sets the notions related to the problem that this paper aims at solving, i.e., how adaptation by analogical extrapolation and revision-based adaptation can meet. For this purpose, the notions and notations used are presented, together with the introduction of a running example in the cooking domain.

### 2.1    A Quick Refresher About Propositional Logic

The formalism for representing cases and domain knowledge in this paper is propositional logic. Let $\mathcal{V}$ be a finite set of symbols, called variables. A formula is either a variable or an expression of one of the forms $\top$, $\bot$, $\neg\varphi$, $\varphi_1 \wedge \varphi_2$, $\varphi_1 \vee \varphi_2$, $\varphi_1 \rightarrow \varphi_2$, and $\varphi_1 \leftrightarrow \varphi_2$ where $\varphi$, $\varphi_1$ and $\varphi_2$ are formulas.

An interpretation $\mathcal{I}$ is a mapping from $\mathcal{V}$ to $\{0, 1\}$ where 0 and 1 denote the Boolean values "false" and "true". The set of all interpretations is denoted by $\Omega$. An interpretation $\mathcal{I} \in \Omega$ is extended on every formulas as follows: $\mathcal{I}(\top) = 1$, $\mathcal{I}(\bot) = 0$, $\mathcal{I}(\neg\varphi) = \textbf{not}\,\mathcal{I}(\varphi)$, $\mathcal{I}(\varphi_1 \wedge \varphi_2) = \mathcal{I}(\varphi_1)\textbf{ and }\mathcal{I}(\varphi_2)$, $\mathcal{I}(\varphi_1 \vee \varphi_2) = \mathcal{I}(\varphi_1)\textbf{ or }\mathcal{I}(\varphi_2)$, $\mathcal{I}(\varphi_1 \rightarrow \varphi_2) = \mathcal{I}(\neg\varphi_1 \vee \varphi_2)$ and $\mathcal{I}(\varphi_1 \leftrightarrow \varphi_2) = \mathcal{I}((\varphi_1 \rightarrow \varphi_2) \wedge (\varphi_2 \rightarrow \varphi_1))$ where $\textbf{not}$, $\textbf{and}$ and $\textbf{or}$ are the classical Boolean operations. A model of a formula $\varphi$ is an interpretation $\mathcal{I}$ such that $\mathcal{I}(\varphi) = 1$ and the set of models of $\varphi$ is denoted by $\mathcal{M}(\varphi)$. A formula $\varphi_1$ entails a formula $\varphi_2$, denoted by $\varphi_1 \models \varphi_2$, if $\mathcal{M}(\varphi_1) \subseteq \mathcal{M}(\varphi_2)$. The formulas $\varphi_1$ and $\varphi_2$ are equivalent, denoted by $\varphi_1 \equiv \varphi_2$, if $\mathcal{M}(\varphi_1) = \mathcal{M}(\varphi_2)$. A formula $\varphi$ is *consistent* if $\mathcal{M}(\varphi) \neq \emptyset$.

Let $\Phi = \{\varphi_1, \varphi_2, \ldots, \varphi_p\}$ be a finite set of formulas. $\bigvee \Phi$ denotes $\varphi_1 \vee \varphi_2 \vee \ldots \vee \varphi_p$. $\bigwedge \Phi$ denotes $\varphi_1 \wedge \varphi_2 \wedge \ldots \wedge \varphi_p$.

In the paper, for the sake of simplicity, the Boolean notations and the propositional notations are sometimes used together (0 and 1 instead of $\top$ and $\bot$ as well as use of propositional connectives between Boolean values).

## 2.2  Notions and Notations Related to CBR

CBR aims at solving problems thanks to a case base, where a case is the representation of a problem-solving episode. Let $\mathcal{P}$ and $\mathcal{S}$ be the space of problems and solutions: a *problem* $\mathbf{x}$ (resp., a *solution* $\mathbf{y}$) is, by definition, an element of $\mathcal{P}$ (resp., of $\mathcal{S}$). A relation on $\mathcal{P} \times \mathcal{S}$ is assumed to exist that is read "has for solution" but is usually not completely known by the CBR system. A *case* is an ordered pair $\texttt{case} = (\mathbf{x}, \mathbf{y}) \in \mathcal{P} \times \mathcal{S}$ such that $\mathbf{x}$ has for solution $\mathbf{y}$. A *source case* $\texttt{case}^s = (\mathbf{x}^s, \mathbf{y}^s)$ is an element of the case base, which is denoted by $\texttt{CB}$. The current problem under solving is called the *target problem*, denoted by $\mathbf{x}^{\mathrm{tgt}}$.

The classical CBR process model is based on the steps of retrieval and adaptation [24], also known as retrieve and reuse [1]. Other steps follow adaptation, but are not considered in this paper. Retrieval selects one or several source case(s) for the purpose of solving $\mathbf{x}^{\mathrm{tgt}}$. Adaptation aims at using the retrieved case(s) for proposing a solution $\mathbf{y}^{\mathrm{tgt}}$ to $\mathbf{x}^{\mathrm{tgt}}$. A *single case adaptation* is an adaptation of a sole retrieved case, otherwise, it is a *multiple case adaptation*.

The *domain knowledge* is a knowledge base $\texttt{DK}$ that can be understood as a set of integrity constraints: a problem $\mathbf{x}$ (resp., a solution $\mathbf{y}$ or a case $(\mathbf{x}, \mathbf{y})$) that is inconsistent with $\texttt{DK}$ is known to be not licit.

It is assumed in this paper that a clear separation of problems and solutions is made. This is not always true: for some applications of CBR, each case constitutes a whole, and a target problem is considered as an incomplete case, thus the problem-solution separation can be made at adaptation time. Since this paper is primarily concerned with adaptation, this assumption is not a big restriction. Therefore, in the propositional setting, $\mathcal{V}$ is partitioned into $\mathcal{V}_\mathcal{P}$ and $\mathcal{V}_\mathcal{S}$ and the variables occurring in a problem (resp., in a solution) are elements of $\mathcal{V}_\mathcal{P}$ (resp., of $\mathcal{V}_\mathcal{S}$). The next section illustrates this idea.

Furthermore, it is assumed that every source case $(\mathbf{x}^s, \mathbf{y}^s)$ and the target problem are fully described: given the domain knowledge, the truth value of each variable $a \in \mathcal{V}_\mathcal{P}$ is known for $\mathbf{x}^s$ and $\mathbf{x}^{\mathrm{tgt}}$ (i.e., $\texttt{DK} \wedge \mathbf{x}^s \models a$ or $\texttt{DK} \wedge \mathbf{x}^s \models \neg a$

and the same for $x^{tgt}$) and a similar constraint holds for $y^s$. A case $(x^s, y^s)$ is represented in propositional logic by a single formula $x^s \wedge y^s$.

According to the classical knowledge model of CBR (see, e.g., [23]), the CBR knowledge base consists in four knowledge containers. Two of them have already been mentioned: the case base CB and the domain knowledge DK. The two others are the retrieval knowledge and the adaptation knowledge, used during the retrieval and adaptation steps.

A final remark can be made here about the notion of similarity in CBR: it is usually said that retrieval aims at finding cases *similar* to the target problem. Now, this notion of similarity should not necessarily be understood as an approximate equality: it may go beyond this. In fact, $(x^s, y^s)$ can be considered similar to $x^{tgt}$ even if $x^s$ and $x^{tgt}$ descriptions are very different, considering that the adaptation of $(x^s, y^s)$ provides a plausible solution to $x^{tgt}$. This is related to the adaptation-guided retrieval principle [25].

### 2.3 Specification of the Running Example

The example used throughout this paper is in the cooking domain. A recipe is represented simply by the type of ingredients it contains. Variables are food names representing recipe classes. For example, fruit represents the class of recipes having at least one fruit as ingredient. The domain knowledge is:

$$DK = \{\text{pineapple} \rightarrow \text{fruit}, \quad \text{cream} \rightarrow \text{sauce}, \quad \text{pesto} \rightarrow \text{sauce},$$
$$\text{St Pierre} \rightarrow \text{fish}, \quad \text{salmon} \rightarrow \text{fish}, \quad \neg(\text{fruit} \wedge \text{pesto})\}$$

The last formula of DK states that a recipe must not have at the same time pesto (that contains garlic) and any fruit (fruit taken in the cooking sense of the term).

The source cases considered in this example are:

$$\text{case}^a = \text{St Pierre} \wedge \text{cream} \wedge \text{vanilla} \wedge \text{curry} \wedge \textit{Nothing else} \quad (1)$$

$$\text{case}^b = \text{salmon} \wedge \text{cream} \wedge \text{vanilla} \wedge \text{pesto} \wedge \textit{Nothing else} \quad (2)$$

$$\text{case}^s = \text{St Pierre} \wedge \text{cream} \wedge \text{pineapple} \wedge \textit{Nothing else} \quad (3)$$

where, for a formula $\varphi$, $\varphi \wedge$ *Nothing else* denotes the formula $\varphi \wedge \Gamma$ where $\Gamma$ is the conjunction of the negative literals $\neg a$ such that $a \in \mathcal{V}$ and $DK \wedge \varphi \not\models a$. For example, $\text{case}^s$ contains the literal $\neg\text{salmon}$.

The target problem is the request of a recipe with fish, pesto but no vanilla:

$$x^{tgt} = \text{fish} \wedge \text{pesto} \wedge \neg\text{vanilla} \quad (4)$$

Thus, for the running example, $\mathcal{V}_\mathcal{P} = \{\text{fish}, \text{pesto}, \text{vanilla}\}$ and $\mathcal{V}_\mathcal{S} = \mathcal{V} \setminus \mathcal{V}_\mathcal{P}$.

### 2.4 Analogical Proportions and CBR

Analogical proportions are statements of the form "$a$ is to $b$ as $c$ is $d$", denoted $ccabcd$. Their origin dates back to Aristotle [3] (at least), and was inspired by a parallel with (geometric) numerical proportions, namely $\frac{a}{b} = \frac{c}{d}$; see [20]. In agreement with this parallel, they are supposed to obey the following postulates: Given a set of items $X$, analogical proportions form a quaternary relation supposed to obey the 3 following postulates (e.g., [13]): for $a, b, c, d \in X$,

1. $a : b :: a : b$ (*reflexivity*);
2. if $a : b :: c : d$ then $c : d :: a : b$ (*symmetry*);
3. if $a : b :: c : d$ then $a : c :: b : d$ (*central permutation*).

The unique minimal Boolean model [22] obeying these 3 postulates is a quaternary propositional logic connective when $X$ is the Boolean set $\mathbb{B} = \{0, 1\}$ [17]:

$$a : b :: c : d = ((a \wedge \neg b) \leftrightarrow (c \wedge \neg d)) \wedge ((\neg a \wedge b) \leftrightarrow (\neg c \wedge d))$$

It makes explicit that "$a$ differs from $b$ as $c$ differs from $d$ (and vice-versa)". It is easy to check that this formula is only valid for the 6 following valuations $0 : 0 :: 0 : 0$, $\quad 1 : 1 :: 1 : 1$, $\quad 0 : 1 :: 0 : 1$, $\quad 1 : 0 :: 1 : 0$, $\quad 0 : 0 :: 1 : 1$, $\quad$ and $1 : 1 :: 0 : 0$.

It can be seen that 1 and 0 play a symmetrical role, which makes the definition *code-independent*. This is formally expressed with the negation operator as: if $a : b :: c : d$ then $\neg a : \neg b :: \neg c : \neg d$. To deal with items, for instance cases, represented by *vectors* of Boolean values, the analogical proportion definition is extended componentwise from $X$ to $X^n$:

$$a : b :: c : d \quad \text{iff} \quad \text{for all } i \in \{1, \ldots, n\}, a_i : b_i :: c_i : d_i$$

This is the basis of an inference principle, first proposed in [18] for nominal values, that can be stated as follows:

$$\frac{\forall i \in \{1, \ldots, n\}, \quad a_i : b_i :: c_i : d_i \text{ holds}}{\forall j \in \{n+1, \ldots, m\}, \quad a_j : b_j :: c_j : d_j \text{ holds}}$$

As can be seen, knowledge from some components of source vectors is transferred to their remaining components, implicitly assuming that the values of the $n$ first components determine the values of the others.

This requires to find ? such that $a : b :: c : ?$ holds. The solution may not exist (e.g., for $0 : 1 :: 1 : ?$). It is solvable if and only if $a = b$ or $a = c$ in the Boolean case. Then, the *unique* solution is given by $? = c$ if $a = b$ and $? = b$ if $a = c$. Thus, we have the following property in the Boolean case

$$a : a :: b : ? \quad \text{if and only if} \quad ? = b \tag{5}$$

which is sometimes taken as a supplementary postulate and is not a consequence of the 3 postulates. This is the basis of the analogical *extrapolation* between cases proposed in [16].

Given a pair of vectors describing problems $(\mathbf{x}^a, \mathbf{x}^b)$, with $n$ components, their comparison yields a partition of the $n$ features in two subsets: the subset of features $\mathcal{E}_{(\mathbf{x}^a, \mathbf{x}^b)}$ for which the values of $\mathbf{x}^a$ and $\mathbf{x}^b$ are equal and the subset $\mathcal{D}_{(\mathbf{x}^a, \mathbf{x}^b)}$ for which they are different. Let us consider two pairs of vectors describing problems $(\mathbf{x}^a, \mathbf{x}^b)$ and $(\mathbf{x}^c, \mathbf{x}^d)$ such that $\mathcal{E}_{(\mathbf{x}^a, \mathbf{x}^b)} = \mathcal{E}_{(\mathbf{x}^c, \mathbf{x}^d)}$ and such that $\forall i \in \mathcal{D}_{(\mathbf{x}^a, \mathbf{x}^b)}, \mathbf{x}^{ai} = \mathbf{x}^{ci}$ and $\mathbf{x}^{bi} = \mathbf{x}^{di}$. Then $\mathbf{x}^a : \mathbf{x}^b :: \mathbf{x}^c : \mathbf{x}^d$ holds true, since

| | fish | pesto | vanilla | sauce | fruit | pineapple | St Pierre | salmon | cream | curry |
|---|---|---|---|---|---|---|---|---|---|---|
| case$^a$ | 1 | 0 | 1 | 1 | 0 | 0 | 1 | 0 | 1 | 1 |
| case$^b$ | 1 | 1 | 1 | 1 | 0 | 0 | 0 | 1 | 1 | 0 |
| case$^s$ | 1 | 0 | 0 | 1 | 1 | 1 | 1 | 0 | 1 | 0 |
| x$^{tgt}$ | 1 | 1 | 0 | ? | ? | ? | ? | ? | ? | ? |
| case$^{tgt}_{extrap}$ | 1 | 1 | 0 | 1 | 1 | 1 | 0 | 1 | 1 | ? |

**Fig. 1.** Analogical inference in the running example.

(i) for each $j \in \mathcal{E}_{(x^a,x^b)} = \mathcal{E}_{(x^c,x^d)}$, $(x^{a_j}, x^{b_j}, x^{c_j}, x^{d_j})$ is of the form $(u, u, v, v)$ or $(u, u, u, u)$ for some values $u$ and $v$ with $u \neq v$ and $u, v \in \{0, 1\}$,

(ii) for each $k \in \mathcal{D}_{(x^a,x^b)} = \mathcal{D}_{(x^c,x^d)}$, $(x^{a_k}, x^{b_k}, x^{c_k}, x^{d_k})$ is of the form $(u, v, u, v)$.

The idea of looking at pairs of cases can be related to the reading of a pair $((x^a, y^a), (x^b, y^b))$ as a virtual rule expressing either that the change from $x^a$ to $x^b$ induces the change from $y^a$ to $y^b$, whatever the problem context (encoded by the features where $x^a$ and $x^b$ are equal), or that the change from $x^a$ to $x^b$ does not modify the solution (in case $y^a = y^b$).

However such virtual rules may have exception in the training set. Indeed there may exist $(x^a, x^b)$, $(x^c, x^d)$ and $(x^{a'}, x^{b'})$ such that

- $x^a : x^b :: x^c : x^d$ and $x^{a'} : x^{b'} :: x^c : x^d$ hold true. It means that $\mathcal{D}_{(x^a,x^b)} = \mathcal{D}_{(x^c,x^d)} = \mathcal{D}_{(x^{a'},x^{b'})}$ and the changes from $x^a$ to $x^b$, from $x^c$ to $x^d$, from $x^{a'}$ to $x^{b'}$ are the same.
- $y^a : y^b :: y^c : y^d$ holds true.
- $y^{a'} : y^{b'} :: y^{c'} : y^{d'}$ *does not hold* for some feature $i$.

This may happen for instance when $y^{a_i} \neq y^{b_i}$ while $y^{a'_i} = y^{b'_i}$. In such a case, the two virtual rules associated to $((x^a, y^{a_i}), ((x^b, y^{b_i}))$ and to $((x^{a'}, y^{a'_i}), (x^{b'}, y^{b'_i}))$ disagree. Then $y^{a_i} = y^{c_i}$ and $y^{b_i} = y^{d_i}$, but the solution of the equation $y^{a'_i} : y^{b'_i} :: y^{c_i} : ?y_i$ is $?y_i = y^{c_i} \neq y^{d_i}$.

Thus the rate of exceptions of the virtual rule associated to a pair $((x^a, y^{a_i}), (x^b, y^{b_i}))$ is an indication of the interest of the pair for analogical inference. This is called the *competence* of the case pair [15]. Note that each rule pertains to a particular feature used in the description of the solutions. Indeed it is not always so that there is a unique rule that computes the adaptation of $y^a$ into $y^b$ from the same problem context.

Let us apply analogical inference to the running example. It is easy to check that its propositional expression gives birth to the table on Fig. 1. Considering the 3 cases (case$^a$, case$^b$, case$^s$), it appears that $x^a : x^b :: x^s : x^{tgt}$ holds true for the features fish, pesto and vanilla. Moreover, as an analogical equation $x^a : x^b :: x^s : x^{tgt}$ can be solved for features sauce, fruit, pineapple, St Pierre, salmon, and cream. Thus, we obtain

$$\text{case}^{\text{tgt}}_{\text{extrap}} = \text{x}^{\text{tgt}} \land \text{y}^{\text{tgt}}_{\text{extrap}}$$
$$\text{with } \text{y}^{\text{tgt}}_{\text{extrap}} \equiv \text{sauce} \land \text{fruit} \land \text{pineapple} \tag{6}$$
$$\land \ \lnot \text{St Pierre} \land \text{salmon} \land \text{cream}$$

Some remarks are worth mentioning:

1. The analogical equation cannot be solved for feature curry, so nothing is inferred regarding its presence or absence from cases (case$^a$, case$^b$, case$^s$). In a genuine example (where the case base would be richer), there may exist a triplet of cases enabling us to conclude on curry.
2. However, even if an exact resolution for curry does not exist, it would be possible to minimize an analogical dissimilarity measure $AD(a,b,c,d)$ for computing an "approximate" solution. $AD(a,b,c,d)$ is equal to the minimal number of flips for moving from $(a,b,c,d)$ to a 4-tuple corresponding to an analogical proportion: $AD(a,b,c,d)$ is maximal (and is equal to 2) if $(1,0,0,1)$ (or $(0,1,1,0)$) [17]. So here the approximate solution would be 0 (i.e., no curry) since $AD(1,0,0,0) = 1$.
3. In general, many triplets of cases can be applied to a given x$^{\text{tgt}}$. Remember that only the triplets built on the most competent pairs are used (when they lead to different conclusions a vote should take place).
4. As can be seen in the computation of case$^{\text{tgt}}$, a modification of case$^s$ takes place: namely, in the context fish and ¬vanilla with addition of pesto, case$^s$ is adapted by changing St Pierre into salmon.
5. In this example, some features are linked by implications. It should be noted that if for two mutually exclusive features $i$ and $j$ x$^{a_i}$ : x$^{b_i}$ :: x$^{c_i}$ : x$^{d_i}$ and x$^{a_j}$ : x$^{b_j}$ :: x$^{c_j}$ : x$^{d_j}$ hold, this entail that it holds as well for a feature $k$ for which $i$ and $j$ are sub-classes, as can be seen on the example (for salmon, St Pierre, and fish). Thus, the analogical extrapolation makes no independence assumption between attributes.
6. One can observe that the roles of case$^b$ and case$^s$ could be exchanged, since analogical proportions are stable by central permutation.

## 2.5   Belief Revision and CBR

This section summarizes the approach to adaptation based on belief revision as introduced in [14] and further developed in [4].

*Belief Revision.* Let us consider an agent having a set of beliefs $\psi$ and that is confronted to another set of beliefs $\mu$ that are supposed to have priority over $\psi$. $\psi$ and $\mu$ are assumed in this paper to be represented in propositional logic. The question raised by belief revision is how the beliefs of the agent evolve by incorporation of $\mu$. When the new set of beliefs are not in contradiction with the old ones—i.e., $\psi \land \mu$ is consistent—then the revision gives simply this conjunction $\psi \land \mu$. Else, according to the minimal change principle of the so-called AGM theory (named after the authors of the paper [2]), belief revision consists in making a "minimal change" of $\psi$ into $\psi'$ so that $\psi' \land \mu$ is consistent, and then, the result of the revision is denoted by $\psi \dot{+} \mu = \psi' \wedge \mu$. Now, the notion of change minimality is not uniquely defined and depends on how change is assessed, so many *revision operators* $\dot{+}$ exist. The AGM theory proposes a set of postulates that an operator $\dot{+}$ should respect. They have been formulated in propositional logic by [12] (for any formulas $\varphi$, $\psi$, $\varphi_1$, $\varphi_2$, $\psi_1$, $\psi_2$ and $\chi$):

($\dot{+}$1) If $\mu$ is consistent then $\psi \dot{+} \mu$ is consistent.
($\dot{+}$2) If $\psi \wedge \mu$ is consistent then $\psi \dot{+} \mu \equiv \psi \wedge \mu$.
($\dot{+}$3) $\psi \dot{+} \mu \models \mu$.
($\dot{+}$4) If $\psi_1 \equiv \psi_2$ and $\mu_1 \equiv \mu_2$ then $\psi_1 \dot{+} \mu_1 \equiv \psi_2 \dot{+} \mu_2$.
($\dot{+}$5) If $(\psi \dot{+} \mu) \wedge \chi$ are consistent then $\psi \dot{+} (\mu \wedge \chi) \models (\psi \dot{+} \mu) \wedge \chi$.
($\dot{+}$6) $(\psi \dot{+} \mu) \wedge \chi \models \psi \dot{+} (\mu \wedge \chi)$.

($\dot{+}$1) states that the agent aims at having consistent beliefs (unless an inconsistent set of beliefs $\mu$ is accepted). ($\dot{+}$2) is linked with the minimal change principle: if $\psi$ is consistent with $\mu$ then the minimal change $\psi \mapsto \psi'$ is $\psi' = \psi$ (i.e., no change). ($\dot{+}$3) is related to the fact that $\mu$ has priority over $\psi$, i.e., the only belief changes are made on $\psi$: $\mu$ is unchanged. ($\dot{+}$4) states that revision respects the principle of independence to syntax (substituting a formula by an equivalent formula should not affect the result of the inference, up to equivalence). It can be shown that the conjunction of postulates ($\dot{+}$5) and ($\dot{+}$6) is equivalence to the following assertion: if $(\psi \dot{+} \mu) \wedge \chi$ is consistent then $\psi \dot{+} (\mu \wedge \chi) \equiv (\psi \dot{+} \mu) \wedge \chi$. In other words, if there is no need to further modify the beliefs after revision of $\psi$ by $\mu$ in order to incorporate $\chi$, then this additional modification is not performed and the new beliefs $\chi$ are simply added to the beliefs $\psi \dot{+} \mu$. This expresses the minimal change principle: when no further change is needed to restore consistency, then no such change is executed.

Despite these postulates, the set of belief revision operators is still wide, depending on the way change is assessed. In particular, it can be assessed thanks to a similarity measure between interpretations, giving birth to a family of revision operators presented below.

*Similarity-based Revision Operators.* A similarity measure on a set $S$ is defined in this paper as a function $\mathtt{sim} : S \times S \to [0,1]$ such that $\mathtt{sim}(a,b) = 1$ iff $a = b$ (for $a, b \in S$). A belief revision operator $\dot{+}^{\mathtt{sim}}$ satisfying the AGM postulates can be defined for every similarity measure $\mathtt{sim}$ on $\Omega$, the set of interpretations:

$$\text{with } \mathtt{sim}^* = \max \{ \mathtt{sim}(\mathcal{I}, \mathcal{J}) \mid \mathcal{I} \in \mathcal{M}(\psi) \text{ and } \mathcal{J} \in \mathcal{M}(\mu) \}$$
$$\mathcal{M}(\psi \dot{+}^{\mathtt{sim}} \mu) = \left\{ \mathcal{J} \in \mathcal{M}(\mu) \,\middle|\, \max_{\mathcal{I} \in \mathcal{M}(\psi)} \mathtt{sim}(\mathcal{I}, \mathcal{J}) = \mathtt{sim}^* \right\} \tag{7}$$

In other terms, the models of $\psi \dot{+}^{\mathtt{sim}} \mu$ are the models of $\mu$ that are the most similar to models of $\psi$. This defines $\dot{+}^{\mathtt{sim}}$ only up to logical equivalence, which is not a problem: any formula $\varrho$ such that $\varrho \equiv \psi \dot{+}^{\mathtt{sim}} \mu$ constitutes the $\dot{+}^{\mathtt{sim}}$-revision of $\psi$ by $\mu$ according to the principle of independence to syntax.

Given a distance function $\mathtt{dist}$ on $\Omega$, a similarity measure can be defined by $\mathtt{sim}(\mathcal{I}, \mathcal{J}) = 1/(1 + \mathtt{dist}(\mathcal{I}, \mathcal{J}))$ for $\mathcal{I}, \mathcal{J} \in \Omega$. In particular, let $H$ be the Hamming distance between interpretations: $H(\mathcal{I}, \mathcal{J})$ is the number of variables

$a$ such that $\mathcal{I}(a) \neq \mathcal{J}(a)$. Let $\mathtt{sim}_H$ be the similarity measure associated with $H$. The revision operator $\dotplus^{\mathtt{sim}_H}$ is the so-called Dalal revision operator [5], denoted by $\dotplus_{\mathrm{Dalal}}$ in the following. The Hamming distance weights every variable equally and considers variables independently: it can be seen as an edit distance on interpretations based on the "flip" edit operation that alters one variable at a time (turning 0 into 1 and conversely). For this reason, when there is no knowledge about how the change has to be measured, the Hamming distance being "neutral" is used for measuring this change. That is why, $\dotplus_{\mathrm{Dalal}}$ is used as a non informed operator, with an empty knowledge about change.

*Revision-based Adaptation* is an approach to single case adaptation based on a revision operator $\dotplus$. The intuition is that the modification of the retrieved case $\mathbf{x}^s \wedge \mathbf{y}^s$ in order to have a proposed solution to the target problem $\mathbf{x}^{\mathrm{tgt}}$ is performed by $\dotplus$. Both the retrieved case and the target problem are interpreted with the domain knowledge, hence the revision to be performed is that of $\mathrm{DK} \wedge \mathbf{x}^s \wedge \mathbf{y}^s$ by $\mathrm{DK} \wedge \mathbf{x}^{\mathrm{tgt}}$. The result of this revision is a formula $\varrho$ entailing $\mathbf{x}^{\mathrm{tgt}}$ (according to postulate $(\dotplus 3)$), thus $\varrho$ is equivalent to a formula $\mathbf{x}^{\mathrm{tgt}} \wedge \mathbf{y}^{\mathrm{tgt}}$ where all variables of $\mathbf{y}^{\mathrm{tgt}}$ belong to $\mathcal{V}_S$. $\mathbf{y}^{\mathrm{tgt}}$ is the proposed solution of $\mathbf{x}^{\mathrm{tgt}}$. Formally, this can be written as follows:

$$(\mathrm{DK} \wedge \mathbf{x}^s \wedge \mathbf{y}^s) \quad \dotplus \quad (\mathrm{DK} \wedge \mathbf{x}^{\mathrm{tgt}}) \quad \equiv \quad \mathbf{x}^{\mathrm{tgt}} \wedge \mathbf{y}^{\mathrm{tgt}} \tag{8}$$

In order to apply revision-based adaptation, a revision operator has to be chosen. This choice is linked on how the change is assessed, that is, using the CBR terminology, the adaptation knowledge $\mathrm{AK}$. So, when no adaptation knowledge is available ($\mathrm{AK} = \emptyset$), the Dalal revision operator is used.

It is noteworthy that the solution $\mathbf{y}^{\mathrm{tgt}}$ provided by revision-based adaptation is necessarily consistent provided that $\mathrm{DK} \wedge \mathbf{x}^{\mathrm{tgt}}$ is, but is not necessarily fully described: $\mathcal{M}(\mathrm{DK} \wedge \mathbf{x}^{\mathrm{tgt}} \wedge \mathbf{y}^{\mathrm{tgt}})$ may contain several interpretations. In such a situation, this means that the revision-based adaptation asserts that there exists a plausible solution $\mathbf{y}$ to $\mathbf{x}^{\mathrm{tgt}}$ that verifies $\mathbf{y} \models \mathbf{y}^{\mathrm{tgt}}$. In the extreme situation, $\mathbf{y}^{\mathrm{tgt}} \equiv \top$, meaning that the revision-based adaptation gives no information on a potential solution of $\mathbf{x}^{\mathrm{tgt}}$.

The running example can be solved using revision-based adaptation. The retrieved case is $\mathbf{case}^s$ and the target problem is $\mathbf{x}^{\mathrm{tgt}}$, defined by (3) and (4). Let us consider this adaptation using $\dotplus_{\mathrm{Dalal}}$. It can be shown that the result $\mathbf{y}^{\mathrm{tgt}}$ of this revision is

$$\begin{aligned} \mathbf{y}^{\mathrm{tgt}}_{\mathrm{Dalal}} \equiv\ & \mathrm{St\ Pierre} \wedge \mathrm{cream} \wedge \mathrm{sauce} \\ & \wedge \neg\mathrm{fruit} \wedge \neg\mathrm{pineapple} \wedge \neg\mathrm{salmon} \wedge \neg\mathrm{curry} \end{aligned} \tag{9}$$

So, this adaptation consists in removing from $\mathbf{case}^s = \mathbf{x}^s \wedge \mathbf{y}^s$ the fruits because their presence would be inconsistent with $\mathrm{DK} \wedge \mathbf{x}^{\mathrm{tgt}}$.

## 3   Bridging Extrapolation and Revision-Based Adaptation

The two approaches to case adaptation presented above—the one based on analogical extrapolation and the one based on belief revision—appear to be quite different: the first one is a multiple case adaptation approach whereas the second one is a single case one (relying respectively on the retrieval of source cases by triplets and by singletons). Nevertheless, the goal of this section is to show how they can meet. First, the approach based on extrapolation is reformulated as a single case adaptation. Second, a revision operator based on source case pairs is defined and it is shown how, under some circumstances, the two approaches to adaptation coincide. This makes it possible to define an approach to adaptation based on both extrapolation and revision that takes into account, on the one hand, the case base and the case pair competence, and, on the other hand, the domain knowledge.

## 3.1   Reformulating Adaptation by Extrapolation as a Single Case Adaptation

In the above presentation of analogical extrapolation, it has been considered that a triplet of cases $(\mathtt{case}^a, \mathtt{case}^b, \mathtt{case}^c)$ is retrieved and then reused in order to solve $\mathbf{x}^{\mathtt{tgt}}$. Now, this can be reformulated in a new way by considering that only $\mathtt{case}^c$ is retrieved, and the other ones, $\mathtt{case}^a$ and $\mathtt{case}^b$, are selected during the adaptation process itself. This "symmetry breaking" has two advantages. First, it can be used for the purpose of an efficient implementation (this issue can be related to the issue of implementing extrapolation algorithms presented in [16]). Second, it makes it possible to match the two approaches to adaptation; for this reason, the retrieved case $\mathtt{case}^c$ is renamed $\mathtt{case}^s = (\mathbf{x}^s, \mathbf{y}^s)$, to better match the notations of single case adaptation.

Therefore, the reformulation of analogical extrapolation as a single case adaptation is as follows:

**Input:** the case to be adapted $(\mathbf{x}^s, \mathbf{y}^s)$, the target problem $\mathbf{x}^{\mathtt{tgt}}$, the case base CB, the preference relation between pairs of cases
**Output:** a set of proposed solutions Y to $\mathbf{x}^{\mathtt{tgt}}$
**1.** Let CandidateCasePairs be the set of $(\mathtt{case}^a, \mathtt{case}^b) \in$ CB $\times$ CB such that $\mathbf{x}^a : \mathbf{x}^b :: \mathbf{x}^s : \mathbf{x}^{\mathtt{tgt}}$ holds and the analogical equation $\mathbf{y}^a : \mathbf{y}^b :: \mathbf{y}^s : ?\mathbf{y}$ is solvable.
**2.** Let BestCandidateCasePairs be the set of most competent case pairs among CandidateCasePairs.
**3.** Let $Y = \left\{ \mathbf{y} \,\middle|\, \begin{array}{l} \mathbf{y} \text{ is the solution of } \mathbf{y}^a : \mathbf{y}^b :: \mathbf{y}^s : ?\mathbf{y} \\ \text{for } (\mathtt{case}^a, \mathtt{case}^b) \in \mathtt{BestCandidateCasePairs} \end{array} \right\}.$
**4.** Y is returned as a set of candidate solutions to $\mathbf{x}^{\mathtt{tgt}}$.

## 3.2   A Revision Operator Based on Competence of Case Pairs

A similarity measure $\mathtt{sim}_{\mathtt{comp}}$ on $\Omega$ that is based on pairs of cases and case pair competence can be defined under some assumptions, hence the revision operator $\dotplus^{\mathtt{sim}_{\mathtt{comp}}}$. It is noteworthy that for the competence preorders presented in [15], these assumptions hold.

The first assumptions is that the competence preorder between pairs of cases can be defined thanks to a *competence level*, i.e., a function compLvl that maps a pair of source cases to a value in $[0, 1]$ such that $(\mathtt{case}^1, \mathtt{case}^2)$ is deemed to be strictly more competent than $(\text{case}^3, \text{case}^4)$ iff $\text{compLvl}(\text{case}^1, \text{case}^2) > \text{compLvl}(\text{case}^3, \text{case}^4)$.

The second assumption just translates the fact that two pairs of cases being in analogy have the same competence, hence the same competence level:

$$\textbf{if } x^a : x^b :: x^c : x^d \textbf{ and } y^a : y^b :: y^c : y^d$$
$$\textbf{then } \text{compLvl}(\text{case}^a, \text{case}^b) = \text{compLvl}(\text{case}^c, \text{case}^d) \qquad (10)$$

The third assumption is that the maximum level of competence is 1 and is reached only by pairs $(\text{case}^a, \text{case}^a)$, for $\text{case}^a \in \text{CB}$. This third assumption can be justified by the fact that if $x^a : x^a :: x^s : x^{tgt}$ then $x^s = x^{tgt}$ (according to (5)) and thus that the analogical equation $y^a : y^a :: y^s : ?y$ has exactly one solution $?y = y^s$ that solves $x^{tgt} = x^s$.

The fourth assumption is that the level of competence is minimal for $(\text{case}^a, \text{case}^b)$ iff this case pair is in analogy with no case pair $(\text{case}^c, \text{case}^d) \in \text{CB}^2$ (which involves that $\{\text{case}^a, \text{case}^b\} \not\subseteq \text{CB}$).

A one-to-one correspondence $\text{caseOf}$ between $\Omega$ and the set of fully described cases can be defined, for $\mathcal{I} \in \Omega$, by $\text{caseOf}(\mathcal{I}) = (x, y)$ is such that $\mathcal{M}(x \wedge y) = \{\mathcal{I}\}$. A similarity measure $\text{sim}_{\text{comp}}$ can be defined using this correspondence and the competence level function:

$$\text{sim}_{\text{comp}}(\mathcal{I}, \mathcal{J}) = \text{compLvl}(\text{caseOf}(\mathcal{I}), \text{caseOf}(\mathcal{J})) \qquad (\text{for } \mathcal{I}, \mathcal{J} \in \Omega)$$

The meeting between the two approaches of adaptation is expressed by the following result (given a source case $(x^s, y^s)$ and a target problem $x^{tgt}$):

$$\begin{array}{l} \textbf{if } y^{tgt} \text{ is the result of revision-based adaptation with } \dotplus = \dotplus^{\text{sim}_{\text{comp}}}, \\ \quad \text{the domain knowledge is empty } (\text{DK} = \top), \\ \quad Y \text{ is the set of solutions obtained by analogical extrapolation} \\ \quad \text{and } Y \neq \emptyset \\ \textbf{then } y^{tgt} \equiv \bigvee Y \end{array} \qquad (11)$$

In other terms, in absence of any domain knowledge, revision-based adaptation using the revision operator $\dotplus^{\text{sim}_{\text{comp}}}$ based on case pair competence gives the same result as analogical extrapolation, unless this latter gives an empty set of solution. Furthermore, it can be shown that if this set is empty, then $y^{tgt} \equiv \top$, i.e., in this case, revision-based adaptation gives no information on the solution.

The proof of (11) mainly consists in applying the definitions: the similarity measure $\text{sim}_{\text{comp}}$ has been chosen so that $\dotplus^{\text{sim}_{\text{comp}}}$-adaptation matches analogical extrapolation when DK is empty.

## 3.3  An Approach to Adaptation Based on Extrapolation and Revision

Now, consider the running example with $\dotplus^{\text{sim}_{\text{comp}}}$-adaptation and, first, with $\text{DK} = \emptyset$. Because of the variable curry, analogical extrapolation gives no solution[1]

---

[1] More precisely, the analogical equation $y^a : y^b :: y^s : y$ has no solution: it is solvable feature by feature on every feature except curry, so a solution can be proposed for every other features.

($Y = \emptyset$) and in this situation, $y^{tgt} \equiv \top$ (no information on the solution is proposed). If the curry variable is discarded from $\mathcal{V}$ then, according to (11), the proposed solution verifies:

$$y^{tgt} = y^{tgt}_{\dot{+}^{sim_{comp}} - \text{adaptation without curry}} \equiv y^{tgt}_{extrap} \tag{12}$$

Now, in order to design an approach that considers all the variables, even the ones such as curry that does not match exactly extrapolation, a similarity measure $sim_{E\&D}$ can be defined that combines $sim_{comp}$ and $sim_H$ (for a given $\alpha$, $0 < \alpha < 1$, and $\mathcal{I}, \mathcal{J} \in \Omega$):

$$sim_{E\&D}(\mathcal{I}, \mathcal{J}) = (1 - \alpha)sim_{comp}(\mathcal{I}, \mathcal{J}) + \alpha sim_H(\mathcal{I}, \mathcal{J}) \tag{13}$$

hence the revision operator $\dot{+}^{sim_{E\&D}}$ (E&D for "extrapolation and Dalal"). Then, for a small enough $\alpha$, $\dot{+}^{sim_{E\&D}}$-adaptation consists in making a minimal number of variable flips on $case^s$ to make extrapolation possible and then in applying it (in fact, it is sufficient that $\alpha < (2|\mathcal{V}|)^{-1}$. ). In particular, if analogical extrapolation applies (on every variables) then no flip is necessary and $\dot{+}^{sim_{E\&D}}$-adaptation coincides with $\dot{+}^{sim_{comp}}$-adaptation (that performs analogical extrapolation).

Applying $\dot{+}^{sim_{comp}}$-adaptation on the running example gives:

$$y^{tgt}_{E\&D} \equiv \mathtt{sauce} \wedge \mathtt{fruit} \wedge \mathtt{pineapple} \atop \wedge \neg\mathtt{St\,Pierre} \wedge \mathtt{salmon} \wedge \mathtt{cream} \wedge \neg\mathtt{curry} \tag{14}$$

The analogical equation $1 : 0 :: 0 : ?y_{curry}$ for finding the value of the feature curry of $y^{tgt}$ by extrapolation of ($case^a, case^b, case^s$) has no solution, so a flip of this feature for $case^s$ gives $case^{s\prime}$ and the triplet ($case^a, case^b, case^{s\prime}$) can be used by extrapolation to solve $x^{tgt}$ in $y^{tgt} = y^{tgt}_{E\&D}$, since $1 : 0 :: 1 : ?y_{curry}$ has a unique solution $?y_{curry} = 0$ (hence $\neg$curry in the proposed solution of $x^{tgt}$).

Now, this adaptation has not taken into account the domain knowledge and, in fact, $DK \wedge x^{tgt} \wedge y^{tgt}_{E\&D}$ is inconsistent, with the DK of the running example (because of the pesto-fruit conflict). Therefore, the proposed approach to adaptation consists in doing a $\dot{+}^{sim_{comp}}$-adaptation taking into account DK as in (8). With the running example, this gives:

$$y^{tgt}_{E\&D\ w/DK} \equiv \mathtt{sauce} \wedge \neg\mathtt{fruit} \wedge \neg\mathtt{pineapple} \atop \wedge \neg\mathtt{St\,Pierre} \wedge \mathtt{salmon} \wedge \mathtt{cream} \wedge \neg\mathtt{curry} \tag{15}$$

which consists in removing fruits from $y^{tgt}_{E\&D}$.

### 3.4   Synthesis

Figure 2 describes the cases and the target problem presented above, as well as the proposed cases $case^s = x^{tgt} \wedge y^{tgt}$ after the different adaptation processes.

This example illustrates how the strengths of two approaches to adaptation can be combined:

| #§ | | fish | pesto | vanilla | sauce | fruit | pineapple | St Pierre | salmon | cream | curry |
|---|---|---|---|---|---|---|---|---|---|---|---|
| 2.3 | $case^a$ | 1 | 0 | 1 | 1 | 0 | 0 | 1 | 0 | 1 | 1 |
| | $case^b$ | 1 | 1 | 1 | 1 | 0 | 0 | 0 | 1 | 1 | 0 |
| | $case^s$ | 1 | 0 | 0 | 1 | 1 | 1 | 1 | 0 | 1 | 0 |
| | $x^{tgt}$ | 1 | 1 | 0 | ? | ? | ? | ? | ? | ? | ? |
| 2.4 | $case^{tgt}_{extrap}$ | 1 | 1 | 0 | 1 | 1 | 1 | 0 | 1 | 1 | ? |
| 2.5 | $case^{tgt}_{Dalal}$ | 1 | 1 | 0 | 1 | 0 | 0 | 1 | 0 | 1 | 0 |
| 3.3 | $case^{tgt}_{E\&D}$ | 1 | 1 | 0 | 1 | 1 | 1 | 0 | 1 | 1 | 0 |
| 3.3 | $case^{tgt}_{E\&D\ w/DK}$ | 1 | 1 | 0 | 1 | 0 | 0 | 0 | 1 | 1 | 0 |

**Fig. 2.** The running example used throughout the paper: problem setting and outcomes $case^{tgt} = x^{tgt} \wedge y^{tgt}$ of adaptation processes presented in the paper. The column #§ indicates the number of the relevant section.

- Analogical extrapolation's strength is to exploit the variations within the case base (variations that can be seen as specific adaptation rules): the variation "Saint-Pierre to salmon" from $case^a$ to $case^b$ is applied to $case^s$ in a similar context.
- $\dotplus_{Dalal}$-adaptation's strength is to take into account the domain knowledge in order to adjust the retrieved case to propose a solution to the target problem: the pesto being incompatible with fruits, pineaple is removed from the recipe.[2]
- $\dotplus^{sim_{comp}}$-adaptation combines the two adaptation approaches: it applies analogical extrapolation on each feature for which it is both possible and consistent with the domain knowledge, and adjust the other features in the $\dotplus_{Dalal}$-adaptation way.

## 4   Related Work and Final Remarks

This paper has considered two very different approaches to case adaptation—analogical extrapolation and revision-based adaptation—and has investigated the issue of how they can meet. It has been shown that, under some circumstances (propositional setting, no domain knowledge, etc.), they coincide (cf. (11)) and that the approach can be extended when domain knowledge is added and/or when analogical proportions holds only for some solution features. The idea is that the case pairs and their competences—used in analogical extrapolation—can be used for "re-shaping the adaptation space" by making more similar the source case and the target problem, and that this similarity is used by the revision operator.

*Related Work.* There is a rich literature on belief revision following the seminal work of Alchourrón, Gärdenfors and Makinson [2], and its expression in a propositional logic setting [12]. However, the idea of applying belief revision to CBR,

---

[2] It is noteworthy that $\dotplus_{Dalal}$-adaptation may do more than removing positive facts as it can substitute a class by a sibling class in the taxonomy (see, e.g., [14]).

as restated in the preliminaries of this paper, can be found only in few works; see [4] in particular.

The idea of an analogical inference pattern based on analogical proportions dates back to [18]. Its application to CBR is suggested in [21], and more systematically investigated in [16].

There has been no work bridging analogical extrapolation and belief revision until now. Keeping in mind that belief revision and nonmonotonic logic are two sides of the same coin in some sense [10], we may however mention a discussion [19] contrasting nonmonotonic reasoning and analogical reasoning, but also providing pathways between them.

*Future Work.* This paper has shown how two models of adaptation can meet, but the question remain of the practical usefulness of this meeting, beyond the running example. For this purpose, a first line of future work is to conduct an experiment comparing the different approaches presented there.

The revision operators $\dot{+}^{sim_{comp}}$ and $\dot{+}^{sim_{comp}}$ may change with the addition of a new case in the case base, e.g., the case $(x^{tgt}, y^{tgt})$ when this new case is validated. Indeed, the competence level of case pairs is computed on the basis of **CB**. The evolution of revision operators over time is an issue related to iterated revision (see, e.g., [7]), so, it would be interesting to study $\dot{+}^{sim_{comp}}$ and $\dot{+}^{sim_{comp}}$ at the light of the postulates of iterated revision postulates.

Lastly, it is known that belief revision can be encoded in possibilistic logic [9], since belief revision relies at the semantic level on epistemic entrenchment relations [11], which are nothing but qualitative necessity relations in the sense of possibility theory [8]. How to process the different kinds of revision/adaptation considered in this paper in the possibilistic setting is another topic for future research.

# References

1. Aamodt, A., Plaza, E.: Case-based reasoning: foundational issues, methodological variations, and system approaches. AI Commun. **7**(1), 39–59 (1994)
2. Alchourrón, C.E., Gärdenfors, P., Makinson, D.: On the logic of theory change: partial meet functions for contraction and revision. J. Symbolic Logic **50**, 510–530 (1985)
3. Aristotle: Nicomachean Ethics. University of Chicago Press (2011). Translated by Bartlett, R.C., Collins, S.D
4. Cojan, J., Lieber, J.: Applying belief revision to case-based reasoning. In: Prade, H., Richard, G. (eds.) Computational Approaches to Analogical Reasoning: Current Trends. SCI, vol. 548, pp. 133–161. Springer, Heidelberg (2014). https://doi.org/10.1007/978-3-642-54516-0_6
5. Dalal, M.: Investigations into a theory of knowledge base revision: preliminary report. In: Proceedings of the Seventh National Conference on Artificial Intelligence (AAAI), pp. 475–479 (1988)
6. d'Aquin, M., Badra, F., Lafrogne, S., Lieber, J., Napoli, A., Szathmary, L.: Case base mining for adaptation knowledge acquisition. In: Veloso, M.M. (ed.) IJCAI 2007, Proceedings of the 20th International Joint Conference on Artificial Intelligence, Hyderabad, pp. 750–755, 6–12 January (2007)

7. Darwiche, A., Pearl, J.: On the logic of iterated belief revision. Artif. Intell. **89**(1–2), 1–29 (1997)
8. Dubois, D., Prade, H.: Epistemic entrenchment and possibilistic logic. Artif. Intell. **50**(2), 223–239 (1991)
9. Dubois, D., Prade, H.: Possibilistic logic - an overview. In: Siekmann, J.H. (ed.) Computational Logic. Volume 9 of Handbook of the History of Logic, 283–342. Elsevier (2014)
10. Gärdenfors, P.: Belief revision and nonmonotonic logic: two sides of the same coin? In: Proceedings of the 9th European Conference on Artificial Intelligence (ECAI 1990), Stockholm, pp. 768–773 (1990)
11. Gärdenfors, P., Makinson, D.: Revisions of knowledge systems using epistemic entrenchment. In: Vardi, M.Y. (ed.) Proceedings 2nd Conference on Theoretical Aspects of Reasoning about Knowledge, Pacific Grove, Morgan Kaufmann, pp. 83–95 (1988)
12. Katsuno, H., Mendelzon, A.: Propositional knowledge base revision and minimal change. Artif. Intell. **52**(3), 263–294 (1991)
13. Lepage, Y.: Analogy and formal languages. Electr. Notes Theor. Comput. Sci. **53**, 180–191 (2001)
14. Lieber, J.: Application of the revision theory to adaptation in case-based reasoning: the conservative adaptation. In: Weber, R.O., Richter, M.M. (eds.) ICCBR 2007. LNCS (LNAI), vol. 4626, pp. 239–253. Springer, Heidelberg (2007). https://doi.org/10.1007/978-3-540-74141-1_17
15. Lieber, J., Nauer, E., Prade, H.: Improving analogical extrapolation using case pair competence. In: Case-Based Reasoning Research and Development, 27th International Conference (ICCBR-2019), Otzenhausen, France (2019)
16. Lieber, J., Nauer, E., Prade, H., Richard, G.: Making the best of cases by approximation, interpolation and extrapolation. In: Cox, M.T., Funk, P., Begum, S. (eds.) ICCBR 2018. LNCS (LNAI), vol. 11156, pp. 580–596. Springer, Cham (2018). https://doi.org/10.1007/978-3-030-01081-2_38
17. Miclet, L., Prade, H.: Handling analogical proportions in classical logic and fuzzy logics settings. In: Sossai, C., Chemello, G. (eds.) ECSQARU 2009. LNCS (LNAI), vol. 5590, pp. 638–650. Springer, Heidelberg (2009). https://doi.org/10.1007/978-3-642-02906-6_55
18. Pirrelli, V., Yvon, F.: Analogy in the lexicon: a probe into analogy-based machine learning of language. In: Proceedings of the 6th International Symposium on Human Communication. Santiago de Cuba, p. 6 (1999)
19. Prade, H., Richard, G.: Cataloguing/analogizing: a nonmonotonic view. Int. J. Intell. Syst. **26**(12), 1176–1195 (2011)
20. Prade, H., Richard, G.: From analogical proportion to logical proportions. Log. Univers. **7**(4), 441–505 (2013)
21. Prade, H., Richard, G.: Analogical proportions and analogical reasoning - an introduction. In: Aha, D.W., Lieber, J. (eds.) ICCBR 2017. LNCS (LNAI), vol. 10339, pp. 16–32. Springer, Cham (2017). https://doi.org/10.1007/978-3-319-61030-6_2
22. Prade, H., Richard, G.: Analogical proportions: from equality to inequality. Int. J. Approximate Reasoning **101**, 234–254 (2018)
23. Richter, M.M., Weber, R.O.: Case representations. In: Case-Based Reasoning, pp. 87–111. Springer, Heidelberg (2013). https://doi.org/10.1007/978-3-642-40167-1_5
24. Riesbeck, C.K., Schank, R.C.: Inside case-based reasoning. Lawrence Erlbaum Associates Inc. Hillsdale, New Jersey (1989)
25. Smyth, B., Keane, M.T.: Using adaptation knowledge to retrieve and adapt design cases. Knowl.-Based Syst. **9**(2), 127–135 (1996)

# Inferring Case-Based Reasoners' Knowledge to Enhance Interactivity

Pierre-Alexandre Murena[1,2(✉)] and Marie Al-Ghossein[1,3]

[1] Helsinki Institute for Information Technology HIIT, Helsinki, Finland
[2] Department of Computer Science, Aalto University, Espoo, Finland
`pierre-alexandre.murena@aalto.fi`
[3] Department of Computer Science, University of Helsinki, Helsinki, Finland
`marie.al-ghossein@helsinki.fi`

**Abstract.** When interacting with a human user, an artificial intelligence needs to have a clear model of the human's behaviour to make the correct decisions, be it recommending items, helping the user in a task or teaching a language. In this paper, we explore the feasibility of modelling the human as a case-based reasoning agent through the question of how to infer the state of a CBR agent from interaction data. We identify the main parameters to be inferred, and propose a Bayesian belief update as a possible way to infer both the parameters of the agent and the content of their case base. We illustrate our ideas with the simple application of an agent learning grammar rules throughout a sequence of observations.

**Keywords:** User modelling · Machine learning for CBR · Bayesian Inference for CBR

## 1 Introduction

Many applications strongly rely on the interactivity between a human user and an Artificial Intelligence (AI). In such applications, a human agent performs actions to complete a specific task in cooperation with an AI agent which guides them along the way, either by providing advice, corrections or by intervening directly in the environment [5]. Intelligent Tutoring Systems (ITS) [1] are an example of such applications, where an AI proposes specific learning materials to help a human learner acquire a specific concept.

Despite their differences, all these applications share an important feature: since they involve the collaboration between two agents, the human user and the AI, they require both agents to have a good understanding of their collaborator [4,18]. From the perspective of the AI, this is done in practice by providing the AI agent with a model of the human user. In the case of ITS, such a model could describe what the learner knows [6,19] or how they acquire knowledge [16]. Alternatively, in model-based recommender systems, a user profile is used to represent their tastes and preferences, based on which items will be recommended by the AI agent.

© Springer Nature Switzerland AG 2021
A. A. Sánchez-Ruiz and M. W. Floyd (Eds.): ICCBR 2021, LNAI 12877, pp. 171–185, 2021.
https://doi.org/10.1007/978-3-030-86957-1_12

Case-Based Reasoning (CBR) has been involved in a long tradition of contributions to the field of interactive systems, some of which made in the domain of education [7], by suggesting how an AI could optimally interact with a user. This paper takes a rather different position: We propose a novel interactive framework that models the human user as a CBR agent, having thus principles from CBR dictate how the user acquires and reuses knowledge from previous observations. Using such a user model enables taking into account different effects that go along with the learning experience, such as the memorization, forgetting, and adaptation of previous observations.

Modeling the user as a CBR agent raises various technical challenges, including the question of how to infer the characteristics of the user from their behavior, in particular when these characteristics are not stationary and evolve throughout the interaction. When the case base of the user is known, it does not seem challenging to infer the other characteristics, such as the similarity metric used for retrieval or the parameters of the adaptation [17]. The main difficulty arises when the content of the case base is unknown to the AI agent. In this paper, we propose to alleviate this uncertainty using Bayesian belief update for a joint inference of the content of the case base and of the CBR characteristics. Although this methodology shows good performances, we also discuss that it would be illusory to expect a full inference, since some CBR configurations cannot be distinguished only based on their outcomes.

The remainder of this paper is organized as follows. In Sect. 2, we introduce a formalization of the interactive process using Partially Observable Markov Decision Processes (POMDP). This formalization introduces the user model as a latent variable and highlights the need for an AI agent to infer the parameters of this model. Section 3 discusses how a CBR agent can be used to model the human user. We then identify the parameters of this CBR agent that need to be inferred during the interaction. The inference itself is described in Sect. 4: after presenting the general principle, we develop a simplified case where the CBR is assumed to be deterministic and we discuss the algorithmic implementation of this procedure. These principles are then applied to specific applications, the results of which are presented in Sect. 5. We conclude the paper with a discussion on the perspectives offered by the presented techniques.

## 2   Problem Statement: Interaction with a CBR Agent

Let $\mathcal{P}$ be a problem space and $\mathcal{S}$ be a solution space. We call a *case* a tuple $(x, y) \in \mathcal{P} \times \mathcal{S}$. The problem of a CBR agent is to infer a plausible solution $y^{tgt} \in \mathcal{S}$ to a problem $x^{tgt} \in \mathcal{P}$.

We consider an agent, the *user*, taking some decisions based on the observation of a sequence of cases. Given a case base $CB$, the sequential process can be described as follows: the user observes a problem $x_t \in \mathcal{P}$ and takes a decision $\omega_t$ in reaction to this problem. We note that this decision $\omega_t$ is not necessarily equal to the estimated solution $\hat{y}_t$ to the problem $x_t$, but is related to it. Then, the user may eventually observe the true solution $y_t$.

In the context of Intelligent Tutoring Systems, a teacher can aim to teach the human learner a grammar rule by showing a sequence of examples. At each step, the teacher suggests a problem ($x_t$) and the user suggests a corresponding solution ($\hat{y}_t$). In this example, the teacher observes directly the estimated solution ($\omega_t = y_t$). However, other more sophisticated applications require a strict distinction between $\omega_t$ and $y_t$. For instance, in a medical context [9], a case can be given by the medical observations of a patient ($x_t$) and the medical diagnosis ($y_t$). However, based on their diagnosis, a physician will take a decision related to the suitable prescription ($\omega_t$). In an interactive system with an AI assistant, only the prescription would be visible to the AI which should then be able to infer the reasoning process of the physician, for their diagnosis and prescription.

A strong hypothesis made by our work is that the mapping $x_t \mapsto \hat{y}_t$ is computed by the agent based on CBR. This hypothesis will be exploited further in Sect. 4 when estimating the parameters of the decision-making.

Whereas the introduced framework focused only on one agent, the user, its more general setting includes additionally the AI agent that may be responsible for selecting the problems. One way to formalize the decision-making of such an artificial agent in interaction with the human user is offered by the Partially Observable Markov Decision Processes (POMDPs), which have been used in various interaction applications such as teaching [12], dialogues [20] or human-robot interaction [2]. A POMDP is defined as a tuple $(S, A, R, T, \Omega, O)$, where $S$ is the set of possible states, $A$ the set of actions (in our context, the cases to present), $R$ a reward function (describing what the AI aims to achieve), $T :$ $S \times A \times S \to [0, 1]$ the state-transition ($T(s_1, a, s_2)$ measures the probability of transition from state $s_1$ to state $s_2$ by playing action $a$), $\Omega$ the set of observations and $O : S \times A \times \Omega \to [0, 1]$ the observation probability ($O(s, a, \omega)$ measures the probability of observing $\omega$ when action $a$ is played in state $s$).

In our context, the state $s$ corresponds to the description of the parameters of the user, which affect their own decision-making. With this POMDP formalization, the user's decision-making is described by the observation probability function $O$, which assesses the probability of the user in a state $s$ to take decision $\omega_t$ based on the problem $x_t$ selected by the AI.

In the following, we denote by $s^{(t)}$ the user state at time $t$. This description is given as a vector containing all the parameters necessary for a representation of the user. For a given $i$, we note $s_i^{(t)}$ to refer to the $i$-th component of the vector, and $s_{-i}^{(t)}$ to designate the vector of all components $s_j$ for $j \neq i$.

An important challenge when solving POMDPs is that the parameters of the user cannot be directly observed by the AI, and some may evolve during the interactive process (e.g. the content of the case base). These changes in the state are described by the transition probability $T$. To alleviate this issue, it is important for the system to be able to infer the value of the states in an online manner, while keeping track of the uncertainty. The remainder of this paper will propose a description of how to define the relevant states when the user bases their decision on CBR, and how the parameters can be inferred in practice.

## 3    Modeling the User as a Case-Based Reasoner

To guarantee an optimal interaction with the user, it is necessary that the definition of the state yields a *forward* model of the user: Given $s^{(t)}$, it must be possible to simulate the user future behavior. In this paper, we assume that the user makes decisions following a CBR. This requires in particular the description of how the user memorizes and reuses previous cases to solve new problems.

**Table 1.** Summary of parameters to infer by the teacher

| Knowledge container | Parameters to infer |
| --- | --- |
| Case base | Content of the case base |
| | Parameters of case retention |
| | Parameters of the forgetting model |
| Domain knowledge | Background knowledge of domain constraints |
| Similarity knowledge | Similarity measure |
| | Parameters of the similarity measure |
| Adaptation container | Algorithm used for adaptation |
| | Parameters of the algorithm |
| | Rules (for a rule-based algorithm) |

Using the definition of the knowledge containers for CBR [14], we propose to split the user's model into four components:

(1) *The case base*, denoted by $CB^U$ is the collection of memorized cases. It is updated upon time by adding or removing elements from the collection. During an interaction, it is important to consider how new cases are added to the case base, but also how cases are removed from the case base. In particular, when considering human users, removing a case can be motivated by a conscious desire to update the case base, but also by unconscious phenomena such as forgetting [11]. The AI agent must be able to have an estimation of the content of the case base in order to choose the most appropriate actions. This requires in practice to infer the parameters of the memorization and forgetting phenomenon.

(2) *The domain knowledge* provides a set of rules dictated by the domain and which constrain the search for a solution to the given problems. These rules can be understood as the background knowledge that the user may or may not have. When the teacher is able to identify potential domain knowledge, it needs to infer whether the user does have it. In case the user does not, the AI can adapt its actions to make such rules understandable, or, in practical applications, the AI may provide explicit explanations [3].

(3) *The similarity knowledge* describes the factors used to assess the similarity between cases, and is used in particular when retrieving cases from $CB^U$ to

solve a new problem $x_t$. For this knowledge container, the inference focuses mostly on the similarity measure used by the learner. In case a finite number of similarity measures can be used, the inference consists in finding out which one is actually preferred by the learner. In more advanced cases, the inference can also focus on parameters of the similarity measure such as the weight coefficients [13].

(4) *The adaptation container* encompasses information that is used for adapting the solution of a retrieved case $(x, y) \in CB^U$ to a new problem $x_t$. In a rule-based system, the adaptation container contains the rules used to perform the adaptation. The inference of the adaptation container requires then to infer which of these rules are used by the user. In a more general case, the rules are replaced by general parameters and/or algorithms for the adaptation.

A summary of the user model parameters to infer is provided in Table 1. In the context of this paper, we will ignore the parameters related to case retention and case forgetting and the inference techniques proposed in the next section cannot apply directly to them. The inference of these parameters will have to be studied in future works.

## 4  Inference of the CBR Parameters

### 4.1  General Principle

A common way to deal with the fact that the states in POMDPs are unobserved, is to evaluate the states using a Bayesian belief update. It can be shown that, in this case, the POMDP is equivalent to a *belief-MDP*. The idea is to estimate the parameters in two steps. First, we estimate the posterior of the state $s^{(t-1)}$ after interaction $t - 1$, using the information obtained at time $t$:

$$p(s^{(t-1)}|x_t, \omega_t) \propto p(\omega_t|x_t, s^{(t-1)})p(s^{(t-1)}) \tag{1}$$

where $p(s^{(t-1)})$ is the prior over the state. The value taken as a prior for the next interaction is obtained by applying the transition function $T$ of the POMDP:

$$p(s^{(t)}) = \mathbb{E}_{p(s^{(t-1)}|x_t, \omega_t)} \left[ T(s^{(t-1)}, (x_t, y_t), s^{(t)}) \right] \tag{2}$$

Although this formulation is the soundest, it is difficult to use in practice when inferring the parameters of a CBR agent, because of the very large dimension of the state space, which must contain all possible case bases. It is applicable though when the case-base of the learner is known.

As a solution, we propose in the following to use marginal distributions over each parameter independently instead of the full joint distribution. This simplification, which is used for computational reasons, yields a loss in terms of the richness of potential correlations between parameters of the model. However, depending on the problem, it is possible to consider some groups of variables together to keep track of some correlations.

When considering the marginals, we consider the update of each of the components $s_i$ of the state $s$. The belief update is then given by:

$$p(s_i^{(t-1)}|x_t, \omega_t) \propto \mathbb{E}_{s_{-i}^{(t-1)}}\left[p(\omega_t|x_t, s_i^{(t-1)}, s_{-i}^{(t-1)})\right] p(s_i^{(t-1)}) \tag{3}$$

where the expected value over the components $s_{-i}^{(t-1)}$ is computed based on the probabilities $p(s_{-i}^{(t-1)})$. For simplicity purposes, we will write the terms $s_i^{(t-1)}, s_{-i}^{(t-1)}$ simply as $s^{(t-1)}$, which is technically correct but loses the intuition that the term $s_i^{(t-1)}$ corresponds to the quantity being updated and $s_{-i}^{(t-1)}$ to the variables of the expected value.

Since the observation is not directly produced by the CBR agent, we decompose the likelihood $p(\omega_t|x_t, s^{(t-1)})$ into two terms, accounting for (i) the result $\hat{y}_t$ of the CBR and (ii) how this result is used to yield observation $\omega_t$:

$$p\left(\omega_t|x_t, s^{(t-1)}\right) =$$
$$\sum_{\hat{y}} \underbrace{p\left(\omega_t|x_t, s^{(t-1)}, CBR(x_t, s^{(t-1)}) = \hat{y}\right)}_{\text{choice of the response given the result of the CBR}} \underbrace{p\left(CBR(x_t, s^{(t-1)}) = \hat{y}\right)}_{\text{result of the CBR}} \tag{4}$$

where the notation $CBR(x_t, s^{(t-1)})$ designates the result of the CBR for problem $x_t$ with $s^{(t-1)}$ as parameters (including the content of the case base).

## 4.2   Inference of the Parameters for a Deterministic CBR

We consider as an illustration the specific case where the learner is a deterministic CBR, i.e. that the retrieval and adaptation are both deterministic functions. In addition, we assume that the learner's output $\omega_t$ is the result of the adaptation, which implies that:

$$p\left(\omega_t|x_t, s^{(t-1)}, CBR(x_t, s^{(t-1)}) = \hat{y}\right) = \mathbb{I}(\omega_t = \hat{y}) \tag{5}$$

where $\mathbb{I}(x) = 1$ if $x$ is true, and $\mathbb{I}(x) = 0$ otherwise. In this context, it can be shown that $p\left(\omega_t|x_t, s^{(t-1)}\right) = \mathbb{I}\left(\omega_t = CBR(x_t, s^{(t-1)})\right)$, and eventually:

$$p(s_i^{(t-1)}|x_t, \omega_t) \propto p_{s_{-i}^{(t-1)}}\left(\omega_t = CBR(x_t, (s_i^{(t-1)}, s_{-i}^{(t-1)}))\right) p\left(s_i^{(t-1)}\right) \tag{6}$$

The CBR process can be divided here into two main steps: the retrieval, denoted by $Ret(x_t)$, which outputs the closest case(s) to $x_t$, and the adaptation, denoted by $Ad(x_t, \mathcal{R})$, which consists in adapting the retrieved cases $\mathcal{R}$ to solve problem $x_t$. It can then be observed that:

$$p_{s_{-i}^{(t-1)}}\left(\omega_t = CBR(x_t, (s_i^{(t-1)}, s_{-i}^{(t-1)}))\right)$$
$$= \sum_{\mathcal{R} \subset CB^U} p_{s_{-i}^{(t-1)}}\left(\omega_t = Ad(x_t, \mathcal{R})\Big|s_i^{(t-1)}\right) p_{s_{-i}^{(t-1)}}\left(Ret(x_t) = \mathcal{R}\Big|s_i^{(t-1)}\right) \tag{7}$$

We note that, in Eq. 7, the sum over all possible results of the retrieval will be in practice reduced to those yielding a correct result during the adaptation (otherwise, the probability that $\omega_t = Ad(x_t, \mathcal{R})$ is 0).

### 4.3   Probability of Retrieval for kNN

In the case where the retrieval is operated by a k Nearest Neighbor algorithm, the probability $p_{s_{-i}^{(t-1)}}(Ret(x_t) = \mathcal{R})$ can be evaluated as follows.

We first consider that the similarity metric and its parameters are fully known. For simplicity and without loss of generality, we assume in the following that the cases are ordered by decreasing similarity to $x$ (in particular, $(x_1, y_1)$ is the most similar to $x$). In practice, this can be obtained using a permutation $\sigma$ reordering the cases. Then the probability that $kNN(x_t)$ outputs $(i_1, \ldots, i_k)$, where $i_1 < \ldots < i_k$, is given by:

$$p\big(kNN(x_t) = (i_1, \ldots, i_k)\big) = \prod_{j \in (i_1, \ldots, i_k)} p(\lambda_j^{(t)} = 1) \prod_{\substack{j=1 \\ j \notin (i_1, \ldots, i_k)}}^{i_k} p(\lambda_j^{(t)} = 0) \quad (8)$$

where $\lambda_i^{(t)} \in \mathbb{B}$ indicate whether case $(x_i, y_i)$ belongs to the user's case base. Note however that the $\lambda_i^{(t)}$ are components of the vector $s^{(t)}$.

When there is uncertainty over the similarity metric and/or its parameters, we can obtain the probability of retrieval by using the law of total probability over these values. Note that the probability in Eq. 7 is computed over the variables $s_{-i}^{(t-1)}$ only, variable $s_i^{(t-1)}$ being fixed and corresponding to the variable being updated. The computational complexity of computing this probability under uncertainty depends on the number of similarity measures and parameters to consider. In particular, this operation requires additional attention in continuous parameter spaces.

### 4.4   Discussion on the Inference Process

The Bayesian inference described in this section is very general and applicable to any situation where the behaviour of the CBR system can be modelled. In particular, we showed how Eq. 4 can be used to assess situations where the exact output of the reasoning is not observed. In terms of implementation however, it is noticeable that the presented techniques can quickly become computationally very expensive, as soon as the number of parameters of the models increases. The variable separation suggested in Eq. 3 goes into the direction of lowering the dimension of the state space, but this dimension is obviously not the only cause of complexity. For instance, Eq. 7 requires to sum over all possible retrieval results, the number of which grows exponentially with the number of cases to retrieve. In the experimental section, we will consider the simple case of 1 neighbor only, in order to keep reasonable space exploration. For future works however, more advanced inference and approximation techniques will be needed, in particular Monte-Carlo techniques or likelihood-free inference.

# 5   Application: Teaching Word Inflection

## 5.1   Presentation of the Application

As an illustrative example, we consider the application of teaching a grammar rule to a learner. In order to teach a new grammatical concept in a foreign language, a commonly used method is to present some examples to the learner, as well as exercises during which the learner aims to solve a series of problems. The purpose of the application considered here is to mimic this teaching procedure.

We focus on the simple case of word inflection, which is the transformation of a word, called *stem*, into an alternative form, called *inflection*. Such transformations are typical of conjugation or declension. In the simplest case, a single rule applies to all stems, but there exist multiple word classes with their specific transformations, and the learner must be able to memorize all these transformations and to know when to apply them. As a typical example, the Institute for the Languages of Finland identifies 51 different declension groups in the Finnish language, which differ mostly in a change of radical. For instance, although the genitive case is obtained by suffixing a -n to the radical, the formation of the radical from the stem varies from one group to the other. We cite here a few examples following the schema (Stem, Radical, Genitive case): ("kissa", "kissa-", "kissan"), ("korpi", "korve-", "korven"), ("rakkaus", "rakkaude-", "rakkauden"), ("Sibelius", "Sibeliukse-", "Sibeliuksen").

When considering this scenario, a case $(x, y)$ is given by the stem (for instance "kissa") and the corresponding genitive form (here "kissan"). The goal of the teacher, as described in Sect. 2, is then to propose an optimal sequence of cases to the learner (which can be seen as exercises).

The learner model we propose is a CBR framework based on the notion of Kolmogorov complexity [8], inspired by the work of Murena et al. (2020) [10] on morphological analogies. Kolmogorov complexity [8] is a theoretical tool measuring how complex the generation of a string is. Intuitively, the character string "0000000000" is less complex than "0110111010" because it can be generated by a simple program. More formally, the complexity of a binary string $x \in \mathbb{B}^*$, denoted $K_M(x)$, is defined as the length of the shortest program, on a reference Turing machine $M$, that outputs $x$.

This definition relies on the choice of a reference Turing machine $M$; theoretical results show that this is not a real issue because of invariance properties, and most applications, including the one of interest here, fix a simple machine to make $K(.)$ computable. For the case of analogies on words, Murena et al. (2020) [10] introduce a simple description language based mostly on the concatenation of character strings. The programs allow the definition of functions with variables, which can be used for instance to assess repetitions of patterns. Although the choice of this language is a parameter *in se*, we consider it as fixed and optimal. Inferring the optimal description language could be an interesting and challenging future direction.

## 5.2   Implementation of a Case-Based Reasoning Learner

We now propose a full description of the CBR learner we use in the context of this application.

*Case Base.* The case base $CB$ is a set of tuples $(x, y)$, where $x$ is the stem and $y$ the inflected form. We consider only finite case bases. We consider a probabilistic retention, where a case is retained with a given probability (see Sect. 5.3, Experiment 3). As stated before, we do not consider the inference of the parameters of the case retention.

*Domain Knowledge.* The domain knowledge is given as a set of rules which determine the validity of a solution and/or affect the adaptation. We consider here the understanding $h \in \mathbb{B}$ of the *vowel harmony* rule in Finnish, which states that the groups of vowels a/o/u and ä/ö/y cannot coexist in a word; according to it, the solution of the analogical equation "maa:maalla::pää:$x$" will be corrected from "päälla" into "päällä".

*Similarity Knowledge.* The retrieval is highly dependent on a distance function between existing cases and a new problem: $d : \mathcal{X} \times \mathcal{Y} \times X$. We identify three main candidate functions, all based on complexity. The first candidate exploits the idea that adaptation knowledge can play a role in the retrieval phase [15]:

$$d_0(a : b, c) = \min_d K(a : b::c : d) - K(a : b) \tag{9}$$

where $K(a : b)$ removes the impact of the complexity of the source case. Distance $d_1(a : b, c)$ is similar, but has $K(a)$ as a regularizer. The third considered distance measures how close the structures of $a$ and $c$ are: $d_1(a : b, c) = K(a::c) - K(a)$

The retrieval phase is then implemented as a $k$-nearest neighbors procedure, where the neighbors are defined according to the chosen distance function. The domain knowledge is then given by two parameters: $d \in \{d_0, d_1, d_2\}$, the chosen distance function, and $k$, the number of neighbors, chosen to be equal to 1 in this paper. The adoption of higher values of $k$ will be explored in future work.

*Adaptation Knowledge.* The retrieved case $\{(a, b)\}$ is reused for solving the new problem $c$ by solving the analogical equations $a : b::c : x$ , using the algorithm proposed in [10], which states that the solution $x$ of the analogical equation minimizes the complexity $K(a : b::c : x)$. This algorithm is non-parametric.

*Discussion.* Altogether, these four knowledge containers fully define the learner's CBR model. We notice that the only free parameters considered in this application are the understanding of vowel harmony ($h \in \mathbb{B}$) and the distance function used for the retrieval ($d \in \{d_1, d_2\}$). Other parameters (for instance $k$ the number of neighbors) could be considered in more sophisticated models. In addition to the inference of these parameters, the teacher must also infer the content of the case base.

## 5.3   Empirical Evaluation

This section presents different experiments for evaluating the process of inference of the CBR parameters, proposed in Sect. 4, in the context of the application of teaching word inflections. We carried out three sets of experiments focusing on different aspects of the inference.

In all of these experiments, the specific task considered is the one of teaching to derive the inessive case of a Finnish word given its nominative case. The list of Finnish words considered is extracted from the one provided by the Institute for Languages in Finland (Kotus)[1]. This list of words also includes characteristics of each word, in particular its group that dictates in part how the radical is formed based on the nominative case. The inessive case was automatically scraped from the Wiktionary dictionary[2]. In the experiments, we considered only words belonging to the 48th type, which contains a large diversity of stem-to-radical transformations.

The main idea of the experiments is to simulate the interaction between a learner, modeled as a CBR agent with fixed parameters $s_{true}$, and an AI agent trying to infer these parameters, over a number of steps. The true CBR model is used to simulate the user's answers and the evaluation of the parameter inference is based on how close the estimated parameters are from their true values. In addition, we also evaluate the ability of the estimated CBR model to reproduce the true behavior of the user. To measure this ability, we introduce a score metric that is measured at each step $t$ and defined as follows:

$$score^{(t)} = \mathbb{E}_{s^{(t)}} \left[ \frac{1}{|CB_{test}|} \sum_{(x,y) \in CB_{test}} \mathbb{I}\left( CBR(x, s^{(t)}) = CBR(x, s_{true})) \right) \right]$$
$$(10)$$

where $CB_{test}$ is a test case base that is introduced for the sole purpose of evaluating the capacity of reproducing the user behavior on a new set of problems.

**Experiment 1: Parameter Evaluation with a Fixed Case Base.** The first set of experiments focuses on the special case of parameter inference when the case base of the user is fixed and does not evolve throughout the interaction. This setting would remove any potential impact of the dynamic character of parameters on the inference process as described by Eq. 2, which would itself be the subject of Experiment 3.

Under this condition, we denote by $CB^U$ the fixed case base of the user that is itself a subset of a larger (also fixed) case base, denoted by $CB$ and containing all possible cases that the user may have observed or learned. We set the size of $CB^U$ to 30 and that of $CB$ to 100. We consider an interaction session of 50 steps, during which the user does not retain any observation but only provides answers to the problems based on its content and parameters. The experiment

---

[1] www.kotus.fi. The link to the list of Finnish words: kaino.kotus.fi/sanat/nykysuomi.
[2] www.wiktionary.org.

is run 20 times and we sample at each run a different $CB^U$ and $CB$ from the complete list of words described above. The parameters of the true user CBR model are set as follows: $h = 0$ and $d = d_2$. The different priors are taken as uniform distributions over the set of related parameters.

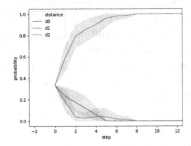

(a) Estimation of the distance used for retrieval.

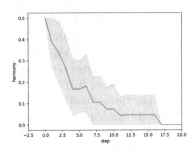

(b) Estimation of the understanding of vowel harmony.

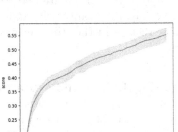

(c) Evolution of the predicted score compared to the real user.

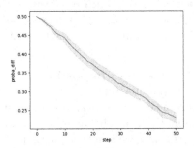

(d) Average error in the prediction of the word probabilities.

**Fig. 1.** Results of Experiment 1, considering the case where the user case base is fixed and measuring the quality of parameter estimation.

As an indication in terms of computational time, running such an experiment (including the 20 runs) takes up to one hour on a machine with one processor Intel Core i5 2.3 GHz and 8G of RAM. Such an experiment also includes measuring the score at each step, which is a costly operation, and does not involve advanced optimization or parallelization of execution.

*Results.* Figure 1 shows the results obtained for this experiment. Figure 1a displays the evolution of the probability of each of the potential distance measures over the number of steps. The estimation of the distance reaches the true value after a few number of steps and the figure only shows the first few steps of the interactive process, after which the values relatively stabilize. Following a similar idea, Fig. 1b shows the estimation of the $h$ parameter related to the understanding of the vowel harmony concept. Its value drops over the number of steps until reaching zero, also showing that it is able to reach the desired value after observing the user solutions. This shows that, even in the absence of a complete certainty over the case base of the user, it remains possible to estimate the parameters of the CBR model.

Figure 1c presents the score metric (Eq. 10) evaluated at each step of the experiment on a fixed $CB_{test}$ of size 100, sampled at the beginning of the experiment from the complete list of words. The score increases over the number of steps, starting from a score of 0 as no proper estimation of the user model has been done at $t = 0$ and the AI agent cannot reproduce the user behavior. The increase of the score metric throughout the experiment shows that the behavior of the estimated user model gets closer to the one of the true user model, which suggests that the estimation quality improves. This idea can also be derived from Fig. 1d where the curve plotting the average difference between the estimated probability of a word from the case base and its true probability, decreases over time. However, and even after a large number of steps, this error does not reach 0: it can be observed that some cases in $CB$ are given a probability of about 0.5. This phenomenon can be explained by an impossibility to discriminate between different words, which are seen by the inference as having a completely similar role, and therefore as completely indiscernible. We discuss further this question of indiscernibility in the next set of experiments.

**Experiment 2: Impossibility of Differentiating Indiscernible States.**
The parameter inference takes as evidence the answers given by the user to a problem set. As mentioned above, it seems that some sets of parameters could exhibit the same behavior (same answers) from the user's side. In this set of experiments, we aim to show that two equivalent states cannot be discernible by the inference process.

We consider the two words "kaura" and "käyrä", having the inessive case as "kaurassa" and "käyrässä" respectively. We focus on the three following states: $s_1 = (kaura \in CB^U, käyrä \notin CB^U, h = 1)$, $s_2 = (kaura \notin CB^U, käyrä \in CB^U, h = 1)$, and $s_3 = (kaura \in CB^U, käyrä \in CB^U, h = 0)$. Since $s_1$ and $s_2$ both incorporate vowel harmony, it can be verified that they hold the same information in terms of how to derive the inessive case from the nominative case, and will therefore provide similar answers to problems.

To compare the probability of each of these states given the user answers,

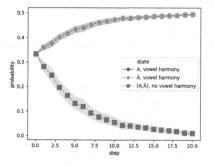

**Fig. 2.** Results from Experiment 2. The states $s_1$ and $s_2$ (blue and orange) yield a similar user's behaviour, and are therefore indiscernible. (Color figure online)

we simulate the behavior of a user having a set of parameters equivalent to $s_1$ on a series of 20 interactions (Note that similar results are obtained with $s_2$).

*Results.* Figure 2 shows the evolution of the probability of each state over the number of steps. It can be seen from the plot that the two states $s_0$ and $s_1$ have the same probability: It is not possible to differentiate between them or favor one over the other only based on the user answers. The probability of state $s_3$ decreases over time until reaching 0: if the user were in state $s_3$, they would retrieve any of the two cases from the case base and would adapt it to form a potentially incorrect answer.

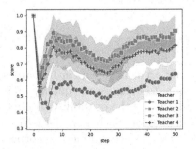 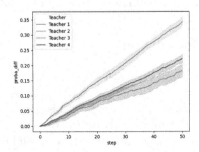

(a) Evolution of the estimated score compared to the real user.

(b) Average error in the prediction of the word probabilities.

**Fig. 3.** Results of Experiment 3: A learner acquires new data during the interaction with a teacher. The teacher estimates the case base, with the following assumptions on retention: (1) Retain with $p = 0.2$; (2) Retain with $p = 0.6$; (3) Retain with $p = 0.95$; (4) retain with $p = 0.5$ when predicting the correct answer and with $p = 0.8$ otherwise.

**Experiment 3: Parameter Inference for a Dynamic Case Base.** We complement the results provided in Experiment 1 by showing that an inference is possible even in a context of a sequential interaction. We mimic a teaching interaction between the AI and the user, during which the teacher displays a sequence of problems from a case base $CB$ of size 50. The learner proposes an answer and observes the actual solution. The presented case is then retained with a probability which depends on the learner's answer: $p = 0.8$ when the answer is incorrect, and $p = 0.5$ otherwise. To infer the learner's CBR model, the teacher exploits a fixed transition dynamics. We compare four possible dynamics: three dynamics having a fixed probabilistic retention (with probabilities $p = 0.2$, $p = 0.6$ and $p = 0.95$) and one having the same dynamics as the learner's. The experiments are led in the same conditions as those of Experiment 1 (20 runs, fixed test base of 100 cases for the score).

*Results.* The experimental results show that the inference of the distance and understanding of vowel harmony is unchanged when using the four transition models. We thus omit to include the corresponding plots. The results presented in Fig. 3 show however that the content of the inferred case base, and consequently the prediction score, are directly affected by the choice of the transition dynamics. In particular, we observe that underestimating the probability of retention (Teacher 1, in blue) causes lower prediction capabilities and more errors on the case base. The teachers with larger probabilities of retention (Teachers 2 and 3, in orange and green) have identical inference of the parameters, which is in particular better than the estimation based on the exact retention model (Teacher 4, in red). These observations highlight the importance of having a good estimation of the transition dynamics. Although this aspect has been ignored in this paper, it is a fundamental and unavoidable future work.

## 6   Conclusion

When interacting with other agents, be it other artificial agents or human users, an AI must be able to understand its teammate to enhance the quality and efficiency of the cooperation. In this paper, we discussed the possibility to use CBR as a paradigm underlying the other agent's behavior. Such a model is particularly interesting when interacting with human users, since it directly incorporates the fact that humans constantly memorize and reuse knowledge from previous experiences. However, it introduces the important challenge of identifying the parameters of such a CBR model based on the observed behavior.

Our first contribution is to clearly identify the dimensions of interest in a CBR model that would need to be inferred (see Table 1). In particular, we discussed that a major but unavoidable challenge is to infer the content of the case base, i.e. what the user knows. This is challenging because of the number of possible configurations for the case base. A second contribution is to demonstrate the feasibility of such an operation: using basic probabilistic tools, we could propose simple algorithms for the inference of the parameters of a CBR agent. For the application of word declension, we succeeded in inferring the parameters used by a CBR user for both the retrieval and the adaptation, when considering fixed and dynamical case bases. However, we also showed that this has limitations: the inference cannot differentiate between different states that exhibit equivalent behaviours, and all the fixed parameters have to be chosen with care. Future research is needed to be able to infer the parameters of case retention, which none of the methods described in our paper can tackle. Furthermore, more advanced techniques will have to be implemented to enable the inference of more complicated models: in particular, *likelihood-free inference* techniques could be valuable tools for approximating more realistic CBR models of human reasoning.

## References

1. Anderson, J.R., Boyle, C.F., Reiser, B.J.: Intelligent tutoring systems. Science **228**(4698), 456–462 (1985)
2. Broz, F., Nourbakhsh, I.R., Simmons, R.G.: Planning for human-robot interaction using time-state aggregated POMDPs. In: AAAI, vol. 8, pp. 1339–1344 (2008)

3. Celikok, M.M., Murena, P.A., Kaski, S.: Teaching to learn: sequential teaching of agents with inner states. arXiv preprint arXiv:2009.06227 (2020)
4. Chakraborti, T., Kambhampati, S., Scheutz, M., Zhang, Y.: AI challenges in human-robot cognitive teaming. arXiv abs/1707.04775 (2017)
5. Dafoe, A., Bachrach, Y., Hadfield, G., Horvitz, E., Larson, K., Graepel, T.: Cooperative AI: machines must learn to find common ground. Nature (593) 33–36 (2021)
6. Elsom-Cook, M.: Student modelling in intelligent tutoring systems. Artif. Intell. Rev. **7**(3–4), 227–240 (1993)
7. Kolodner, J.L., Cox, M.T., González-Calero, P.A.: Case-based reasoning-inspired approaches to education. Knowl. Eng. Rev. **20**(3), 299–304 (2005)
8. Li, M., Vitányi, P., et al.: An Introduction to Kolmogorov Complexity and its Applications, vol. 3. Springer, Heidelberg (2008)
9. Marling, C., Whitehouse, P.: Case-based reasoning in the care of Alzheimer's disease patients. In: Aha, D.W., Watson, I. (eds.) ICCBR 2001. LNCS (LNAI), vol. 2080, pp. 702–715. Springer, Heidelberg (2001). https://doi.org/10.1007/3-540-44593-5_50
10. Murena, P.A., Al-Ghossein, M., Dessalles, J.L., Cornuéjols, A.: Solving analogies on words based on minimal complexity transformations. In: International Joint Conference on Artificial Intelligence, IJCAI (2020)
11. Nioche, A., Murena, P.A., de la Torre-Ortiz, C., Oulasvirta, A.: Improving artificial teachers by considering how people learn and forget. arXiv preprint arXiv:2102.04174 (2021)
12. Rafferty, A.N., Brunskill, E., Griffiths, T.L., Shafto, P.: Faster teaching via POMDP planning. Cogn. Sci. **40**(6), 1290–1332 (2016)
13. Richter, M.M.: Similarity. In: Perner, P. (ed.) Case-Based Reasoning on Images and Signals, pp. 25–90. Springer, Heidelberg (2008). https://doi.org/10.1007/978-3-540-73180-1_2
14. Richter, M.M., Michael, M.: Knowledge Containers. Readings in Case-Based Reasoning. Morgan Kaufmann Publishers (2003)
15. Smyth, B., Keane, M.T.: Using adaptation knowledge to retrieve and adapt design cases. Knowl.-Based Syst. **9**(2), 127–135 (1996)
16. Sottilare, R.A., Graesser, A., Hu, X., Holden, H.: Design Recommendations for Intelligent Tutoring Systems: Volume 1-Learner Modeling, vol. 1. US Army Research Laboratory (2013)
17. Stahl, A.: Learning similarity measures: a formal view based on a generalized CBR model. In: Muñoz-Ávila, H., Ricci, F. (eds.) ICCBR 2005. LNCS (LNAI), vol. 3620, pp. 507–521. Springer, Heidelberg (2005). https://doi.org/10.1007/11536406_39
18. Tabrez, A., Luebbers, M.B., Hayes, B.: A survey of mental modeling techniques in human-robot teaming. Curr. Robot. Rep. 1–9 (2020)
19. Woolf, B.P.: Building Intelligent Interactive Tutors: Student-Centered Strategies for Revolutionizing E-learning. Morgan Kaufmann (2010)
20. Young, S., Gašić, M., Thomson, B., Williams, J.D.: POMDP-based statistical spoken dialog systems: a review. Proc. IEEE **101**(5), 1160–1179 (2013)

# A Case-Based Approach for the Selection of Explanation Algorithms in Image Classification

Juan A. Recio-García[1]([⊠]) [iD], Humberto Parejas-Llanovarced[1],
Mauricio G. Orozco-del-Castillo[2,3], and Esteban E. Brito-Borges[2,3]

[1] Department of Software Engineering and Artificial Intelligence,
Instituto de Tecnologías del Conocimiento, Universidad Complutense de Madrid,
Madrid, Spain
{jareciog,hparejas}@ucm.es
[2] Department of Systems and Computing, Tecnológico Nacional de México/IT de
Mérida, Merida, Mexico
mauricio.orozco@itmerida.edu.mx
[3] AAAIMX Student Chapter at Yucatan, Mexico (AAAIMX), Association
for the Advancement of Artificial Intelligence, Merida, Mexico
esteban.brito@aaaimx.org

**Abstract.** Research on eXplainable AI (XAI) is continuously proposing novel approaches for the explanation of image classification models, where we can find both model-dependent and model-independent strategies. However, it is unclear how to choose the best explanation approach for a given image, as these novel XAI approaches are radically different. In this paper, we propose a CBR solution to the problem of choosing the best alternative for the explanation of an image classifier. The case base reflects the human perception of the quality of the explanations generated with different image explanation methods. Then, this experience is reused to select the best explanation approach for a given image.

**Keywords:** Image explanations · User experience · Case-based explanations

## 1 Introduction

With the success of Machine Learning (ML), interpretability for ML systems has become an active focus of research. XAI research tries to solve several questions related to the increasing need for interpretable models, such as: How should interpretable models be designed? What to explain? When to explain?

Supported by the Horizon 2020 Future and Emerging Technologies (FET) programme of the European Union through the ERA-NET (CHIST-ERA-19-XAI-008 - PCI2020-120720-2) and the Spanish Committee of Economy and Competitiveness (TIN2017-87330-R).

A. A. Sánchez-Ruiz and M. W. Floyd (Eds.): ICCBR 2021, LNAI 12877, pp. 186–200, 2021.
https://doi.org/10.1007/978-3-030-86957-1_13

If we focus on the explanation of deep learning image classifiers, we can find several proposals in the literature [10,23]. There are two major approaches: model-dependent models that analyze the internal behavior of the classifier to provide explanations, and model-independent models where the classification process is considered as a black-box. Concretely, in this paper we analyze and compare four different XAI methods for the explanation of deep neural networks classifiers: Integrated Gradients and XRAI (eXplanation with Ranked Area Integrals) belonging to the group of model-dependent explainers; and LIME (Local Interpretable Model-Agnostic Explanations) and Anchors that are relevant model-agnostic explanation methods.

This work is based on two basic hypotheses: (i) the performance of the explanation method depends on the nature and features of the source image, and (ii) there is no algorithmic solution to determine which explanation method is most suitable for a concrete image. This way, given an image classification to be explained, we can only use previous explanation experiences to select the most suitable XAI method. These hypotheses follow previous research in CBR applied to XAI that has pointed out the importance of taking advantage of the human knowledge to generate and evaluate explanations [14,17].

For example, in our previous work [11] we presented a case-based reasoning method that takes advantage of human knowledge to generate explanations. Concretely, we defined and evaluated a CBR solution to the problem of configuring the well-known LIME algorithm, that attempts to understand a global black-box classification model by perturbing the input of data samples. However, this method applies a generic setup for any image, that leads to inadequate explanations. The CBR solution is based on a case base of images and their associated "optimal" LIME configurations according to the users. From this case base, we implement the CBR-LIME method where, given a new query image, similar images are retrieved, and their corresponding configurations are reused to generate an explanation through the LIME algorithm.

Generalizing this idea, in this paper we propose a CBR solution to the problem of selecting the best explanation approach for an image classifier. The case base reflects the human perception of the quality of the explanations generated with different XAI approaches. Then, given a query image, several similarity metrics are applied to find the most suitable explanation approach. Here, similarity metrics play a very relevant role as we can compare images using different points of view: feature vectors, pixel-to-pixel, structural similarity or color distribution.

This paper is organized as follows: Sect. 2 introduces the XAI algorithms. Section 3 describes the CBR process and the case base elicitation process. In Sect. 4 we demonstrate the benefits of our approach using both off-line and on-line evaluations. Concluding remarks are discussed in Sect. 5.

## 2   Background

CBR can provide a methodology to reuse experiences and generate explanations for different AI techniques and domains of applications. Therefore, we can find

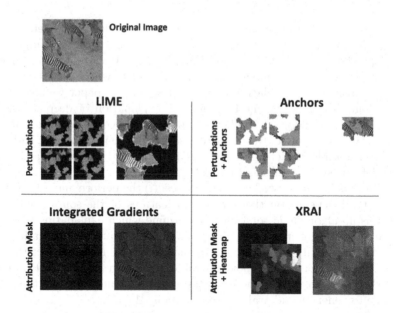

**Fig. 1.** Visual description of the explanation process followed by LIME, Anchors, IG and XRAI.

several initiatives in the CBR literature to explain AI systems. Some relevant early works can be found in the review by [7]. Recently there is a relevant body of work on CBR applied to the explanation of black-box models. Most of these works are post-hoc explanation systems, where CBR follows the model-agnostic approach to explain black-box models [2,5,8,21].

Leaving aside CBR as an explanation method per se, the goal of this paper is to apply CBR to find the most suitable explanation model given an image classification. Here, there are several XAI models for the explanation of image classifiers to be considered. Local surrogate models, such as LIME or Anchors [12,13], focus on explaining individual predictions instead of the whole global model. They are model-agnostic models based on perturbation mechanisms. Another popular local surrogate model similar to LIME is SHAP [9]. It is based on the game theory concept of Shapley values and explains the prediction of an instance by computing the contribution of each feature to the prediction. We can also find alternative approaches such as Integrated Gradients (based on Shapley values too) and XRAI (eXplanation with Ranked Area Integrals) that aim to explain the relationship between model's predictions in terms of its features. These methods are model-dependent and belong to the group of backward propagation mechanisms. The backward propagation starts with the layer that is producing the given target, e.g. certain classification, and estimates the contribution of neurons in the layer previous to that target.

Figure 1 shows graphically the process followed by LIME, Anchors, IG and XRAI to provide explanations for images. Next, we provide further details of these XAI methods.

## 2.1 LIME

LIME focuses on training local surrogate models to explain individual predictions given by a global black-box prediction model. In a general way, it analyses the behavior of the global prediction model through the perturbation of the input data.

In order to figure out what features of the input are contributing to the prediction, it perturbs the input data around its neighborhood and evaluates how the model behaves. Then, it trains an interpretable local model that weighs these perturbed data points by their proximity to the original input. This local model should be a good and explainable local approximation of the black-box model. Mathematically, it is formulated as follows [12]:

$$explanation(x) = \underset{g \in G}{\arg\min} \, L(f, g, \Pi_x) + \Omega(g). \tag{1}$$

This equation defines an explanation as a model $g \in G$, where $G$ is a class of potentially interpretable models, such as linear models or decision trees. The goal is to minimize the loss function $L$ that measures how close the explanation is to the prediction of the original model $f$ given a proximity measure $\Pi_x$. This proximity measure defines the size of the neighborhood around the predicted instance $x$ that is used to obtain the explanation. Additionally, it is necessary to minimize the complexity (as opposed to interpretability) of the explanation $g \in G$, denoted as $\Omega(g)$.

Regarding the perturbation of the input data, it depends on its type. For tabular data, LIME creates new samples by perturbing each feature individually based on statistical indicators. For text and images, the solution is to remove words or parts of the image (called superpixels). Finally, the interpretable surrogate model used by LIME is linear regression, corresponding to the $\Omega(g)$ function in Eq. 1. Here, the user has to define the number of the top superpixels being considered.

## 2.2 Anchors

The use of linear regression makes LIME unable to explain the model correctly on some scenarios where simple perturbations are not enough. Ideally, the perturbations would be driven by the variation that is observed in the dataset. The same authors proposed a new way to perform model interpretation which is Anchors [13]. Anchors is also a local model-agnostic explanation algorithm that explains individual predictions, i.e., only captures the behavior of the model on a local region of the input space. However, it improves the construction of the perturbation data set around the query. Instead of adding noise to continuous features, hiding parts of the image, to learn a boundary line (or slope) associated to the prediction of the query instance, Anchors improves LIME using a "local region" instead of a slope.

Ribero et al. [13] have demonstrated the usefulness of anchors by applying them to a variety of ML tasks (classification, structured prediction, text generation) on a diverse set of domains (tabular, text, and images). They also ran a user study, where they observed that anchors enable users to predict how a model would behave on unseen instances with much less effort and higher precision as compared to existing techniques for model-agnostic explanation, or no explanations.

### 2.3　Integrated Gradients

Integrated Gradients (IG) is a gradients-based method to efficiently compute feature attributions with the same axiomatic properties as the Shapley value [16]. IG aims to explain the relationship between a model's predictions in terms of its features. It has many use cases including understanding features importance, identifying data skew, and debugging model performance. IG as an interpretability technique is applicable to any differentiable model (e.g., images, text, structured data) that allows it to scale to large networks and feature spaces such as images [19].

In the IG method, the gradient of the prediction output is calculated with respect to the features of the input, along an integral path [3]. First, the gradients are calculated at different intervals of a scaling parameter. For image data, imagine this scaling parameter as a "slider" that is scaling all pixels of the image to black. Integrated gradients can be visualized by aggregating them along the color channel and scaling the pixels in the actual image by them.

The formula for IGs is as follows [18]:

$$IntegratedGradients_i(x) = (x_i - x_i') \times \int_{\alpha=0}^{1} \frac{\partial F(x' + \alpha \times (x - x'))}{\partial x_i} d\alpha, \quad (2)$$

where $i$ represents each feature (pixel) of input $x$, $x'$ is the baseline image (black image), and $\alpha$ is a interpolation constant to perturb features.

### 2.4　XRAI

XRAI (eXplanation with Ranked Area Integrals) is an explanation model based on the IGs method. It is recommended for image models where it is desirable to localize attributions at the region vs. pixel level [3]. XRAI is a region-based saliency method, which first over-segments the image, then iteratively tests the importance of each region, coalescing smaller regions into larger segments based on attribution scores [19]. Attributions are calculated by back-propagating the prediction score through each layer of the network, back to the input features. These methods are in general faster than perturbation-based methods since they usually require a single or constant number of queries to the neural network (independent of the number of input features) [4].

The XRAI method combines the IG method with additional steps to determine which regions of the image contribute the most to a given class prediction [4]. These steps are:

1. Pixel-level attribution: XRAI performs pixel-level attribution for the input image. In this step, XRAI uses the integrated gradients method with a black baseline and a white baseline.

2. Oversegmentation: Independently of pixel-level attribution, XRAI over-segments the image to create a patchwork of small regions. XRAI uses Felzenswalb's graph-based method [1] to create the image segments.
3. Region selection: XRAI aggregates the pixel-level attribution within each segment to determine its attribution density. Using these values, XRAI ranks and orders the segments to determines which areas of the image are most salient, or contribute most strongly to a given class prediction.

Next, we present our CBR process to select the most suitable explanation method for a given image.

## 3   CBR Process Specification

As explanations depend on their utility to the user, it is not possible to find an algorithmic solution to find the most suitable explanation algorithm given an image classification. Therefore, we propose the use of a CBR approach where a case base of instances and their most suitable explanation method is collected and reused to provide explanations. Next, we present the case base elicitation process, several similarity metrics that have been considered, as well as alternative reuse strategies.

### 3.1   Case Base Elicitation

The case base of images has been obtained from the dataset provided by the Visual Genome project [6]. We selected 200 images that were confidently classified by Google's Inception deep convolutional neural network architecture [20] with a predominant class ($precision > 95\%$). The distribution of images according to the predominant class is not uniform, having classes with different number of images, as illustrated in Fig. 2. For every image, we generated four different explanations with the XAI algorithms presented before: LIME, Anchors, IG and XRAI. Then, these four explanations were presented to users, that could select the most suitable explanatory image, as illustrated in Fig. 3. Explanations were randomly shuffled, and the corresponding explanation method is not displayed to the user. Each time the user selects an explanation, a new image and its corresponding explanations are shown until the 200 images have been voted. Concretely, users were asked to select the most specific explanation, meaning that, in case of two similar images, they should choose the one with less image area.

After repeating this process with 15 users we collected a total of 3.000 votes (15 per image) that were used to generate the case base. The description of each case is the image itself –its pixel matrix, $M$– plus the feature's vector returned

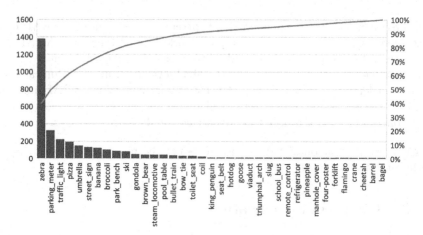

**Fig. 2.** Pareto diagram showing the distribution of images (total and percentage) according to the predominant class predicted by the classifier.

by the classifier $\overrightarrow{f}$. The solution is the number of votes given by the users to each explanation strategy, denoted as $L, A, I, X$. This representation of cases can be formalized as:

$$Case = \langle D, S \rangle$$
$$where$$
$$D = \langle M, \overrightarrow{f} \rangle$$
$$S = \langle L, A, I, X \rangle. \tag{3}$$

The preliminary analysis of the explanation methods voted by the users denoted a higher preference for the backward propagation mechanism, independently of the predominant class in $\overrightarrow{f}$. The distribution of the votes according to the explanation method is shown in Fig. 4. Here, it is specially relevant that the IG method got up to 45% of the votes, followed by XRAI (30%), LIME (18%) and Anchors (7%).

### 3.2 Similarity Metrics

A key element in the CBR process is the similarity metric used for the retrieval of similar images (and their corresponding votes distribution) from the case base. We have defined four different approaches:

**Pixel-to-pixel.** A straightforward method to retrieve similar images is the comparison of the pixel matrix. This similarity metric uses the difference between the pixels of both images:

$$sim(D_q, D_c) = 1 - |M_q - M_c|. \tag{4}$$

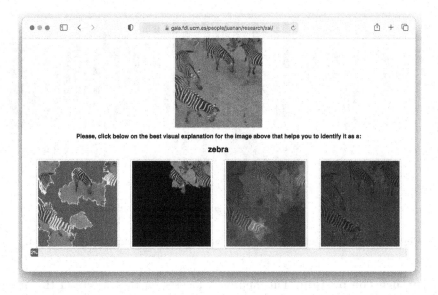

**Fig. 3.** Web application used to vote for the best explanation and generate the case base. The original image and the majoritarian predicted class is shown on the top. Images below are generated through the four XAI methods being considered.

**Histogram correlation.** This similarity metric is based on the correlation between the color histograms of the images. It is defined as follows:

$$sim(D_q, D_c) = 1 - hcorr(h(M_q), h(M_c))$$
$$hcorr(H_1, H_2) = \frac{\sum_i (H_1(i) - \bar{H}_1)(H_2(i) - \bar{H}_2)}{\sqrt{\sum_I (H_1(i) - \bar{H}_1)^2 \sum_i (H_2(I) - \bar{H}_2)^2}} \tag{5}$$

$$where$$
$$\bar{H}_k = \frac{1}{N} \sum_J H_k(J). \tag{6}$$

Here, function $h()$ obtains the color histogram of a pixel matrix with length $N$.

**Structural similarity index (SSIM).** This metric compares the structural changes in the image. It has demonstrated good agreement with human observers in image comparison using reference images [24]. The SSIM index can be viewed as a quality measure of one of the images being compared, provided the other image is regarded as of perfect quality. It combines three comparison measurements between the samples of $x$ and $y$: luminance $l$, contrast $c$ and structure $s$:

$$sim(D_q, D_c) = [l(M_q, M_c)]^\alpha \cdot [c(M_q, M_c)]^\beta \cdot [s(M_q, M_c)]^\gamma \tag{7}$$

Where $\alpha$, $\beta$, and $\gamma$ are parameters that define the relative importance of each component.

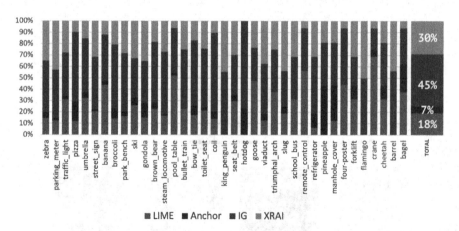

**Fig. 4.** Votes distribution aggregated according to the explanation method.

**Feature comparison.** This approach focuses on the objects in the image that were identified by the classifier being explained. Therefore, this similarity metric is based on the comparison of the feature vectors $\overrightarrow{f}$ by applying a distance metric such as the Euclidean distance:

$$sim(D_q, D_c) = 1 - Eucl\_Dist(\overrightarrow{f}_q, \overrightarrow{f}_c). \tag{8}$$

Then, the $k$ most similar images can be selected. This retrieval process is illustrated in Fig. 5, where the three nearest neighbors for a given image query are retrieved using the strategies defined before.

## 3.3   Reuse Strategies

The following step in the CBR cycle is solution reuse. Here, we need to define the strategy to aggregate the votes received by one or several images and assign the corresponding explanation method. To do so, we need to define the function that returns the class (LIME, Anchors, IG or XRAI) with the largest number of votes that are stored in the solution of a case:

$$mostVoted(S) = \underset{m \in \{l,a,i,x\}}{\arg\max} \; votes(m, S), \tag{9}$$

where function $votes()$ returns the number of votes stored in solution $S$ for a given explanation method $x$. With this function we can define the following strategies to aggregate the solutions of the $k$-nearest neighbors of the query:

**Simple voting.** This strategy returns the majoritarian class in the solutions of the nearest neighbors:

$$sv(S_1, \dots, S_k) = \underset{m \in \{l,a,i,x\}}{\arg\max} \bigcup_{i \in \{1,\dots,k\}} mostVoted(S_i). \tag{10}$$

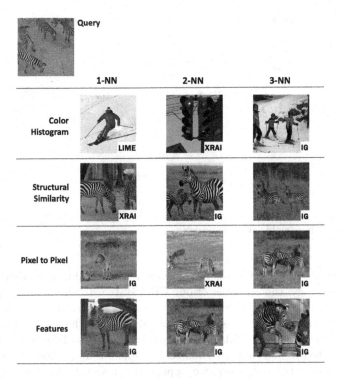

**Fig. 5.** Example of 3NN retrieval using each similarity metric. Nearest neighbors show the explanation method with the largest number of votes given by the users in the bottom-right corner.

**Aggregated voting.** This alternative strategy aggregates the votes received by each explanation method in each solution of the k-NNs and then computes the majority class:

$$av(S_1, \ldots, S_k) = mostVoted(A) \tag{11}$$
$$where$$
$$A = \langle L^+, A^+, I^+, X^+ \rangle$$
$$m^+ = \sum_{i \in \{1,\ldots,k\}} votes(m, S_i). \tag{12}$$

## 4   Evaluation

In order to demonstrate the benefits of our case-based approach to find the most suitable explanation method for an image, we performed an evaluation using cross-validation.

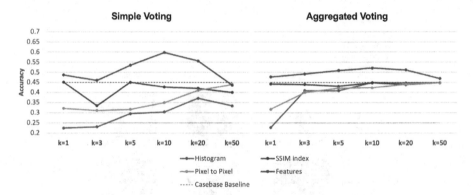

**Fig. 6.** Evaluation result for several values of $k$ using the Simple Voting (left) and Aggregated Voting (right) reuse strategies.

The goal of this evaluation is to compare, using each image similarity metric and reuse strategy, the outcome of our CBR system to the actual explanation method chosen by the users. Here, it is important to note that the distribution of the solutions in the case base is very unbalanced, as the explanation method assigned to each case is based on the majoritarian class. This way, the 62% of the cases in the original case base had IG as explanation method. To reduce this bias and balance the case base, we applied a random subsampling over cases with solution IG until they represent 45% of the total cases. This figure tries to reproduce the original distribution of votes presented in Fig. 4. This way, a dummy classifier that always returns IG as the explanation method for any image will obtain an accuracy of 45%. We can consider that value the baseline of our system, although in a completely balanced case base, this baseline should be 25% as we are considering four classes.

Results, using a 50-times leave-one-out evaluation, are summarized in Fig. 6. We can clearly observe that the best similarity metric is *feature comparison*. It achieves an accuracy around 0.6 for $k = 10$. The *structural similarity index* is the second best similarity metric, although it does not overcome the case base baseline. The two similarity metrics based on the comparison of the pixel matrices, either *pixel to pixel* or through the *color histogram*, report a low performance. Regarding the comparison of the reuse strategies, *aggregated voting* is more stable, specially when $k$ is greater than 3. However, it does not achieve as good results as the *simple voting* strategy. It is also worth noting that, independently of the reuse strategy, the best accuracy is obtained when the number of nearest neighbors is relatively high ($k > 10$). Within this scenario, most of the configurations of the CBR system achieve an accuracy close to 0.6, that in comparison with our case-base baseline (0.45) represents a great improvement. This improvement is even more relevant if we compare with the theoretical baseline of a 4-class classifier (0.25).

Finally, we also simulated the learning process of our CBR system and analyzed its behavior as the case base grows. This analysis is presented in Fig. 7,

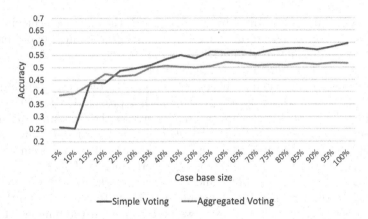

Fig. 7. Analysis of the learning process of the CBR system. Line chart shows the evolution of the accuracy for the optimal configuration from Fig. 6: $k = 10$, *feature-based* similarity and *simple voting* reuse strategy.

where we can observe that the accuracy of the CBR system grows approximately linearly with respect to the size of the case base, specially for the *simple voting* aggregation strategy. It is a clear indicator that the performance of the system could increase if there were more cases available.

## 5    Conclusions and Future Work

This paper presents a case-based reasoning method that takes advantage of human knowledge to select the best explanation model for image classification. Concretely, we have defined and evaluated a CBR solution to the problem of selecting from four explanation methods –LIME, Anchor, IG and XRAI– that represent the two major explanation approaches: model-independent and model-dependent. To evaluate this CBR system we have collected a case base of images and their associated "optimal" explanation models according to the votes of several users. The main conclusion obtained from the elicitation of the case base is that there is a predominant explanation method according to the users' votes: Integrated Gradients. This method belongs to the group of model-dependent explanation models, and surprisingly it clearly beats XRAI, that can be considered as an evolution of IG. From the group of model-agnostic models, we also find a similar conclusion: LIME is preferred over Anchors, although the former is considered as an enhancement of LIME.

From the collected case base, we implement a system where, given a new query image, similar images are retrieved, and their corresponding XAI methods are proposed as the best explanation approach. Here, the behavior of the similarity metrics to retrieve the most similar cases is a key element to be carefully addressed as there are several alternative approaches: features comparison, pixel-to-pixel or structural similarity, or even color correlation. Finally, as the experimental evaluation has demonstrated, the *feature-based similarity*, that compares the features identified by the classifier, is the only one achieving a significant performance. This result corroborates the initial findings of our previous work [11] where *feature-based* similarity was used to find the proper configuration of the LIME method using a CBR approach.

The evaluation presented in this paper compares through cross-validation the images in the case base and the explanation method proposed by the CBR system or voted by the users. To compare the images, we use the four similarity metrics and two alternative reuse strategies: *simple and aggregated voting*. The results of the evaluation demonstrated that our CBR system achieves an accuracy up to 60% that significantly improves the baseline performance of 45%.

This paper leaves many open lines for future work. Firstly, we would like to compare with other novel explanation approaches that are constantly being developed, such as [15,22]. Moreover, we need to extend the number of voting users, to confirm the predominance of the IG method and discard any kind of bias on the voting process. As the process of selecting the best explanation image for a given image and classification may be considered subjective, we could enhance it by asking for preliminary features about the image or class (for example: how do you describe a zebra? or what is the most important feature of a zebra?).

Another line of future work is the analysis of the combination of the similarity metrics, as they have been analyzed in isolation. However, we could define additional metrics that combine, for example, features and structural similarity. We must also analyze the impact of the case base quality in the explanation process regarding cold-start scenarios where no similar images are available in order to find out the minimum similarity threshold required to provide good explanations.

Finally, our evaluation only includes images that are confidently classified by the neural network, so we need to evaluate the impact of multi-class images that combine several elements (for example: an image containing a zebra, a lion and an elephant). Within this scenario, it is unclear how the model-dependent methods will behave as the introspection of the neural network becomes more difficult. On the other side, it is possible that the model-agnostic methods, such as LIME or Anchors, will raise their performance, as it is much easier to identify the regions (superpixels) that led to the classification.

# References

1. Felzenszwalb, P.F., Huttenlocher, D.P.: Efficient graph-based image segmentation. Int. J. Comput. Vis. **59**(2), 167–181 (2004). https://doi.org/10.1023/B:VISI.0000022288.19776.77. http://link.springer.com/10.1023/B:VISI.0000022288.19776.77
2. Gates, L., Kisby, C., Leake, D.: CBR confidence as a basis for confidence in black box systems. In: Bach, K., Marling, C. (eds.) ICCBR 2019. LNCS (LNAI), vol. 11680, pp. 95–109. Springer, Cham (2019). https://doi.org/10.1007/978-3-030-29249-2_7
3. Google LLC: AI Explanations Whitepaper pp. 1–28 (2019). https://storage.googleapis.com/cloud-ai-whitepapers/AI%20Explainability%20Whitepaper.pdf

4. Kapishnikov, A., Bolukbasi, T., Viégas, F., Terry, M.: XRAI: better attributions through regions. In: Proceedings of the IEEE International Conference on Computer Vision, vol. 2019-October, pp. 4947–4956, June 2019. http://arxiv.org/abs/1906.02825

5. Keane, M.T., Kenny, E.M.: How case-based reasoning explains neural networks: a theoretical analysis of XAI using post-hoc explanation-by-example from a survey of ANN-CBR twin-systems. In: Bach, K., Marling, C. (eds.) ICCBR 2019. LNCS (LNAI), vol. 11680, pp. 155–171. Springer, Cham (2019). https://doi.org/10.1007/978-3-030-29249-2_11

6. Krishna, R., et al.: Visual genome: connecting language and vision using crowd-sourced dense image annotations (2016). https://arxiv.org/abs/1602.07332

7. Leake, D.B., McSherry, D.: Introduction to the special issue on explanation in case-based reasoning. Artif. Intell. Rev. 24(2), 103–108 (2005). https://doi.org/10.1007/s10462-005-4606-8

8. Li, O., Liu, H., Chen, C., Rudin, C.: Deep learning for case-based reasoning through prototypes: a neural network that explains its predictions. In: McIlraith, S.A., Weinberger, K.Q. (eds.) Proceedings of the Thirty-Second AAAI Conference on Artificial Intelligence, AAAI-18, pp. 3530–3537. AAAI Press (2018)

9. Lundberg, S.M., Lee, S.I.: A unified approach to interpreting model predictions. In: Guyon, I., et al. (eds.) Advances in Neural Information Processing Systems, vol. 30, pp. 4765–4774. Curran Associates, Inc. (2017)

10. Ras, G., van Gerven, M., Haselager, P.: Explanation methods in deep learning: users, values, concerns and challenges. In: Escalante, H.J., et al. (eds.) Explainable and Interpretable Models in Computer Vision and Machine Learning. TSSCML, pp. 19–36. Springer, Cham (2018). https://doi.org/10.1007/978-3-319-98131-4_2

11. Recio-García, J.A., Dí-az-Agudo, B., Pino-Castilla, V.: CBR-LIME: a case-based reasoning approach to provide specific local interpretable model-agnostic explanations. In: Watson, I., Weber, R. (eds.) ICCBR 2020. LNCS (LNAI), vol. 12311, pp. 179–194. Springer, Cham (2020). https://doi.org/10.1007/978-3-030-58342-2_12

12. Ribeiro, M.T., Singh, S., Guestrin, C.: "Why should i trust you?": explaining the predictions of any classifier. In: Proceedings of the 22nd ACM SIGKDD International Conference on Knowledge Discovery and Data Mining, pp. 1135–1144. Association for Computing Machinery, New York (2016). https://doi.org/10.1145/2939672.2939778

13. Ribeiro, M.T., Singh, S., Guestrin, C.: Anchors: high-precision model-agnostic explanations. In: McIlraith, S.A., Weinberger, K.Q. (eds.) Proceedings of the Thirty-Second AAAI Conference on Artificial Intelligence, AAAI-18, pp. 1527–1535. AAAI Press (2018). https://www.aaai.org/ocs/index.php/AAAI/AAAI18/paper/view/16982

14. Roth-Berghofer, T., Richter, M.M.: On explanation. Künstliche Intell. KI 22(2), 5–7 (2008)

15. Selvaraju, R.R., Cogswell, M., Das, A., Vedantam, R., Parikh, D., Batra, D.: Grad-CAM: visual explanations from deep networks via gradient-based localization. In: 2017 IEEE International Conference on Computer Vision (ICCV), pp. 618–626 (2017). https://doi.org/10.1109/ICCV.2017.74

16. Shapley, L.S., Shubik, M.: The assignment game I: the core. Int. J. Game Theory 1(1), 111–130 (1971). https://doi.org/10.1007/BF01753437. http://link.springer.com/10.1007/BF01753437

17. Sørmo, F., Cassens, J., Aamodt, A.: Explanation in case-based reasoning-perspectives and goals. Artif. Intell. Rev. 24(2), 109–143 (2005). https://doi.org/10.1007/s10462-005-4607-7

18. Sturmfels, P., Lundberg, S., Lee, S.I.: Visualizing the impact of feature attribution baselines. Distill (2020). https://doi.org/10.23915/distill.00022. https://distill.pub/2020/attribution-baselines

19. Sundararajan, M., Taly, A., Yan, Q.: Axiomatic attribution for deep networks. In: 34th International Conference on Machine Learning, ICML 2017, vol. 7, pp. 5109–5118, March 2017. http://arxiv.org/abs/1703.01365

20. Szegedy, C., et al.: Going deeper with convolutions. In: Computer Vision and Pattern Recognition (CVPR) (2015). http://arxiv.org/abs/1409.4842

21. Weber, R.O., Johs, A.J., Li, J., Huang, K.: Investigating textual case-based XAI. In: Cox, M.T., Funk, P., Begum, S. (eds.) ICCBR 2018. LNCS (LNAI), vol. 11156, pp. 431–447. Springer, Cham (2018). https://doi.org/10.1007/978-3-030-01081-2_29

22. Xu, S., Venugopalan, S., Sundararajan, M.: Attribution in scale and space. In: Proceedings of the IEEE/CVF Conference on Computer Vision and Pattern Recognition (CVPR), June 2020

23. Yuan, H., Cai, L., Hu, X., Wang, J., Ji, S.: Interpreting image classifiers by generating discrete masks. IEEE Trans. Pattern Anal. Mach. Intell. 1 (2020). https://doi.org/10.1109/TPAMI.2020.3028783

24. Wang, Z., Bovik, A.C., Sheikh, H.R., Simoncelli, E.P.: Image quality assessment: from error visibility to structural similarity. IEEE Trans. Image Process. **13**(4), 600–612 (2004). https://doi.org/10.1109/TIP.2003.819861

# Towards Richer Realizations
# of Holographic CBR

Renganathan Subramanian[1]([⊠]), Devi Ganesan[1], Deepak P[1,2],
and Sutanu Chakraborti[1]

[1] Indian Institute of Technology, Madras, Chennai 600036, TN, India
ch16b058@smail.iitm.ac.in, {gdevi,sutanuc}@cse.iitm.ac.in
[2] Queen's University Belfast, Belfast, UK
deepaksp@acm.org

**Abstract.** Holographic Case-Based Reasoning is a framework developed to build cognitively appealing case-based reasoners with proactive and interconnected cases. Improved realizations of the Holographic CBR framework are developed using the principles of dynamic memory proposed by Roger Schank and tested on their cognitive appeal, efficiency, and solution quality compared to other relevant systems.

**Keywords:** Case-based reasoning · Cognitive CBR · Holographic systems · Dynamic memory

## 1 Introduction

Case-Based Reasoning (CBR) has its inspiration in Roger Schank's seminal work on Dynamic Memory [13] that aspired to model learning in computers based on how learning happens in humans. Schank also proposed the concepts of scripts, plans, and goals as possible knowledge structures used by humans while understanding a piece of text. Kolodner actively worked on Schank's work on dynamic memory and built a computer program called CYRUS [9] which stands for Computerized Yale Retrieval and Updating System. In particular, CYRUS was an attempt to model the reconstructive nature of human memory. CYRUS can be uniquely contrasted against the current day CBR systems in terms of its rich case representation. CBR systems like CREEK [1], CELIA [16], CHEF [8] were also built with richly inter-connected case structures. However, in the conventional CBR theory, a case is usually represented as a simple problem and solution pair with no provision to accommodate the interconnections/dependencies between cases. In other words, the conventional CBR theory does not have provisions to neatly accommodate the richly inter-connected case representations found in the complex CBR systems of the past. Holographic CBR [7] is an attempt to provide a single conceptual framework that can cover a spectrum of CBR systems with case representations ranging from simple problem-solution pairs to complex inter-connected cases. This is achieved by modifying the case representation to include a solo and a holo component. While the solo component stands

© Springer Nature Switzerland AG 2021
A. A. Sánchez-Ruiz and M. W. Floyd (Eds.): ICCBR 2021, LNAI 12877, pp. 201–215, 2021.
https://doi.org/10.1007/978-3-030-86957-1_14

for the conventional problem-solution pair, the holo component is responsible for acquiring/storing/updating all the interconnections between cases.

The idea of holographic CBR was inspired by the holographic nature of the human brain. A holographic system is whose individual components contain information about the whole system and where the whole system can be reconstructed from each of the parts (with some loss in detail). For example, when a holographic image of an apple is broken down into small pieces, every piece would still be able to reconstruct the full apple image to some extent under appropriate conditions. Similarly, the holonomic brain theory suggests that every part of the brain contains some information about the whole. Motivated by this observation, holographic CBR was proposed with a holo component to capture the connection/relation of a single case to the entire case base. However, the realizations proposed in [7] are simplistic in that they restrict themselves to the mode of knowledge acquisition during the case acquisition/case addition process. In knowledge-rich domains, a holographic reasoner learns the relation of a case to the whole from the domain expert, whereas, in a knowledge-light setting, it attempts to infer the same from the cases already present in the case base using bottom-up learning methods. It is interesting to note that the paradigm of holographic CBR opens an avenue for integrating both top-down and bottom-up approaches in the building of a cognitively appealing case base. While there is significant scope for exploration in this aspect, in this paper, we focus primarily on forming generalized cases in a holographic reasoner during the case acquisition process itself. This involves invoking a failure-driven reminding process combined with bottom-up learning of the connections between cases. We have, however, restricted our work to regression and classification tasks. In the past, there have been works on generalizations and abstraction in CBR [3,11,18]. However, we are interested in developing a robust and cognitively appealing bottom-up approach to the same.

We discuss the Holographic CBR framework in Sect. 2. Section 3 introduces the key ideas realized and the realizations built. We present our results in Sect. 4 and summarize our findings, and discuss the future scope in Sect. 5.

## 2   Holographic Case-Based Reasoning

Traditional CBR systems treat cases as isolated entities. Any changes made to one case do not affect the rest of the case base. This is unlike human memory, where a new experience affects related memories, and information is not localized. This idea stems from the experiments of Lashley on mice [10] and observations of Pribram on accident victims [12]. Even when parts of the brain were removed, an organism could still form a hazy recollection of past experiences instead of completely losing them. This shows the "holographic" nature of human memory, where every part of the system contains information about the entire system. Inspired by this, Holographic CBR [7] treats cases as proactive interconnected entities which actively affect and are affected by any changes to the rest of the case base. Holographic cases develop their own *local* knowledge containers, which helps them understand their problem-solving competence in relation to the rest of the case base. They also proactively interact with and modify the case base. More importantly, this interaction is not necessarily engineered in the reasoner but is learned as the reasoner solves new problems. This Holographic CBR framework has certain key properties.

**The Holographic Case:** Cases in a holographic CBR system are made up of two parts. The *solo component* stores the problem-solution as in traditional CBR systems and represents the individual experience that the case stands for. On the other hand, the *holo component* defines the case's role with respect to the case base and captures diverse forms of relationship of a case with other cases in the case base.

**Holographic Case Addition:** New cases are added to the case base only when the system cannot solve the new case using the existing cases. Thus, the case base grows only when it identifies a knowledge gap. Instead of merely adding the new case to the case base, the system informs the existing cases of the new case's presence. It highlights why the existing case base could not solve the problem and the new case's value-addition and is later used to decide when/how to use the newly added case to solve future problems.

**Holographic Problem Solving:** The system has a coarse knowledge of the competence of the different cases inside it, but problem-solving happens in a decentralized manner. Cases are expert problem solvers in their neighborhoods. The system uses its global similarity knowledge to retrieve a case to solve a new problem. The retrieved case, in turn, uses its holo component to identify if any other cases can solve the problem better and, if found, transfers control to such a better case.

This treatment of CBR has several advantages. The ability of cases to interact with other cases allows us to design helpful ways to use, modify, and reorganize the case base. The presence of explanations for adding a case not only ensures that only useful cases get added but also highlights the added case's novelty. Ganesan et al. [7] built holographic CBR realizations, which demonstrated some of these ideas. However, the realizations restrict themselves to the mode of knowledge acquisition during the case acquisition/case addition process and explored only limited cognitive ideas. We utilize this framework to infuse several dynamic memory ideas, absent in Ganesan et al.'s initial realizations like forming generalizations based on multiple cases, updating links between cases based on usage, etc., into our CBR realizations to make it more cognitively appealing. These ideas draw inspiration from Schank's works on human understanding and the properties of a dynamic memory system.

## 3  Methodology

### 3.1  Key Ideas

In this section, we provide the intuition and justification for the key cognitive ideas implemented. These are then implemented in holographic CBR systems in Sect. 3.2. We use an example of an animal classification task from the UCI Zoo Dataset to motivate these ideas. In this task, each animal (case) has certain binary biological features like - *lays eggs, produces milk, is a predator, airborne,* etc., and belongs to a class which is one of mammal, bird, fish, amphibian, reptile, flight, and non-flight invertebrate. The CBR system is presented with the cases one at a time from which it learns and then classifies new cases.

**Forming Generalizations.** Schank observed that past experiences stored in memory might contain detailed information not relevant to solving a given task [14]. When two experiences are similar, and when their differences do not contribute additional value to problem-solving, it is useful to form generalizations by combining such similar experiences. These generalizations should only retain information which 1) help them solve the task and 2) differentiate them from other non-similar experiences. This makes the system more efficient by focusing only on important information and ignoring unnecessary details. Moreover, it helps in identifying novel information present in new experiences by comparing them with existing generalizations.

For example, whether an animal is a predator or not does not help in the classification. When the system sees several animals with the same class but different predator values, it should identify this unnecessary feature. Similarly, it should be able to find features that have typical values for a certain class. For example, the class *mammals* has *lays eggs* as predominantly false. With this information, the system should form generalizations about mammals that ignore the useless feature and highlights the typical feature. This generalization can immediately capture interesting information in new cases. For example, when faced with a platypus case (which lays eggs but is still a mammal), the system can identify its novelty by comparing it to the mammal generalization.

**Failure-Driven Reminding.** When a CBR system uses a similar past case to solve a new problem, it expects that the solutions to the past and the new problem are similar. When such *expectations* do not match the ground truth (*expectation-failure*), there is a scope for the system to learn. Schank hypothesized that a dynamic memory system should explain such expectation failures and use them to extract valuable information from the new experience and retain it in memory. Thus, we want a system that remembers its past mistakes and their reasons, which it uses to avoid making similar mistakes again. For example, a tuatara (reptile) and a newt (amphibian) share all features except *aquatic* but belong to different classes. So, a CBR system might incorrectly use its memory of tuatara to classify a newt. Once the mistake is identified, the system should realize that the aquatic feature explains the failure. This intuition forms the basis for our *failure-driven links* present in a case's *holo component*. These links are created when a case makes a mistake in solving a new case. They hold explanations identified by the system for failures and connect the two cases. They later help the system avoid mistakes by reminding it of its past mistakes. In our example, if the system retrieves tuatara again to solve a new problem, it checks if the new case is also aquatic. If so, the system remembers its past mistakes and instead transfer problem-solving control to the connected case newt.

**Outcome-Driven Solo and Holo Component Weights.** Cases in holographic CBR have two components - a *solo* component and a *holo* component. However, not all information present in these components would be equally important in solving the CBR task. Also, which information is important might differ from one case to another. For example, a seal (mammal) and a frog (amphibian) differ in four features - *hair, legs, milk,* and *eggs*. But the differing feature *legs* is less important as there are other 4-legged mammals. However, between a honeybee (flight-invertebrate) and a carp (fish), the feature *legs* is important as all flight-invertebrates have legs while no fish do. We have introduced the concept of outcome-driven weights to handle this by which the feature *legs* gets a higher weight in certain cases and a lower weight in certain other cases. These weights indicate the correctness and utility of different parts of information stored in a case. The system progressively learns these weights as it solves new problems.

### 3.2   Holographic CBR Realization Framework

In this section, we describe a framework of Holographic CBR, which implements the concepts described earlier. We use this framework to build and test two systems for a classification task and a regression task.

**Components of the Framework.** We propose two levels of memory units - cases and generalized cases. Generalized cases are made of multiple cases as discussed in Sect. 3.1. These units have a solo component that stores their standalone expertise and a holo component connecting them with other units. Both these components have outcome-driven weights (solo-weights and holo-weights) as discussed in Sect. 3.1 which denote their importance in the unit. When faced with a new problem, these units are retrieved by the system and are used to solve the new problem.

*Cases:* Cases are the storehouses of knowledge from individual training data points. Each case represents one data point from which the system has learned and serves as a primary knowledge source to solve new problems. Each case is stored within a generalized case. A case as shown in Fig. 1 contains the problem definition, solution, and local knowledge in its solo component. It has information about its relative competence with respect to its generalized case in its holo component that is updated as the system learns.

*Generalized Cases (GCs):* Schank introduced Memory Organization Packets which are organizers of individual experiences centered around common contexts or similar themes. Storing experiences within these MOPs would highlight the interesting aspects of the experiences, and if such interesting aspects are absent or irrelevant, the MOPs aid in removing the unnecessary experience. MOPs are also connected to other MOPs based on important differences.

Similar to this, Generalized Cases are combinations of cases with similar problem representations and similar solutions. Every time a GC is retrieved to solve a new case during training and is able to do so, the knowledge in the new case updates the GC, and the new case is stored within the GC. The GC can replace the individual cases if the individual case offers no additional value. Thus, a GC represents a region in the problem space that has similar solutions. A GC, as shown in Fig. 1 contains a generalized problem description which is a combination of the problem descriptions of the individual cases. The GC's relative competence with respect to other GCs is stored in its holo component as failure-driven links. Multiple cases are retained within a GC.

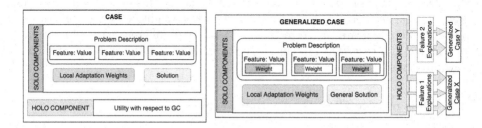

**Fig. 1.** Memory units visualized

*Solo Component:* The solo component of a memory unit contains:

- *Problem Description:* These are features and their corresponding values. In addition, GCs have outcome-driven weights discussed in Sect. 3.1 that indicate the importance of a feature-value combination. For GCs, the value for a real feature is the mean value from the cases stored within it. In contrast, categorical features have multiple values for each feature (the feature's value in each case inside it) with different importance weights for each value.
- *Solution:* The solution to the problem the memory unit represents. For regression tasks, a GC's solution is the mean solution of the cases within it.
- *Local Adaptation Knowledge:* This is present in regression tasks and modifies a memory unit's solution to account for differences between its problem description and that of a new case. Local adaptation knowledge allows different cases to have different adaptation applicable to that point in the problem space. We have used the difference between the values of features in the retrieved and to-be-solved case to perform adaptation. Let the unit retrieved have a feature vector $x = [x_1, \ldots x_n]$ and the new problem have $\tilde{x} = [\tilde{x}_1, \ldots \tilde{x}_n]$. We know $f(x)$ and want to predict $f(\tilde{x})$. We define adaptation weights $w_1^a, \ldots w_n^a$ for each feature which are initialized as 0 and progressively learned (as discussed in Sect. 3.2). We perform adaptation using:

$$f(\tilde{x}) = f(x) + w_1^a(\tilde{x}_1 - x_1) + w_2^a(\tilde{x}_2 - x_2) + \ldots + w_n^a(\tilde{x}_n - x_n) \qquad (1)$$

*Holo Component:* This stores the relative competence of a memory unit with respect to the rest of the case base. In a regression setup, GCs initially store the individual cases within them and compare their adaptation success to that of the GC for future problems. The results of this comparison (how better the case is at solving new problems when compared to its GC, measured in terms of closeness of adapted solution to the ground truth) are stored in each case's holo component and are used to remove cases that do not add additional value over the GC. In classification setups, where the GC and its cases have the same solution and adaptation is not required, such competence is meaningless. Hence, cases update the GCs but are not retained in the case base.

For a GC, the holo component stores the failure-driven links explained in Sect. 3.1 and connects GCs with one another. When a GC is retrieved but cannot solve a new problem, the explanations for the failure identified are stored in these links, and the new problem is added to the case base and connected via this link. These links are later used to transfer problem-solving control from the retrieved GC to another GC linked to it. In addition, these links also have weights denoting the GC's confidence in their explanation.

**Learning Processes in the Framework.** This section describes how the memory units are created/updated during the learning phase. The system is presented with training data points one at a time, and it iteratively *learns by solving.*

---

**Class 1:** Holographic CBR System

---

```
 1  Class HOLOGRAPHIC_SYSTEM:
 2  GC_base //List of GCs in the system
 3  Function ADD_CASE(newCase):
 4    if GC_base is not emtpty then
 5      retGC = Retrieve GC in GC_base with closest solo-weighted distance to newCase
 6      retCG.ADD_CASE(newCase)
 7    else
 8      Create newGC by copying newCase, initialize empty holo component and add to
          GC_base
 9    end
10  end
11  Function SOLVE(newCase):
12    retGC = Retrieve GC in GC_base with closest solo-weighted distance to newCase
13    retGC.SOLVE(newCase)
14  end
15  end
```

---

*Initial Solo-Based Retrieval:* When a new case (QUERY) is encountered, the system compares this problem description with the solo problem description of each of its GCs (GC$_i$) to compute a distance between the two (RET$_d$(GC$_i$, QUERY)). It weighs each feature $f$ by the corresponding outcome-driven solo weight ($w_f^{(S,i)}$). For real features, we use a standardized-Euclidean distance:

$$\text{RET}_d(\text{GC}_i, \text{QUERY}) = \sqrt{\sum_{f \in \text{features}} w_f^{(S,i)} \cdot \frac{\left(x_{f,\text{GC}_i} - x_{f,\text{QUERY}}\right)^2}{\text{Var}(f)}} \qquad (2)$$

And for categorical features we use a modified weighted-Hamming distance:

$$\text{RET}_d(\text{GC}_i, \text{QUERY}) = \sum_{f \in \text{features}} \sum_{x_{f,k} \in \text{GC}_i(f)} w^{(S,i)}_{x_{f,k}} \cdot I(x_{f,k} \neq x_{f,\text{QUERY}}) \quad (3)$$

where $x_f$ is the value taken by the feature $f$ (Categorical features in a GC can have multiple values each of which is represented as $x_{f,k}$), $Var(f)$ is the variance of feature $f$, and $I(\text{condition}) = 1$ if the condition is true and 0 otherwise. The GC with the lowest distance to the new case is retrieved (Class 1, Step 5) and is used to predict the solution to the new problem (Class 2, Step 6), and this solution is validated with the ground truth (expectation validation).

*Formation of Generalized Cases:* The first case encountered by the system is stored within a GC with features, values, and solution equal to this case (Class 1, Step 8). It has equal outcome-driven weights for all features and no holo components. When a GC is later retrieved, and there is an expectation success, the problem descriptions are modified to the means of the problem descriptions of the existing cases within the GC, and the new case (for real features) or new feature-value pairs are added to the description (for categorical features) as shown in Class 3, Step 7. If none of the existing GCs can solve a new case, a new GC is again created by copying the new case in Class 3, Step 19.

*Update of Outcome-Driven Solo Weights:* The outcome-driven solo weight of a feature in a GC is increased if the values for the feature in the GC and the case are close during expectation success (Class 3, Step 8) and are far apart during expectation failures (Class 3, Step 16). For example, if a GC containing frog and newt (amphibians) is retrieved to classify a toad (also an amphibian and hence an expectation success), the weight for the feature *backbone*, that has a matching value of 1 in both the GC and the new case, increases. Similarly, if the same GC is retrieved to classify a flea (flight-invertebrate and hence an expectation failure), the weight for the feature *backbone*, which has mismatching values in the GC and the unsolvable case, increases again. Thus, mismatching during failures and matching during successes increases the feature-value importance and vice-versa. For categorical features, we use the ratio of times the feature had matching values in successes and mismatching values in failures to the number of times the GC was retrieved as the weight for a feature. For real features, we compute the difference in values ($\text{DIFF}_f(\text{GC}_i, \text{QUERY})$) for feature $f$ in the GC ($\text{GC}_i$) and a new case (QUERY) as:

$$\text{DIFF}_f(\text{GC}_i, \text{QUERY}) = \frac{(x_{\text{GC}_i,f} - x_{\text{QUERY},f})^2}{\text{Var}(f)} \quad (4)$$

Here $x_{\text{GC}_i,f}$ and $x_{\text{QUERY},f}$ are the values of the feature in the $i^{th}$ GC and a new case respectively. Since small differences should increase feature weights during expectation success and decrease weights during expectation failure, we update

**Class 2:** Holographic Case

```
1  Class CASE:
2  │ solo_component //Problem definition, solution and adaptation weights
3  │ holo_component //Relative competence with GC
4  │ pointer_to_GC //Parent GC within which the CASE is stored
5  │ Function PREDICT(newCase):
6  │ │ Use adaptation weights to predict solution for newCase
7  │ end
8  │ Function UPDATE_ADAPT(newCase):
9  │ │ Use difference between CASE.predict(newCase) and newCase's solution to update
   │ │ adaptation weights of CASE
10 │ end
11 │ Function UPDATE_HOLO(newCase):
12 │ │ Check whether CASE or CASE.pointer_to_GC is better at predicting newCase and store
   │ │ result in CASE.holo
13 │ │ If CASE is consistently worse, DELETE CASE
14 │ end
15 end
```

**Class 3:** Holographic Generalized Case

```
1  Class GC(CASE):
2  │ solo_component //Problem definition, solution and adaptation weights
3  │ holo_component //Failure driven links connecting to other GCs
4  │ cases //Cases stored within GC pointer_to_system
5  │ Function ADD_CASE(newCase):
6  │ │ if GC.PREDICT(newCase) close to solution of newCase then
7  │ │ │ Update problem description and solution of GC
8  │ │ │ Increase(decrease) solo weights of features with close(far) values in newCase and GC
9  │ │ │ Decrease(increase) holo weights of links with close(far) values in newCase and GC
10 │ │ │ for case in GC.cases do
11 │ │ │ │ case.ADAPT_WEIGHT(newCase)
12 │ │ │ end
13 │ │ │ GC.ADAPT_WEIGHT(newCase)
14 │ │ │ Add newCase to GC.cases
15 │ │ else
16 │ │ │ Decrease(increase) solo weights of features with close(far) values in newCase and GC
17 │ │ │ transfers = Use holo links to find linked GCs with correct linkGC.predict(newCase)
18 │ │ │ if transfers is empty then
19 │ │ │ │ Create newGC by copying newCase, initialize empty holo component, add to GC_base
20 │ │ │ │ Create holo link from GC to newGC for every feature
21 │ │ │ │ Initialize holo weights based on difference in feature values between newGC and GC
22 │ │ │ else
23 │ │ │ │ for linkGC in transfers do
24 │ │ │ │ │ Increase(decrease) holo weight of links between linkGC and GC where the link value
   │ │ │ │ │ and the corresponding feature value in newCase are close(far)
25 │ │ │ │ end
26 │ │ │ end
27 │ │ end
28 │ end
29 │ Function SOLVE(newCase):
30 │ │ if Solo-weighted distance between GC and newCase high then
31 │ │ │ transferGC = Use holo links to find linked GC with minimum holo link distance to
   │ │ │ newCase
32 │ │ │ if Holo distance between newGC and transferGC small then
33 │ │ │ │ transferGC.SOLVE(newCase)
34 │ │ │ end
35 │ │ else
36 │ │ │ retCase = Case in GC.cases closest to newCase
37 │ │ │ RETURN retCase.PREDICT(newCase)
38 │ │ end
39 │ end
40 end
```

the solo weights($w^{(S,i)}$) for feature $f$ in the $i^{\text{th}}$ GC for expectation success as follows. $\eta$ is a learning rate parameter between 0 and 1 to avoid over-fitting.

$$w_f^{(S,i)} \leftarrow w_f^{(S,i)} + \eta^{(S,i)} \frac{\max(d_f) + \min(d_f) - d_f}{\sum_f (\max(d_f) + \min(d_f) - d_f)} \tag{5}$$

$$w_f^{(S,i)} = \frac{w_f^{(S,i))}}{\sum_f w_f^{(S,i)}} \tag{6}$$

For expectation failure we use:

$$w_f^{(S,i)} \leftarrow w_f^{(S,i)} + \eta^{(S)} \frac{d_f}{\sum_f d_f} \qquad w_f^{(S,i)} = \frac{w_f^{(S,i)}}{\sum_f w_f^{(S,i)}} \tag{7}$$

*Formation of Failure-driven Links and Holo Update:* When an expectation fails, the system must explain the failure and create failure-driven holo links (if they do not exist). All feature-values of the unsolved new case are possible explanations for the failure and become links between the initially retrieved GC and a new GC which only has the new case (Class 3, Step 20). However, not all feature-value pairs are equally valid explanations. Features whose values differ significantly between the GC and the case are more likely to be the correct explanations and are weighed more (Class 3, Step 24). However, during an expectation success, the links are not needed, and existing links should not match with the new case. Thus weights of links that match with the new case are reduced, and weights of links that do not match are increased (Class 3, Step 9). Hence, every time a GC is retrieved to solve a new problem, the holo weights of existing failure-driven links are updated/created depending on whether they are useful or not.

For example, a GC made of antelope and buffalo (mammals) might have links to another GC made of crab and lobster (non-flight invertebrates) with features *aquatic:1*, *eggs:1*, and *backbone:0*. When the first GC is retrieved to classify a dolphin (mammal but has *aquatic:1*), there is an expectation success, and the failure-driven links should not be used. Thus, the weight of the link *aquatic:1*, which spuriously matched, goes down while the confidence in *eggs:1* and *backbone:1* as valid failure-driven links goes up.

For categorical features, we assign the holo weight for a failure-driven link as the ratio of the number of times the link matched when needed (expectation failure) or did not match when not needed (expectation success) to the total number of times the GC was retrieved. For real features, we use the difference in feature values between the link and new case defined as LINK_DIFF$_f$(link$_{ij,f}$, QUERY) where link$_{ij,f}$ is the failure-driven link between GCs $i$ and $j$ with feature $f$. This is calculated as:

$$\text{LINK\_DIFF}_f(\text{link}_{ij,f}, \text{QUERY}) = \frac{(\text{link}_{ij,f} - x_{\text{QUERY},f})^2}{\text{Var}(f)} \tag{8}$$

If the holo weights are $w_f^{(H,ij)}$, during expectation failure, we update it using:

$$w_f^{(H,ij)} \leftarrow w_f^{(H,ij)} + \eta^H \frac{\max(d_f) + \min(d_f) - d_f}{\sum_f (\max(d_f) + \min(d_f) - d_f)} \tag{9}$$

$$w_f^{(H,ij)} = \frac{w_f^{(H,ij)}}{\sum_f w_f^{(H,ij)}} \tag{10}$$

as features with small differences in values are more likely to be correct links.

Whereas, during expectation success, we want to decrease weights of features which have small difference in values and update them as:

$$w_f^{(H,ij)} \leftarrow w_f^{(H,ij)} + \eta^H \frac{d_f}{\sum_f d_f} \qquad w_f^{(H,ij)} = \frac{w_f^{(H,ij)}}{\sum_f w_f^{(H,ij)}} \tag{11}$$

*Learning Adaptations:* For regression tasks, the adaptation weights in the solo component need to be learnt. During expectation success, when the difference between problem descriptions of the new case and GC is small, the adaptation weights of the GC and the cases stored within it are updated using the Newton's method for optimization (Class 3, Step 11). If $\hat{f}(x)$ and $f(x)$ are the adapted and true solutions of the new case respectively, we update the adaptation weight vector($w^a$) to minimize the squared difference between these two using:

$$w^a \leftarrow w^a - \left[ (\tilde{x} - x)((\tilde{x} - x))^T \right]^{-1} \left[ \left( \hat{f}(x) - f(x) \right) (\tilde{x} - x) \right] \tag{12}$$

**Solving a New Problem.** We discuss how the system solves a new problem during the prediction phase.

- *GC Retrieval:* The reasoner uses solo-weighted distance to find the closes GC to solve the new problem (Class 1, Step 12). If the distance between the retrieved GC and the new case is greater than a threshold, the system must decide whether to use this GC or transfer control to another linked GC.
- *Failure-driven Reminding:* The reasoner matches failure-driven links to identify potential GCs to transfer problem-solving control (Class 3, Step 31). The sum of matching weights that lead to any GC denotes the usefulness of a transfer. Control is transferred to the linked GC with the maximum confidence if the total weight leading to such a GC is greater than a threshold.
- *Local-Adaptation:* However, if the confidence is low, control is retained with the GC. If the GC has cases stored inside it, the case closest to the new problem is retrieved (Class 1, Step 36), and its solution is adapted using the case's local adaptation. If no cases are stored within the GC, the GC's adaptation knowledge is used to predict the solution.

Thus, the reasoner can use the interconnected and proactive case base of the holographic CBR framework to implement the dynamic memory ideas of outcome-driven weighted retrievals, forming and validating expectations, creating generalizations, and failure-driven reminding to learn and solve problems.

# 4    Results and Interpretations

This section tests our realizations on their efficiency and solution quality. All results are averages over 50 runs where the train-test split and the order of case addition are randomized. We build two systems that solve different tasks using data from the UCI Machine Learning Repository [5]:

- *Zoo Case Base:* This is a classification task that has been discussed before in Sect. 3.1. It has animals from 7 classes represented by 16 categorical biological features.
- *Energy Efficiency Case Base* [17]: This is a regression task to predict the heating load and cooling load requirements of buildings (that is, energy efficiency) using eight real-valued building parameters.

## 4.1    Comparison with Baseline

In this section, we compare our holographic approach with other systems which perform the same tasks to illustrate the improvements obtained by the holographic framework. For the classification task, we compare our system with two machine learning models and two alternate holographic systems:

- *Naive Bayes* [2]: This is a parametric ML model that does not retain experiences but builds a model using the entire training data at once.
- *K-Nearest Neighbors* [6]: This is a non-parametric model which retains the entire training data but uses multiple (k) data points taken together to solve a problem. This is a simplified version of our holographic CBR without weights, failure-driven links, or generalizations.
- *Ganesan et al.'s System:* This previous holographic system does not have generalizations or outcome-driven weights but has expert-given failure-driven links to connect cases.
- *ML Switching Model:* This is a modification made to our realization where the failure-driven links transfer control to an ML model (Naive Bayes) instead of other GCs during expectation failures.

For the regression task, we compare the system with:

- *Ridge Polynomial Regression:* This is a parametric method that, unlike our approach, does not retain experiences but instead builds a model based on all the training data points taken at once.
- *K-Nearest Neighbors:* This non-parametric model uses the average solution of the k-nearest neighbors to predict the new solution.
- *Ganesan et al.'s System:* This previous system has local adaptation with a weighted linear regression model in each case. It does not have control transfers or generalizations and retains all training data points as cases.

The results of the comparison are in Table 1. The improved holographic systems outperform the ML models. The holographic system does not lose out on the solution quality despite its cognitive appeal. KNN, which retains all the data

**Table 1.** Comparison with other systems (averages over 50 runs)

| Classification models | Test accuracy | Regression models | Test RMSE |
|---|---|---|---|
| Current system | 93.771 | Current system | 1.4606 |
| Previous system | 91.523 | Previous system | 1.8272 |
| KNN | 93.226 | KNN | 2.1749 |
| Naive Bayes | 91.649 | Polynomial ridge regression | 3.1693 |
| ML Switching Model | 92.718 | | |

points, can still not outperform our approach, which only retains a fraction of the training data. In our approach, cases are aware of each other's competence and can coordinate better to solve the problem. On the other hand, in (parametric) ML models (Naive Bayes and Ridge Regression), the training data points interact to create the model but lose their individual competencies when a model which might not reflect the ground truth is enforced. Regression builds a model by treating the entire problem space as one while the holographic system has local pockets of knowledge in the local knowledge containers.

Our approach thus finds a middle ground by retaining cases but also allowing them to interact in a holographic fashion. It is able to identify structures in the problem in a bottom-up fashion and exploit it to achieve better performances. Our system also outperforms the previous holographic realizations that treat all components equally important and also miss out on the merits of forming generalizations. This highlights the importance of the generalization mechanism, which identifies regions of similarity in the problem space and combines the knowledge present in multiple similar cases.

### 4.2 Tests for Efficiency

CBR systems suffer from the utility problem [16] where, as the case base grows, the knowledge of the system and its solution quality improve, but the system's efficiency drops. It has been observed that better indexing and case base mainte-nance can handle this trade-off [4]. This section tests how our system handles this trade-off by monitoring its efficiency as the number of training cases increases. The holographic reasoner has additional computation costs over a traditional CBR system due to control transfer using failure-driven links. The number of such transfers is an indication of this additional cost. We track this as a proxy for efficiency along with the test data accuracy/RMSE. The system's test accuracy should ideally increase with more training data without reducing efficiency.

In both the tasks, as shown in Fig. 2a and Fig. 2b, we observe that as the train-ing data increases, the test performance increases without significantly increasing the control transfers. Contrarily, the number of transfers decreases after a point. As the amount of training data crosses a limit, the outcome-driven weights are well-tuned, and as a consequence, the initial retrievals are more accurate. Even when control transfers are used, they arrive at the correct solution faster. This

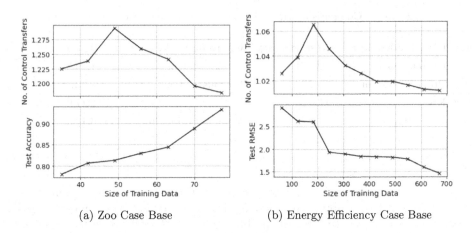

(a) Zoo Case Base          (b) Energy Efficiency Case Base

**Fig. 2.** Tracking efficiency and accuracy with increase in training data

indicates the system's ability to improve both its efficiency and solution quality with increasing training data.

## 5   Conclusions and Future Directions

We have expanded the holographic CBR framework and developed improved realizations that draw insights from popular models of dynamic memory and are cognitively appealing. We have demonstrated holographic CBR's broad scope, which offers an interconnected and proactive case base to build practical systems that can outperform traditional methods in selected tasks both in terms of efficiency and solution quality. With this, we aim to establish holographic CBR as a general-purpose CBR framework using which we can build a myriad of systems with different applications, memory models, amount of domain knowledge, and end goals. In this way, we establish holographic CBR not as a problem-solving tool but rather as a paradigm to design such tools.

We aim to view CBR the way it was envisioned during its initial phases and look past the haze created by practical constraints. By framing CBR as a Memory-based Reasoning Framework and improving its cognitive appeal using insights from models of human understanding, we aim to demonstrate the richness of the CBR framework and its relevance in building better Artificial Intelligence systems [15]. This research work is a step in that direction.

The ideas developed here - forming generalizations, failure-driven reminding, and outcome-driven weights are just a few of the numerous cognitive mechanisms that can be realized. More importantly, the way in which these have been realized in this work is not the only way to do so. Nevertheless, we hope that the results from this work and the ideas presented pave the way for further integrating cognitive memory-based reasoning components with holographic CBR.

# References

1. Aamodt, A.: Knowledge-intensive case-based reasoning in CREEK. In: Funk, P., González Calero, P.A. (eds.) ECCBR 2004. LNCS (LNAI), vol. 3155, pp. 1–15. Springer, Heidelberg (2004). https://doi.org/10.1007/978-3-540-28631-8_1
2. Bayes, T., Price, N.: LII: an essay towards solving a problem in the doctrine of chances. By the late Rev. Mr. Bayes, FRS communicated by Mr. Price, in a letter to John Canton, AMFRS. Philos. Trans. R. Soc. Lond. **53**, 370–418 (1763). https://doi.org/10.1098/rstl.1763.0053
3. Bergmann, R., Vollrath, I.: Generalized cases: representation and steps towards efficient similarity assessment. In: Burgard, W., Cremers, A.B., Cristaller, T. (eds.) KI 1999. LNCS (LNAI), vol. 1701, pp. 195–206. Springer, Heidelberg (1999). https://doi.org/10.1007/3-540-48238-5_16
4. De Mantaras, R.L., et al.: Retrieval, reuse, revision and retention in case-based reasoning. Knowl. Eng. Rev. **20**(3), 215–240 (2005). https://doi.org/10.1017/S0269888906000646
5. Dua, D., Graff, C.: UCI machine learning repository (2017). http://archive.ics.uci.edu/ml
6. Fix, E., Hodges, J.L.: Discriminatory analysis. Nonparametric discrimination: consistency properties. Int. Stat. Rev./Revue Internationale de Statistique **57**(3), 238–247 (1989). https://doi.org/10.2307/1403797
7. Ganesan, D., Chakraborti, S.: Holographic case-based reasoning. In: Watson, I., Weber, R. (eds.) ICCBR 2020. LNCS (LNAI), vol. 12311, pp. 144–159. Springer, Cham (2020). https://doi.org/10.1007/978-3-030-58342-2_10
8. Hammond, K.: Chef: a model of case-based planning. In: AAAI (1986)
9. Kolodner, J.L.: Reconstructive memory: a computer model*. Cogn. Sci. **7**(4), 281–328 (1983). https://doi.org/10.1207/s15516709cog0704_2
10. Lashley, K.S.: In search of the engram, pp. 454–482. Physiological Mechanisms in Animal Behavior. Academic Press, Oxford (1950)
11. Maximini, K., Maximini, R., Bergmann, R.: An investigation of generalized cases. In: Ashley, K.D., Bridge, D.G. (eds.) ICCBR 2003. LNCS (LNAI), vol. 2689, pp. 261–275. Springer, Heidelberg (2003). https://doi.org/10.1007/3-540-45006-8_22
12. Pribram, K.H.: Brain and Perception: Holonomy and Structure in Figural Processing. Lawrence Erlbaum Associates, Hillsdale (1991)
13. Schank, R.: Dynamic memory - a theory of reminding and learning in computers and people (1983)
14. Schank, R.C.: Language and memory*. Cogn. Sci. **4**(3), 243–284 (1980). https://doi.org/10.1207/s15516709cog0403_2
15. Slade, S.: Case-based reasoning: a research paradigm. AI Mag. **12**(1), 42 (1991). https://doi.org/10.1609/aimag.v12i1.883
16. Smyth, B., Cunningham, P.: The utility problem analysed. In: Smith, I., Faltings, B. (eds.) EWCBR 1996. LNCS, vol. 1168, pp. 392–399. Springer, Heidelberg (1996). https://doi.org/10.1007/BFb0020625
17. Tsanas, A., Xifara, A.: Accurate quantitative estimation of energy performance of residential buildings using statistical machine learning tools. Energy Build. **49**, 560–567 (2012). https://doi.org/10.1016/j.enbuild.2012.03.003
18. Zito-Wolf, R., Alterman, R.: Multicases: a case-based representation for procedural knowledge. In: Proceedings of the 14th Annual Conference of the Cognitive Science Society, pp. 331–336. Lawrence Erlbaum (1992)

# Handling Climate Change Using Counterfactuals: Using Counterfactuals in Data Augmentation to Predict Crop Growth in an Uncertain Climate Future

Mohammed Temraz[1,2], Eoin M. Kenny[2], Elodie Ruelle[3], Laurence Shalloo[3], Barry Smyth[1], and Mark T. Keane[1,2(✉)]

[1] Insight Centre for Data Analytics, University College Dublin, Dublin, Ireland
{barry.smyth,mark.keane}@ucd.ie
[2] VistaMilk SFI Research Centre, University College Dublin, Dublin, Ireland
{mohammed.temraz,eoin.kenny}@ucdconnect.ie
[3] VistaMilk SFI Research Centre, Teagasc, Animal and Grassland Research, Fermoy, Ireland
{elodie.ruelle,laurence.shalloo}@teagasc.ie

**Abstract.** Climate change poses a major challenge to humanity, especially in its impact on agriculture, a challenge that a responsible AI should meet. In this paper, we examine a CBR system (PBI-CBR) designed to aid sustainable dairy farming by supporting grassland management, through accurate crop growth prediction. As climate changes, PBI-CBR's historical cases become less useful in predicting future grass growth. Hence, we extend PBI-CBR using data augmentation, to specifically handle disruptive climate events, using a counterfactual method (from XAI). Study 1 shows that historical, extreme climate-events (climate outlier cases) tend to be used by PBI-CBR to predict grass growth during climate disrupted periods. Study 2 shows that synthetic outliers, generated as counterfactuals on an outlier-boundary, improve the predictive accuracy of PBI-CBR, during the drought of 2018. This study also shows that an case-based counterfactual method does better than a benchmark, constraint-guided method.

**Keywords:** Climate change · Counterfactual · Data augmentation · Grass

## 1 Introduction

Climate change is arguably the single, biggest challenge facing the world today. The United Nations "AI for Good" platform promotes AI technologies to meet this challenge and the UN's Sustainability Goals [1]. But, how can we predict an uncertain future using historical data that may no longer apply; how can we build predictive systems that can handle the "concept drift" created by climate change, a drift that may make past training-data irrelevant. In this paper, we explore one attempt to meet such challenges in supporting a sustainable smart agriculture. We show how an AI system, PBI-CBR [2, 3], that aids dairy farmers in sustainable grass management, can better handle crop growth

© Springer Nature Switzerland AG 2021
A. A. Sánchez-Ruiz and M. W. Floyd (Eds.): ICCBR 2021, LNAI 12877, pp. 216–231, 2021.
https://doi.org/10.1007/978-3-030-86957-1_15

prediction in the face of disruptive climate events. Specifically, we explore the novel use of counterfactual techniques to augment training data, to improve future predictions during climate-disrupted periods. The intuition is that a case-based counterfactual technique [4] can generate new cases by adapting historical cases to better handle climate change; these methods "re-combine" historical cases to produce new synthetic cases that are "offset" from past cases to better predict the future. This counterfactual technique is tested against actual grass-growth data for 2018, in Ireland, a year of disrupted weather, causing a forage crisis across the dairy sector in Europe [5].

In the next section, we detail the grass-growth prediction problem and the case-based reasoning (CBR) system developed to handle it for farmers (PBI-CBR [2, 3] see Sect. 1.1). Then, we consider the relevant literature from CBR, counterfactual techniques in XAI and the novel use of counterfactuals in data augmentation (see Sect. 1.2). Finally, we close this introduction by considering the research questions addressed and the novelties that arise (see Sect. 1.3). We then report two major studies that: (a) determine how PBI-CBR currently handles the prediction of climate-disruptive events (such as those in 2018; see Sect. 2), (b) comparatively test the PBI-CBR system using two different counterfactual methods – an instance-guided and constraint-guided one – that generate synthetic cases differently for this prediction problem (see Sect. 3).

### 1.1 The Problem: Grass Growth Prediction for Sustainable Dairy Farming

While some climate activists have argued that many sectors of agriculture should simply be abandoned - as humanity moves from a meat-based diet to a vegetarian (or indeed vegan) one - the short-term feasibility of such radical changes is questionable. *Agroecology* may be more feasible, where farming systems are changed to embrace more sustainable practices [6]. In the dairy sector, such a move could be achieved by adopting *pasture-based dairy systems* where animals are predominantly fed on grass outdoors (i.e., on pastures) rather than on meal/supplements indoors [6]; these pasture-based systems have lower carbon costs (e.g., feed is not transported over long distances), and grassland can also be used as a carbon sink. However, such agricultural practices hinge on the development of a precision agriculture to support sustainability; in the dairy sector, one initiative relies on the accurate prediction of grass growth to help farmers estimate feed budgets for dairy herds [6–10].

**Grass Budgeting & Sustainability.** Accurate grass budgeting sits at the heart of this sustainable, dairy alternative which, in turn, hinges on farmers accurately predicting the grass growth on their farm in coming weeks [2, 8, 9]. When grass growth is predicted accurately a dairy farmer can (i) improve grass utilization, thus reducing reliance on meal/supplements (reducing carbon costs), (ii) reduce fertilizer use (and potential nitrate pollution), and (iii) extend of the grazing season (reducing greenhouse gas emissions, see [7]). The Irish dairy sector mainly operates a pasture-based system with national sustainability goals [7]. To support these efforts, an online grassland management system aids farmers in this task, the PastureBase Ireland (PBI) system [9].

**PastureBase Ireland.** Since 2013, Ireland's national agricultural research organization, *Teagasc*, have provided PastureBase Ireland (PBI, https://pasturebase.teagasc.ie) as a grassland management system for Irish dairy farmers [9]. PBI has 6,000+ users of the ~18,000 dairy farmers in Ireland. The PBI database has weekly records of *grass covers* for individual farms from 2013 to the present (here, a *cover* refers to the amount of grass available from each paddock/field on a farm for a given day). The current PBI system provides a farmer with a model of their farm (i.e., the paddock sizes) and the current herd-size, to help them estimate feed budgets for a week ahead. At present in PBI, grass growth is calculated by comparing the grass cover of the current week with the previous week's cover. In the future, PBI will provide predictions of grass growth; traditionally, using *mechanistic models* such as the *Moorepark St. Gilles Grass Growth* model (MoSt [8]). Currently, these models make region-level predictions based on weather and farm variables and farm-level predictions for selected farms. The present paper is a collaboration exploring AI techniques, PBI-CBR, in this problem domain [2].

**Predicting Grass Growth Using PBI-CBR.** PBI-CBR [2, 3] applies CBR to grass-growth prediction using historical data from the PBI system, that has been entered by farmers about their own farms; this data has cases recording the time-of-year, farm-id, current-grass-cover (i.e., dried grass biomass above 4 cm grass height) and 3 weather parameters (i.e., rainfall, temperature, and solar-radiation; see Fig. 1). PBI-CBR uses a $k$-NN to predict grass-growth-rates for the following week using its cases. However, the historical data is very noisy; some cases have missing data, different farms have different numbers of cases, and some are manifestly incorrect (e.g., impossible growth rates from data-entry errors). PBI-CBR cleans this data using a novel method – called *Bayesian Case-Exclusion* – where cases that are predictive-outliers are excluded using a separate gold-standard distribution of grass growth [8]. This cleaned case-base spanning several years (2013–2016) makes accurate predictions for grass growth in future years (optimally, for $k = 25$–$40$) as well providing *post-hoc* explanatory cases from the same/similar farm. In this paper, we examine PBI-CBR's grass-growth predictions for atypical, disruptive climate events. For instance, the summer of 2018 was unusually hot with low rainfall across Europe. Grass tends to grow faster as temperatures rise (up to 25 °C), but the absence of soil-water can interrupt growth and lead to burnt plants (at >30 °C). So, in the Irish summer of 2018, when grass-growth rates typically are at their highest, growth fell back to near zero causing a feed crisis in the dairy sector.

### 1.2   Related Work: Counterfactuals from XAI to Data Augmentation

This paper focuses on the use of counterfactual methods for data augmentation as a solution to improving grass growth prediction in the context of climate change. However, to date, counterfactual methods are usually only used in explainable AI (XAI) rather than in data augmentation (for reviews see [11, 12]). In example-based post-hoc explanation strategies, counterfactual explanations have become very popular and are argued to be superior to factual explanations [13]. Imagine you have applied for a bank loan and are refused by an automated AI system, a *counterfactual explanation* might say "if you requested a loan that was 10% lower, you would have got the loan". In the last two years, counterfactual methods have received huge attention in XAI. We review this XAI work and the few papers that use counterfactuals for data augmentation.

```
-----------------------------------      -----------------------------------
             Normal Case                               Extreme Case
-----------------------------------      -----------------------------------
Farm ID: 6332                            Farm ID: 6314
Cover CoverID: 73113                     Cover CoverID: 133875
Cover Date: 04/07/2016                   Cover Date: 02/07/2018
GrowthRateFarm: 50.0                     GrowthRateFarm: 21.0
Week: 27                                 Week: 27
Month: 7                                 Month: 7
-----------------------                  -----------------------
Parameter    | Actual value | The mean   Parameter    | Actual value | The mean
-----------------------                  -----------------------
Rain              1.2          1.8       Rain              0.0          1.1
Temperature      19.0         19.1       Temperature      27.1         21.8
Solar Radiation 1209.0      1454.2       Solar Radiation 2959.0      1946.4
-----------------------                  -----------------------
```

**Fig. 1.** Examples of PBI-CBR's cases from two different farms for week-27 in 2016 and 2018; a "normal" case where weather values are close the mean and an "extreme" outlier case where temperature and solar radiation are very high relative to the mean (i.e., a climate-disrupted event)

**Counterfactual Generation in XAI.** In CBR, counterfactuals have been traditionally cast as Nearest Unlike Neighbours (NUNs; [14–16]); namely, the closest neighbouring case to a test case, just over a decision boundary in the dataset. Keane and Smyth [4] re-christened NUNs as *native counterfactuals*, to distinguish them from the *synthetic counterfactuals* generated by current XAI counterfactual methods. Wachter et al.'s [17] seminal paper cast synthetic counterfactual generation as a constraint optimization problem using gradient descent over a space of blindly-perturbed datapoints; using a loss function to find the "best counterfactual", balancing the proximity of the counterfactual case to the test case against its closeness to the decision boundary. So, this method aims to generate the "closest possible world" to the test case, in which the counterfactual case is minimally different and sparse (i.e., there are few feature differences between test and counterfactual). In the XAI literature, this method has been extensively used and extended with additional constraints (diversity, causality, feature-importance) and other generative methods (such as, using genetic algorithms, GANs, VARs; see [12] for a full review). Mothilal et al.'s [18] *Diverse Counterfactual Explanations* (DiCE) extends this method to include *diversity* constraints; so that for given $p$, the set of counterfactuals produced minimizes the distance within the set, while maximizing the range of features changed across the set. DiCE seems to generate counterfactuals that are valid, diverse, and sparse. Interestingly, [18] also proposed the notion of "substitutability" as an evaluation method for counterfactual XAI; namely, that if a set of counterfactuals were good, one could substitute them for the original dataset to achieve equivalent predictive performance. In the present tests, we use DiCE as it has become a benchmark-method for tests of counterfactual generation (e.g., see [19]).

However, Keane and Smyth [4] proposed a very different *case-based counterfactual method* that exploits known counterfactual relationships in the dataset. Their *instance-guided method* finds the test case's nearest-neighbour that takes part in a so-called *explanation case (xc,* which we label as *cf).* An explanation case is a pair of mutually-counterfactual cases which differ by at most 2 features. The test case and the counterfactual case from this nearest *cf* are used to generate a new "good" counterfactual for the test case, by combining the test-case features with the (at most) 2 difference-features from the *cf's* counterfactual. In the loan scenario, imagine historical cases that form a native counterfactual about customer-A who was *refused* a *$5k* loan (a female accountant earning $50k a year, who is *2.5 years* in her current job) and customer-B who was *granted* a *$4k* loan (a female accountant on $50k a year, who is *3 years* in her current job). Assume customer-C (a female accountant earning $50K and 2.5 years in her job) has also been *refused* a $5k loan. In this scenario, the native counterfactual suggests a counterfactual explanation saying "if you wait 6-months and re-apply for a lower loan (of say $4k), you will be *granted* the loan". So, customer-C's nearest neighbour in the dataset is customer-A (who has the same refusal outcome), but because there is a close counterfactual case, customer-B (with a different outcome), a counterfactual scenario for customer-C can be generated, using the difference-features found (time-in-job, loan-requested). In Study 2 reported here, this method is used in data augmentation tests and compared to DiCE (see Sect. 3.1 for a full algorithmic description).

**Counterfactual in Data Augmentation.** Beyond XAI, our hypothesis is that counterfactual methods could also play a role in data augmentation, that generated synthetic counterfactual cases could improve the predictive accuracy of a model. Though there are now 100s of papers on counterfactuals in XAI, only a handful of papers consider their use in data augmentation [22–25]. Recall, that Mothilal et al. [18] proposed *substitutability* as a way to evaluate counterfactual XAI methods; namely, that a good set of generated counterfactuals should be able to substitute for the original dataset. Hasan [22] explicitly tested this idea using DiCE, to determine if an augmented dataset using DiCE's counterfactuals could act as a proxy dataset; however, the improvements found were minimal. [23] consider the problem of *dataset shift*, where there is a divergence between the context in which a model was trained and tested; they use the notion of "counterfactual risk" to diagnose this problem using causal models. However, this work does not use the XAI counterfactual methods that have been extensively tested; hence, this work's status, reproducibility and/or generality is unclear. So, in the current tests, we use two proven counterfactual methods from the XAI-literature (i.e., [4, 18]).

### 1.3 Research Questions and Novelties

In this paper, we test whether the counterfactual methods, developed in XAI, can be applied to the challenging concept-drift problems associated with climate-change; specifically, in the context of grass growth prediction for dairy farmers (using the PBI-CBR model). So, we determine whether generated synthetic, counterfactual cases can be used to improve prediction during periods of climate disruption (focusing on 2018). However, before we can consider whether counterfactual data augmentation might work, there are several prior steps that need to be considered. First, we need to understand how the PBI-CBR model currently handles grass-growth prediction when it encounters climate-disrupted events (as test cases); a reasonable hypothesis might be that it uses historical-cases capturing past climate-disruptive events. However, this begs the non-trivial question of how one might define "past climate-disruptive events". Hence, we perform two major studies, one that aims to understand how PBI-CBR actually predicts grass growth for climate-disruptive events (Study 1; Sect. 2) and one that comparatively tests whether counterfactual data augmentation methods can improve PBI-CBR's performance on such climate-disruptive events (Study 2; Sect. 3). So, these studies aim to answer 3 research questions:

*RQ1*: How does PBI-CBR currently handle grass growth prediction involving climate-disruptive events?
*RQ2*: Can PBI-CBR's prediction of climate-disruptive events be improved by counterfactual data-augmentation methods?
*RQ3*: And, if prediction is improved by counterfactuals, which counterfactual methods work best?

As we shall see, several significant novelties arise from the answers to these research questions (see Sect. 4). In the following sections, we describe the studies carried out.

## 2   Study 1: Predicting Climate Disruption with PBI-CBR

In this first study, we analyze PBI-CBR's grass-growth predictions when it encounters climate disruptive events (RQ1). However, before we can assess its performance, we need to define cases that potentially reflect climate-disruptive events (see Sect. 2.1). Then, armed with this definition, we perform two experiments. The first experiment determines whether historical extreme-climate cases in the PBI-CBR case-base tend to be used to predict growth rates when extreme-climate test-cases are encountered (using 2018 as a test year; see Expt. 1a in Sect. 2.2). This may seem like an obvious test but it is not. The grass-growth dataset is very noisy, as the cases come from end-user data-entry on the PBI website; while PBI-CBR automatically cleans the original dataset, it is still not clear whether "outlier" cases are "true outliers" representing actual extreme-weather events on a farm or just invalid data-points created by data-entry errors (e.g., in temperature or growth data). The second experiment considers the effects of varying $k$ in the model on these results (see Expt. 1b in Sect. 2.3).

Both experiments used PBI dataset drawn from 6,000 + farms over 6 years 2013–2018 (N = 70,091)[1]; divided as follows 2013 (N = 5,205), 2014 (N = 6,852), 2015 (N = 9,695), 2016 (N = 14,777), 2017 (N = 18,611), and 2018 (N = 14,951). In general, the number of cases increases each year as more farmers joined the PBI system. So, in both experiments, the years 2013–2016 (N = 36,529) were used as training data and 2018 (N = 14,951) was used as the test data (2017 was also run but not reported); 2018 had many extreme climate events with high-temperatures, high solar-radiation and low summer rainfall that caused a feed-crisis for the sector (as grass growth is inhibited by low soil moisture and damage by solar radiation). As such, it is a real-life, test-case of the climate challenges now facing agriculture. However, we first need to define which cases are likely to be the ones that reflect extreme-climate events.

### 2.1   Defining a Class Boundary for Climate Outlier Cases

To run our tests on PBI-CBR we need some definition of which cases might reflect extreme-weather events (n.b., extreme values could just be data-entry errors). Here, we

---

[1] Note, this is after pre-processing to remove noisy cases (originally, N = 138,970).

used a statistical approach to define, what we call, *climate outlier cases;* that is, cases that appear to capture extreme weather events by virtue of having high/low extreme values for either temperature, rainfall, or solar radiation. As weather data follows a normal distribution $[X \sim (\mu, \sigma^2)]$ for each week, we defined *climate outliers* as cases with values that are >2 standard deviations above/below the mean for a given week. So, this filter was applied to all the cases for a given week (e.g., week 12) aggregated over all the years (2013–18) in the dataset (an in-year weekly-average produces broadly similar results). More formally, weather parameters high/low outliers are defined as:

$$High\,Outliers = X_i > \mu + 2\sigma$$
$$Low\,Outliers = X_i < \mu - 2\sigma$$

when $X_i$ is an observation, $\mu$ and $\sigma$ are the mean and standard deviation for a given week. Figure 1 shows two sample cases, a "normal" farm case with typical weather features for the week-27 of 2016 and a "extreme" climate-outlier case for the same week of 2018. Figure 2a shows the distribution of *temperature* values for each week across 2013–2018 (with box plots) and Fig. 2b shows the high and low outliers found for each week in this combined dataset. Note, how there are many high-temperature outliers in summer weeks and many low-temperature ones in winter weeks.

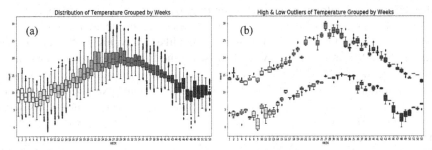

**Fig. 2.** The distribution of temperature values (with box plots) (a) for all cases by weeks of the year (from 2013–2018) with (b) high and low outliers separated out.

When we remove these climate outliers from the overall PBI-CBR dataset (N = 70,091) for 2013–2018, we find 7,324 *unique outliers*[2]. Most climate outliers reflect rainfall extremes (44%, N = 3,500), with others reflecting extremes of temperature (38%, N = 2,997) and solar radiation (18%, N = 1,414). The percentage of outlier cases in each year is fairly constant, though frequencies increase across years (in-year %'s shown): 2013 (16%, N = 836), 2014 (10%, N = 707), 2015 (10%, N = 1,008), 2016 (9%, N = 1,259), 2017 (10%, N = 1,778), 2018 (12%, N = 1,736). In these experiments, we used 2013–2016 as the training set, testing it mainly against 2018 (we found equivalent results for 2017, though as it was more "normal", the effects were less pronounced). So, in these experiments the 2013–16 PBI-CBR case-base had N = 36,529 cases, when all cases are included, and N = 32,719 cases when the climate outliers were excluded.

---

[2] A *unique outlier* is a case with an extreme value on any of its weather features.

**Table 1.** Frequencies of training outliers used to predict test outliers in 2018 (Expt.1a)

|  | Training (*Outliers*) | Training (*Non-Outliers*) |
|---|---|---|
| *Test-Outliers* (N = 1,736) | 1,534 *(88.4%)* | 202 *(11.6%)* |
| *Test-Non-Outliers* (N = 1,248) | 144 *(11.5%)* | 1,104 *(88.5%)* |

## 2.2 Experiment 1a: The Contribution of Climate Outliers to Predictions

This experiment determines whether historical extreme-climate cases in the PBI-CBR dataset tend to be used to predict growth rates when extreme-climate problem-cases are encountered (using 2018 as a test year).

**Setup & Method.** This experiment ran a version of PBI-CBR (for the years 2013–2016) with and without its climate outliers (as defined above); so, we compared the (a) *original* system with all training outliers *included* (PBI-CBR$^O$; N = 36,529) and (b) PBI-CBR$^{EX}$, a version of the system with all training climate-outliers *excluded* (N = 32,719). For all tests $k = 30$, the value found to deliver the highest accuracy in previous tests of PBI-CBR [2]. The measure used was the Absolute Error (AE) found for each test case in a given year (measured in kg/DM/ha), where the *AE = |actual-grass-growth - predicted-grass-growth|*. *Mean Absolute Error* (MAE) is the aggregate measure over all test-cases for a given condition.

**Results & Discussion.** The results showed that the presence of climate-outliers in PBI-CBR training set significantly improved the performance of the system. The absolute-errors across the 2018 test-set showed that PBI-CBR$^O$ (MAE = 20.20 kg/DM/ha) performed reliably better than PBI-CBR$^{EX}$ (MAE = 20.35 kg/DM/ha) which excluded the climate outliers, $t(14950) = 3.58, p < 0.001$, one-tailed[3]. While these MAE differences may not look large, they could be quite significant for a given farm. Remember the measure here is kg/DM/ha (kilograms of dried grass/matter per hectare), so a 0.50 kg error could be a lot of grass, as it is multiplied by the size of the farm for each weekday. Importantly, we also determined which training-cases were being used to make predictions for the 2018 test-cases to determine whether PBI-CBR$^O$ succeeds by using past extreme-climate events to handle new extreme-climate events. Specifically, that in PBI-CBR$^O$ the *climate-outliers in the training set* are used to make predictions for *climate-outliers in the test set*. Note, the test-cases were all the outlier cases (N = 1,736) in 2018 and all the non-outlier cases with "good" predictions in 2018 (N = 1,248); where a *good prediction* was one with an *AE* equal to or better than the *MAE* for all test cases in that year. Table 1 shows that in solving 2018 test-cases, there is a marked tendency for *climate-outlier test-cases* to be solved by *climate-outlier training-cases* (~88% of the time). This result confirms the intuition that PBI-CBR is succeeding by flexibly assembling similar cases in atypical, local regions of the problem space to make better predictions. Recall, this

---

[3] Similar results were found for tests of 2017, though less marked, as that year has fewer disruptive events: PBI-CBR$^O$ (MAE = 18.58 kg/DM/ha) did better than PBI-CBR$^{EX}$ (MAE = 18.62 kg /DM/ha) without the climate outliers, $t(18610) = 1.9, p < 0.05$, one-tailed.

result is based on $k = 30$ for PBI-CBR, so in the second experiment we varied $k$, to get a sensitivity analysis of this result.

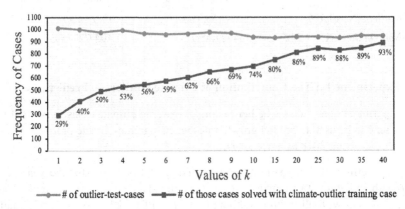

**Fig. 3.** Expt. 1b results showing the number of outlier-test-cases and frequency (and %) of outlier-test-cases solved by outlier-training-cases for different values of $k$ (for 2018)

### 2.3   Experiment 1b: Role of Training Outliers at Values of $k$

Expt.1a was run using the optimal $k = 30$, with predictions being made by averaging the grass-growth values over all cases in $k$. However, it would be good to know at what $k$-value these training-outlier-cases begin to play a role in solving test-outlier-cases. If these training outliers appear in solving test-cases at low values of $k$, then it means these cases are being readily recruited to solve test-cases (n.b., predicted values of grass growth are based on mean of the cases in $k$). So, in this experiment $k$ was varied and role of training-climate-outliers in predictions was noted.

**Setup & Method.** Using the 2018 test-set, ~1,000 test-cases with "good predictions" were tested for every value of $k = 1$–40; where a *good prediction* was one with an *AE* equal to or better than the *MAE* for all test cases in that year (n.b., differs for each $k$).

**Results & Discussion.** Figure 3 shows the results of varying $k$ on the occurrence of outlier-training-cases that solve outlier-test-cases. Stated simply, it shows that by $k = 4$, climate-outlier training-cases are contributing to predictions in >50% of climate-outlier test-cases showing that these key past cases are being used. So, having established that climate-outlier cases *are* used to make better predictions for disruptive climate events, in the next study we consider whether counterfactual methods can be used to generate new outlier-cases, to augment the dataset, and improve prediction even further.

## 3   Study 2: Predicting Climate Disruption with Counterfactuals

In Study 1, we saw that PBI-CBR's grass-growth predictions benefit from the use of historical extreme-climate cases to deal with future extreme-climate test-cases (answering RQ1). Notably, other unreported experiments, showed that the outlier cases used by

PBI-CBR to predict climate-extreme events, were sparse *two-difference* native counterfactuals. So, these outliers typically have two feature-value differences that "change" a "normal" case into an outlier, counterfactual-case; for instance, a normal case for farm-x in week-12 with moderate sunshine and growth is counterfactually "changed" to an outlier case with very-high solar-radiation and a very-low grass growth (as grass has been burnt off). This finding identifies the key outlier cases used to realise performance success seen in Study 1 (and is used in our algorithm).

In Study 2, we determine whether counterfactual methods from XAI have a role to play in data augmentation (RQ2 and RQ3). So, we explore the idea that counterfactual methods can be used to populate a case-base with new, synthetic cases that improve predictive performance. Specifically, in PBI-CBR, whether counterfactual methods can find new, synthetic *outlier cases* that improve predictive performance for *extreme-climate test-cases* in the future. So, in this study, we compare the performance of PBI-CBR using its native counterfactuals as outlier cases, as a baseline, against PBI-CBR using synthetic, counterfactually-generated outlier-cases. Note, this test pits native-counterfactual outliers against counterfactually-generated outliers to assess whether the artificial datapoints can "beat" naturally-occurring outlier cases.

Study 2 also performs comparative tests of two counterfactual algorithms from the XAI literature: Mothilal et al.'s [18] DiCE and Keane & Smyth's [4] case-based method. These two methods take quite different approaches. DiCE randomly generates a space of perturbed cases and then finds the best counterfactuals based on balancing proximity and diversity constraints [17, 18]. In contrast, the case-based method adapts cases from the original dataset; it finds a nearest neighbour to the test case, involved in a "good" counterfactual (i.e., a good native counterfactual) and then adapts the test case using the feature-differences found in this native counterfactual (see Fig. 4). In the next subsection, we detail our variant of this algorithm for data augmentation; it differs in how it selects test-cases and how it uses a statistically-defined boundary between "normal" and "extreme" climate cases (see Sect. 3.1, Figs. 1 and 4).

### 3.1 A Case-Based Counterfactual Augmentation Algorithm (CFA)

The *Counterfactual Augmentation* (CFA) method generates synthetic counterfactual cases in three main steps: (i) "good" counterfactual pairs, $cf(x, x')$, are initially computed over the whole case-base, $X$, (ii) given a test case, $p$, a nearest neighbour case, $x$, is retrieved from the set of counterfactual pairs, $cf(x, x')$, and (iii) then, a new synthetic counterfactual case, $p'$, is produced by adapting the original test-case, $p$, using feature-difference values from $x'$. More formally:

*Definitions:*

- Normal (non-outlier) case = $x_i$ $(x1, x2, x3, \ldots, x_i)$, where $x_i \in X$
- Counterfactual (outlier) case = $x_i'$ $(x1', x2', x3', \ldots, x_i')$, where $x_i' \in X$
- CF pair $cf(x, x') \Leftrightarrow target(x_i) \neq target\left(x_i'\right)$
- $K$-nearest neighbors = $k$-NN
- Difference between two cases = $Diff$

Step 1   **Identify native counterfactual (CF) pairs, $cf(x, x')$:** CFA first finds all possible "good" counterfactual pairs $cf(x, x')$ that already exist in a case-base, $X$ (pairing a normal case and its outlier counterfactual). These *native counterfactuals, cf(x, x')*, pair cases either side of $2\sigma$ *climate boundary*. Each of these native pairs has a set of *match-features* and a set of *difference-features*, where the differences determine the class change (e.g., the counterfactual case may a high temperature value relative to the normal case, resulting in a different grass-growth outcome; see Fig. 1)

Step 2   **For a test case, $p$, find its nearest neighbour, $x$, from the CF pairs:** Given a test case, $p$, CFA uses a $k$-NN to find its nearest neighbour, $x$, from the set of native counterfactual pairs, $cf(x, x')$. The test case, $p$, is drawn from those "normal" cases that do not take part in CF-pairs, those unpaired cases that do not occur in $cf(x, x')$

Step 3   **Transfer feature values from $x'$ to $p'$ and from $p$ to $p'$:** Having identified a candidate native, $cf(x, x')$ for the test case, $p$, CFA generates the synthetic counterfactual, $p'$ for $p$, such that:

– For each of the *difference-features* between $x$ and $x'$, take the values from $x'$ into the synthetic counterfactual case, $p'$.
– For each of the *match-features* between $x$ and $x'$, take the values from $p$ into the new counterfactual case, $p'$.

Clearly, the definition of a "good" counterfactual pairing is a critical parameter in this algorithm. On psychological grounds, [4] defined a "good" counterfactual to be one with no more than two feature-differences, taking a strong position on sparsity. Interestingly, subsequent user testing has shown that people prefer counterfactual explanations with 2–3 feature-differences (even over ones with 1 feature-difference [26]). Indeed, in an analysis of the outliers used in Study 1 (not reported here), we found that 2-difference native counterfactuals produced more accurate performance relative to 3-, 4- and 5-difference ones in PBI-CBR. So, the above algorithm, as in [4], uses the 2-difference definition of counterfactual "goodness" in Study 2.

### 3.2   Experiment 2: Using Synthetic Counterfactual Cases to Predict Growth

In the present experiment, PBI-CBR's predictive performance on 2018 is run by comparing it's native-counterfactual dataset (as a baseline), against datasets of synthetic counterfactuals generated by the *Counterfactual Augmentation* (CFA) and *Diverse Counterfactual Explanations* (DiCE) methods. We want to assess how these data augmentation techniques deal with climate-extreme events. So, we report MAE by month (n.b., the "year" consists of the 9 months in which cattle graze). In 2018, the climate disruption occurred in March (as an unusually cold spring), July (very hot summer) and October (a cold autumn; see Fig. 5).

**Setup & Method.** This experiment ran a version of PBI-CBR with three different datasets testing its performance against the climate-disruptive year of 2018 (see Fig. 5). All datasets used the $k$-NN to predict the grass growth-rates (measured in kg/DM/ha),

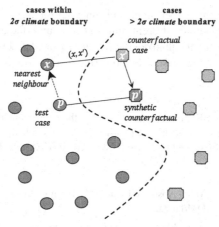

**Fig. 4.** Counterfactual Augmentation (CFA): A test case, $p$, finds a nearest neighbour, $x$, taking part in a "good" native counterfactual in the case-base, $cf(x, x')$, and then uses the difference-features of the counterfactual-case, $x'$, to generate a new synthetic counterfactual-case, $p'$, combining them with the matching-features of the original test case, $p$. The synthetic counterfactual-case, $p'$, is added to the case-base to improve future prediction.

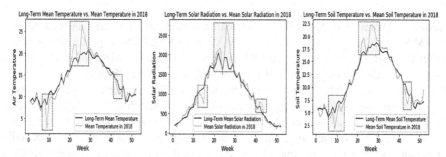

**Fig. 5.** Three graphs compare the long-term means for three weather-variables – air temperature, solar radiation and soil temperature – to the mean values in 2018, outlining the three main climate-disruptive periods in 2018 (i.e., March, July, and October).

with $k = 30$, the value found to deliver the highest accuracy in previous tests. Again, as before, the measure used was the Mean Absolute Error (AE) found over the 2018 test-set based on averaging the *Absolute Error (AE)*, where $AE = |actual\text{-}grass\text{-}growth - predicted\text{-}grass\text{-}growth|$. The three datasets used:

- *Native-CF:* "good" native counterfactuals (i.e., 2 feature-difference ones) from the original PBI-CBR dataset (N = 2,500)
- *DiCE:* the *synthetic counterfactuals* generated by DiCE from finding the best counterfactual for the test-cases in 2013–2016 (N = 2,500)
- *CFA: synthetic counterfactuals* generated by CFA based on adapting native counterfactuals for each of the test-cases in 2013–2016 (N = 2,500)

Originally, we ran this experiment with unequal datasets, CFA (N = 4,028) and DiCE (N = 14,951) generate different numbers of counterfactuals for the 2013–2016 data. The current experiment equalized the counterfactual datasets (to N = 2,500) taking the mean results of 5 random case-selections (the results does not change for these test variants).

**Table 2.** Study 2: PBI-CBR predictions (2018) for three different datasets: (i) good native counterfactuals from the original dataset (*Native-CF*), and synthetic counterfactuals from the (ii) constraint method (DiCE) and (iii) case-based method (CFA); the best results are shown in bold.

|  | Mean Absolute Error (*MAE*) of growth kg DM/ha/day | | | | | | | | |
|---|---|---|---|---|---|---|---|---|---|
|  | Feb | Mar | Apr | May | Jun | July | Aug | Sept | Oct |
| Native-CF | **40.9** | 40.2 | 19.0 | **26.4** | 24.8 | 30.0 | 21.7 | **16.7** | 33.0 |
| DiCE | 41.0 | 35.9 | 30.4 | 48.8 | 30.8 | 25.6 | 31.2 | 25.0 | **22.7** |
| CFA | 41.3 | **31.3** | **17.6** | 30.2 | **21.8** | **23.4** | **19.4** | 17.2 | 25.7 |

**Results & Discussion.** Overall, for the year, absolute-error for 2018 showed Native-CF (MAE = 20.1 kg/DM/ha) and CFA (MAE = 23.8 kg/DM/ha) performing reliably better than DiCE (MAE = 30.1 kg/DM/ha). Table 2 shows the MAE values by month. Overall, the CFA does best in 5/9 months, with the Native-CF doing best in 3 and DiCE just 1 (see Table 2, Figs. 5 and 6); notably, CFA succeeds in periods where the most climate-disruption occurred, the cold spring (March-April), the hot summer (Jun-Aug), and is a close second to DiCE for the cold autumn (October; see Fig. 6). Furthermore, CFA appears to generate better data augmentations than DiCE, especially in the climate-disrupted months. The MAE across the July test-set showed that CFA (MAE = 23.4 kg/DM/ha) performed better than DiCE (MAE = 25.6 kg/DM/ha) and the Native-CF (MAE = 30.0 kg/DM/ha) conditions, $F(2, 4446) = 50.49, p < .001$; with a decrease in the error rate of up to 22%. Also, in March, the MAE decreased from 40.2 (Native-CF) to 31.3 (CFA), $F(2, 2169,) = 65.59, p < .001$; the more extreme the disruption the better CFA seems to perform (see Fig. 6).

# 4   Conclusions: Novelties, Explications and Caveats

The present paper exhibits some of the *promise* that AI, and specifically CBR, offers to the challenge of climate change; specifically, we can see how AI might be applied to climate problems in sustainable, dairy agriculture. It shows that the counterfactual methods developed for XAI can be usefully deployed to augment datasets, with synthetic cases, that improve subsequent predictions in climate-disruptive periods. This result is significant because it shows that these techniques can be used to supplement historical datasets to better predict what could have been "an unpredictable future".

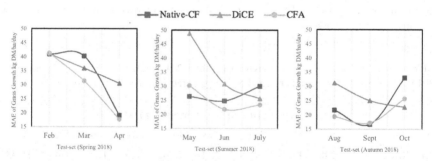

**Fig. 6.** Error in grass-growth predictions (MAE of kg DM/ha/day) in the spring, summer and autumn of 2018 for the Native-CF, CFA and DiCE datasets. Note, the counterfactual methods, CFA and DiCE, consistently do better than the native counterfactuals (Native-CF) in the climate-disrupted months (March, July, October)

**Novelties.** Specifically, we have answered the three research questions posed in the introduction: we have shown that (a) the original PBI-CBR system makes accurate predictions for climate-disrupted periods by relying on historical outlier cases (RQ1), (b) its prediction of crop growth in climate-disruptive events can be improved by counterfactual data-augmentation methods (RQ2), (c) the case-based CFA method performs better on this task than a benchmark optimization method (RQ3). As such, this paper reports several significant novelties: namely, key discoveries on how (a) AI methods for data augmentation can be used to deal with climate change, (b) counterfactual methods can be successfully used for data augmentation, (c) case-based counterfactual techniques can generate useful synthetic datapoints.

**Why Does This Work?** When we first discovered these effects of counterfactual data augmentation, they appeared (to us) to be both exciting and, somewhat magical. We asked ourselves "Why does this work?". How can a generated synthetic datapoint better predict a future event over historical data? There seemed to be no good reason for why it might work? Now, having completed these experiments (and a 100 more not reported here), it is beginning to become clear why this case-based counterfactual method succeeds. CBR is often claimed to be optimal when "local" views of the data are needed to solve problems (as it seems to be here), rather than generalized, "global" functions over the whole dataset (e.g., as in iterative optimization methods, such as neural networks). When we encounter a good native-counterfactual in a dataset, we essentially find a rule (a bit like an adaption rule) that tells us what minimal set of feature-changes move a case over a decision boundary. CFA exploits this implicit-knowledge in the case-base when it adapts the native-counterfactual to produce a synthetic counterfactual case, so these artificial cases are "meaningful offsets" from historical cases (it's like applying a good adaptation rule to generate new synthetic data-point). Notably, DiCE does not do this. DiCE perturbs feature-values and filters results based on broad constraints of proximity and diversity; as such, while it may "hit on" a case that is useful for solving the problem it does not do this in the guided way that CFA works. From another perspective, the present outlier cases here are essentially pivotal cases in competence terms [21] and CFA is effectively generating novel, synthetic pivotal cases that, of course, have a high probability of being useful. These are some of the reasons why we think this case-based data augmentation works.

**Table 3.** PBI-CBR's growth predictions for 9 months of 2018, with (PBI-CBR$^{O+CFA}$) and without (PBI-CBR$^O$) the synthetic counterfactual outliers generated by the CFA method

| | Mean Absolute Error (*MAE*) of growth kg DM/ha/day | | | | | | | | |
|---|---|---|---|---|---|---|---|---|---|
| | Feb | Mar | Apr | May | Jun | July | Aug | Sept | Oct |
| PBI-CBR$^O$ | **22.8** | **16.7** | 17.08 | **21.2** | **23.6** | 30.3 | 19.8 | **16.5** | **16.03** |
| PBI-CBR$^{O+CFA}$ | 23.3 | 17.8 | **17.06** | 21.7 | 23.7 | **29.9** | **19.6** | 16.6 | 16.33 |

**Caveats & Concerns.** However, there are some caveats we should keep in mind about these data augmentation successes. First, we have shown these results in one dataset; so can we be confident they generalize? Temraz and Keane [27] have applied this method to many standard datasets and found similar improvements. Second, note that in Study 2 we performed a carefully controlled study, pitting native counterfactuals against synthetic ones to determine the impacts of the latter. If one was using CFA in the PBI-CBR system, one would presumably add the generated counterfactuals to the original historical dataset and then run that full-dataset on 2018. When we do this, we can see that CFA still delivers improvements, but only in the more extreme months (April, July, August; see Table 3). So, obviously, the relative impacts of these techniques will wax and wane depending on the severity of the climate events encountered. Finally, the CFA method used here could be improved: Smyth and Keane [20] have proposed a more general counterfactual method than CFA, that appears to deliver better explanatory counterfactuals. It remains to be seen whether these are also better augmenting counterfactuals. Indeed, this raises a broader question about whether the explanatory versus data-augmentation requirements on counterfactual methods will, at some stage, diverge as they do appear to be very different use-contexts. But that is, as they say, a question for another day.

# References

1. Rosenzweig, C., Iglesias, A., Yang, X.B., Epstein, P.R., Chivian, E.: Climate Change and U.S. Agriculture. centre for health and the global environment. Harvard Medical School, Boston, MA, USA (2000)
2. Kenny, E.M., et al.: Predicting grass growth for sustainable dairy farming: a CBR system using bayesian case-exclusion and post-hoc, personalized explanation-by-example (XAI). In: Bach, K., Marling, C. (eds.) ICCBR 2019. LNCS (LNAI), vol. 11680, pp. 172–187. Springer, Cham (2019). https://doi.org/10.1007/978-3-030-29249-2_12
3. Kenny, E.M., et al.: Bayesian case-exclusion for sustainable farming. In: IJCAI-20 (2020)
4. Keane, M.T., Smyth, B.: Good counterfactuals and where to find them: a case-based technique for generating counterfactuals for explainable AI (XAI). In: Watson, I., Weber, R. (eds.) ICCBR 2020. LNCS (LNAI), vol. 12311, pp. 163–178. Springer, Cham (2020). https://doi.org/10.1007/978-3-030-58342-2_11

5. EU Parliament Briefing on the EU dairy sector (2018). https://www.europarl.europa.eu/Reg Data/etudes/BRIE/2018/630345/EPRS_BRI(2018)630345_EN.pdf
6. Altieri, M.A.: Agroecology: The Science of Sustainable Agriculture. CRC Press, Boca Raton (2018)
7. Teagasc: The Dairy Carbon Navigator: Improving carbon efficiency on Irish dairy farms
8. Ruelle, E., Hennessy, D., Delaby, L.: Development of the Moorepark St Gilles grass growth model (MoSt GG model). Eur. J. Agron. **99**, 80–91 (2018)
9. Hanrahan, L., et al.: PastureBase Ireland. Comput. Electron. Agric. **136**, 193–201 (2017)
10. Hurtado-Uria, C., Hennessy, D., Shalloo, L., O'Connor, D., Delaby, L.: Relationships between meteorological data and grass growth over time in the south of Ireland. Ir. Geogr. **46**(3), 175–201 (2013)
11. Karimi, A.H., Barthe, G., Schölkopf, B., Valera, I.: A survey of algorithmic recourse: definitions, formulations, solutions, and prospects. arXiv preprint arXiv:2010.04050 (2020)
12. Keane, M.T., Kenny, E.M., Delaney, E., Smyth, B.: If only we had better counterfactual explanations. In: IJCAI-21 (2021)
13. Dodge, J., Liao, Q.V., Zhang, Y., Bellamy, R.K., Dugan, C.: Explaining models. In: IUI-19, pp. 275–285 (2019)
14. Nugent, C., Doyle, D., Cunningham, P.: Gaining insight through case-based explanation. J. Intell. Inf. Syst. **32**(3), 267–295 (2009)
15. McKenna, E., Smyth, B.: Competence-guided case-base editing techniques. In: Blanzieri, E., Portinale, L. (eds.) EWCBR 2000. LNCS, vol. 1898, pp. 186–197. Springer, Heidelberg (2000). https://doi.org/10.1007/3-540-44527-7_17
16. Dasarathy, B.V.: Minimal consistent set (MCS) identification for optimal nearest neighbor decision systems design. IEEE Trans. Syst. Man Cybern. **24**(3), 511–517 (1994)
17. Wachter, S., Mittelstadt, B., Russell, C.: Counterfactual explanations without opening the black box: automated decisions and the GDPR. Harv. J. L. Tech. **31**, 841 (2018)
18. Mothilal, R.K., Sharma, A., Tan, C.: Explaining machine learning classifiers through diverse counterfactual explanations. In: FAT*20, pp. 607–617 (2020)
19. Schleich, M., Geng, Z., Zhang, Y., Suciu, D.: GeCo: quality counterfactual explanations in real time. arXiv preprint arXiv:2101.01292 (2021)
20. Smyth, B., Keane, M.T.: A few good counterfactuals. arXiv preprint:2101.09056 (2021)
21. Smyth, B., Keane, M.T.: Remembering to forget. In: Proceedings of the 14th international Joint Conference on Artificial intelligence (IJCAI-95), pp. 377–382 (1995)
22. Hasan, M.G.M.M.: Use case of counterfactual examples: data augmentation. In: Proceedings of Student Research and Creative Inquiry Day (2020)
23. Subbaswamy, A., Saria, S.: Counterfactual normalization: proactively addressing dataset shift using causal mechanisms. In: UAI-18, pp. 947–957 (2018)
24. Zeng, X., Li, Y., Zhai, Y., Zhang, Y.: Counterfactual generator. In: Proceedings of the Conference on Empirical Methods in Natural Language Processing, pp. 7270–7280 (2020)
25. Pitis, S., Creager, E., Garg, A.: Counterfactual data augmentation using locally factored dynamics. In: Advances in Neural Information Processing Systems (2020)
26. Förster, M., Klier, M., Kluge, K., Sigler, I.: Fostering human agency: a process for the design of user-centric XAI systems. In: ICIS-2020, paper 1963 (2020)
27. Temraz, M., Keane, M.T.: Solving the class imbalance problem using a counterfactual method for data augmentation. Under review (2021)

# A Case-Based Approach to Data-to-Text Generation

Ashish Upadhyay[1]([✉]), Stewart Massie[1], Ritwik Kumar Singh[2], Garima Gupta[2], and Muneendra Ojha[2]

[1] Robert Gordon University, Aberdeen, UK
{a.upadhyay,s.massie}@rgu.ac.uk
[2] International Institute of Information Technology, Naya Raipur, India
{ritwik17100,garima17100,muneendra.ojha}@iiitnr.edu.in

**Abstract.** Traditional Data-to-Text Generation (D2T) systems utilise carefully crafted domain specific rules and templates to generate high quality accurate texts. More recent approaches use neural systems to learn domain rules from the training data to produce very fluent and diverse texts. However, there is a trade-off with rule-based systems producing accurate text but that may lack variation, while learning-based systems produce more diverse texts but often with poorer accuracy. In this paper, we propose a Case-Based approach for D2T that mitigates the impact of this trade-off by dynamically selecting templates from the training corpora. In our approach we develop a novel case-alignment based, feature weighing method that is used to build an effective similarity measure. Extensive experimentation is performed on a sports domain dataset. Through Extractive Evaluation metrics, we demonstrate the benefit of the CBR system over a rule-based baseline and a neural benchmark.

**Keywords:** Data-to-Text · Textual CBR · Feature weighting

## 1 Introduction

Data-to-Text Generation (D2T) is a process that automatically generates textual summary of insights extracted from structured data [9,25]. With business processes often generating huge amount of domain-specific data, which is not easily understandable by humans, there is a growing need to synthesise this data by converting it into textual summaries that are more accessible. There are many real-world applications, from weather or financial reporting [10,14,27] to medical support or sports journalism [4,20,26,31]. D2T is expected to be one of 5 core technologies enabling an economic impact of $5 trillion annually by 2025.

D2T requires two separate problems to be addressed: **content selection**, deciding important content from the input data (implicit or explicit), as in *what to say?*; and **surface realisation**, conveying the selected content into textual summaries, as in *how to say?* Traditional methods use a modular approach

© Springer Nature Switzerland AG 2021
A. A. Sánchez-Ruiz and M. W. Floyd (Eds.): ICCBR 2021, LNAI 12877, pp. 232–247, 2021.
https://doi.org/10.1007/978-3-030-86957-1_16

to divide the generation task into several smaller modules. These modules are based on carefully crafted domain-specific rules and templates [25,27]. Recently, neural based learning approaches have shown promising results by integrating all modules into a single end-to-end architecture and learning domain-specific as well as generation rules from parallel corpora of data and summaries [21,31].

Neural systems demonstrate greater fluency and diversity in generated textual summaries but often *hallucinate* by producing inaccurate information that is not supported by the input data. One of the main reasons for hallucination is that the systems have to learn multiple domain specific rules to make sense of input data as well as learn how to verbalise that data [21]. Rule-based systems however are able to produce high quality texts in terms of accuracy but at the expense of diversity in the generated texts. Although, in real-world data-to-text applications, accuracy is usually much more important than fluency and diversity, thus making rule-based systems state-of-the-art in real-world applications.

We propose a Case-Based Reasoning (CBR) approach that learns content selection and realisation separately to generate accurate and diverse texts. Our model learns to choose important entities from the data and then verbalises them via templates extracted from the training corpus. We run experiments to evaluate our proposed method on a sports domain dataset, SportSett [28] and demonstrate it produces better quality texts than neural systems while also maintaining diversity. The contributions are as follows[1]:

1. introduction of a CBR D2T model with separated planning and realisation;
2. development of a novel feature weighting technique using case-alignment that can be used for weighting features in problems with complex solutions;
3. demonstrating with experiments the benefits of our CBR approach over neural and rule-based systems, as well as our case-alignment feature weighting method over information gain feature weighting.

## 2   Related Works

Natural Language Generation can be divided into two sub-fields based on the input to the systems. The task of generating text from unstructured linguistic data is referred as Text-to-Text generation; while generating text from structured non-linguistic data is known as D2T generation [9]. D2T has been studied for decades. One of the very first systems proposed in 1980s generated textual summaries of financial data [14]. Later other systems were developed to generate weather forecasts [10,27] and medical support documents [20,26]. These were modular systems developed using carefully engineered rules and templates with the help of domain experts dividing the whole task into several sub-tasks. Later in 2000s, statistical learning methods tried combine different modules into a single architecture [3,13]. Some methods also performed just content selection based on statistical methods while utilising rules for surface realisation [4,12]. Advancements in deep learning has boosted the interest in neural-D2T. Several

---

[1] The code can be found at https://github.com/ashishu007/data2text-cbr.

datasets [7,8,31] and systems [22,24,31] have been proposed to utilise neural networks for solving D2T task. Most datasets, are not realistic of real-world challenges as they only verbalise few entities and do not require content selection on the input data. Some datasets, however, such as RotoWire [31] and MLB [22] do reflect real-world challenges and require extensive content selection from input data to verbalise many entities in the textual summaries. Initially, neural systems employed different versions of sequence-to-sequence models to generate text utilising an end-to-end architecture [22,24,31]. These end-to-end systems, despite being able to generate very fluent texts, fared poorly on the accuracy, coherence and structuring of the information in summaries. Pipeline-based systems that separate content-planning from realisation attempt to address some of these issues [6,21].

D2T has also been studied in CBR with systems ranging from weather forecasts to obituary generation [1,2,30], although these systems have been limited to generating smaller texts describing very few entities. As with most CBR systems, similarity is a key component of CBR-D2T systems as effective similarity measures ensure relevant previously solved problems are reused. Feature weighting is one approach to developing an effective similarity measure [11,32], however there are challenges in comparing feature weighting schemes where the problem has a textual solution. Case-alignment measures [15,17] have been employed for feature weighting in classification and regression environments [11]. Our proposed feature-weighting method uses all features to measure the case-alignment of a case-base and utilises this information to assign weights to the features.

## 3   Background

The data in most D2T task is organised on three dimensions. There are multiple **entities** (for example, players or teams in the case of sports domains), which are described by multiple **features** (points or goals scored in a match), all of which belong to an **event** (a match played between two teams). To simplify, a dataset will have multiple events consisting of multiple entities described by multiple features. Datasets such as RotoWire [31], SportSett [28] and MLB [22] have similar properties. Systems trained on these datasets require extensive content selection on the input data and coherent realisation of longer texts which reflects the challenges of real-world applications.

An example summary from a sports domain dataset, SportSett is shown in Fig. 1a with a subset of its corresponding box-score stat in Fig. 1b. The summaries in such problems have different level of complexities. For example, S12 in Fig. 1a identifies that *Reggie Jackson* scored a 'triple-double' in the game which was the continuation of top performance from the last game. There are at least two types of information that are explicitly not available in the input data:

- first, scoring triple-double is identified if the player scores double-digits in three categories - PTS, REB, and AST. To convey this information, either developers need to explicitly design a rule that identifies if a player scored such a thing (in case of a rule-based system) or the system needs to learn

| | |
|---|---|
| S01 | The Philadelphia 76ers (16-52) defeated the Detroit Pistons (24-44) 94-83 on Wednesday in Philadelphia . |
| S02 | The 76ers were able to pull off the win despite Nerlens Noel leaving the game with a right foot contusion after playing just 22 minutes . |
| S03 | He had 11 points on 5-of-10 shooting , four rebounds and three blocks in that time and did not return . |
| S04 | It was Ish Smith who led the way for Philadelphia , as he was moved into the starting point guard role while Isaiah Canaan was moved to the bench . |
| S05 | Smith thrived in the role , recording 15 points on 6-of-12 shooting , eight assists and three steals in 26 minutes . |
| S06 | Canaan struggled coming off the bench , putting up just nine points on 2-of-10 shooting and four assists in 22 minutes . |
| S07 | Jason Richardson shot his way out of a slump , scoring 15 points on 4-of-7 shooting in 25 minutes , as it was just the first time in five games that Richardson scored in double figures . |
| S08 | Conversely , Robert Covington struggled , as he shot just 1-of-6 from the field on his way to three points in 22 minutes off the bench . |
| S09 | It was a quick fall back to reality for the Pistons , as just a day after upsetting the Grizzlies and ending a 10 - game losing streak , they lost to one of the NBA 's worst teams . |
| S10 | They were without Greg Monroe , who sat out his second consecutive game with a strained knee . |
| S11 | Despite the loss , Detroit did have some strong performances . |
| S12 | Reggie Jackson followed up Tuesday night 's big game with a triple-double , putting up 11 points on just 4-of-17 shooting , 11 rebounds and 10 assists in 32 minutes . |
| S13 | Kentavious Caldwell-Pope also followed up his 24-point performance on Tuesday in a strong way , scoring 20 points on 7-of-16 shooting and grabbing eight rebounds in 36 minutes . |
| S14 | Up next , the 76ers will take on the Knicks at home Friday , while the Pistons head home Saturday to take on the Bulls . |

(a)

| Starters | MP | FG | FGA | FG% | 3P | 3PA | 3P% | FT | FTA | FT% | ORB | DRB | TRB | AST | STL | BLK | TOV | PF | PTS | +/- |
|---|---|---|---|---|---|---|---|---|---|---|---|---|---|---|---|---|---|---|---|---|
| Kentavious Caldwell-Pope | 35:41 | 7 | 16 | .438 | 1 | 4 | .250 | 5 | 6 | .833 | 1 | 7 | 8 | 2 | 1 | 0 | 3 | 3 | 20 | -2 |
| Anthony Tolliver | 33:50 | 3 | 5 | .600 | 2 | 4 | .500 | 1 | 2 | .500 | 3 | 2 | 5 | 0 | 0 | 1 | 2 | 2 | 9 | +5 |
| Andre Drummond | 32:17 | 3 | 8 | .375 | 0 | 0 | | 3 | 3 | 1.000 | 7 | 7 | 14 | 0 | 1 | 2 | 1 | 4 | 9 | -3 |
| Reggie Jackson | 32:00 | 4 | 17 | .235 | 1 | 3 | .333 | 2 | 2 | 1.000 | 3 | 8 | 11 | 10 | 0 | 0 | 5 | 3 | 11 | -4 |
| Caron Butler | 17:51 | 3 | 8 | .375 | 1 | 6 | .167 | 0 | 0 | | 0 | 0 | 0 | 0 | 2 | 0 | 0 | 1 | 7 | +1 |

(b)

**Fig. 1.** (a) An example summary from SportSett dataset. (b) Subset of box-score from the same game.

the rules to infer if a player scored such a thing (in case of a learning-based system). The player scoring triple-double is not explicitly given in the data;
– second, the claim that it was the continuation of top performance from the last game by the same player. To convey this information, the system needs access to the data from previous games as well which is not available for the system during run-time.

There are numerous examples of such situations, where several rules are needed to infer the information either from data of same event or of previous events. There can be a large number of different combinations of data that can be mentioned in a summary. Even in other domains, such as finance or weather reporting, an ample portion of summaries discuss the information aggregated over many entities, features or events [14,27]. This is one of the many reasons for the hallucination of neural systems, as they do not just have to learn the rules of language generation but also have to learn the rules to infer contextual information, as well as learn content selection. On the other hand, for a rule-based system, it can be extremely difficult to write rules for discussing important insights from all the possible combinations from the input data.

There is clear trade-off in current state-of-the-art rule-based and neural-based D2T systems. Based on the requirement of the application, the developed system will either be accurate with monotonous, non-fluent texts, if using rule-based system; or capable of generating fancy and diverse texts but with some inaccurate information, if using neural system. Our proposed CBR system tries to reduce the impact of this trade-off between accuracy and fluency of texts by learning to reuse previous sentences based on the entities' feature values within an event. It will provide more accurate texts than neural systems without sacrificing diversity.

## 4    Methodology

The input to a D2T system is a set of records organised by entities and features. Based on these records, a output summary needs to be generated that describes the event to which given entities belong. We assume that a summary is the combination of multiple components organised on higher-level. Typically sports summaries (such as in SportSett [28] and MLB [22]) are organised in four components: **winning team**, a sentence about which team won, with scores; **teams' performance**, few sentences about teams' performance; **players' performance**, several sentences about different players performance from the game; and **next game**, next game fixtures of both teams;

Although, this assumption may be an over-generalisation of the problem, it does apply to a large portion of the SportSett dataset, where more than 90% of the summaries follow similar pattern (with some extra sentences in between). Also, similar components can be identified in other domains [12,27]. The texts in the second and third components (teams' and players' performance) appear to follow the principle of 'similar problems have similar solutions' and thus a CBR approach is used for their generation. The methods described later in the section are mostly centred around these two components. The methodology is briefly shown in the Fig. 2. The first and last components (winning team and next game) use manual rules to select an appropriate template from a bank of around 10 templates. The template bank is created by selecting a few standard sentences from the training set.

### 4.1    Case-Base Creation

First of all, we create separate case-bases for different components (for players' and teams' performance). The case-base creation is a semi-automated process where first textual summaries are broken into sentences and similar sentences are clustered into the same group. Clusters related to the different components are identified manually and then used to extract the templates [13].

**Semantic Clustering.** We first extract all the sentences from the training set summaries and then abstract them based on their named-entities and pos-tags using the method described in [29]. The abstracting process uses open-source NLP libraries spaCy and neuralcoref combined with some domain-specific rules. Through this process, a sentence from the dataset '*The Atlanta Hawks (41-9) beat the Washington Wizards (31-19) 105-96 on Wednesday.*', is transformed to '*PROPN-ORG (X-Y) beat the PROPN-ORG (X-Y) X-Y on NOUN- DATE.*'.

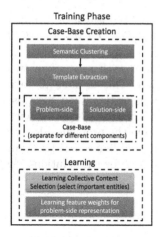
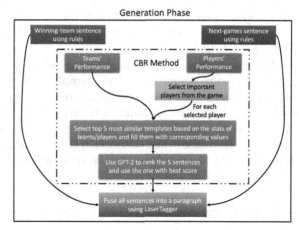

**Fig. 2.** Our methodology for CBR-D2T

These abstract sentences are then embedded into a 786 dimensional vector using DistilRoBERTa Language Model[2]. We plot the embedded sentences on a 3-dimensional space using the UMAP algorithm [18] and count around 50 clusters based on the plot's view. Then a K-Means clustering algorithm is used to cluster the embedded sentences into 50 similar groups. A manual process is then used to combine the similar clusters and assign them a label identifying the concept cluster items represent. This way we reduce from 50 clusters to 31 clusters, out of which four represent sentences from the teams' component and ten from the players' component (rest 17 contain sentences with more complex facts difficult to classify into just team or player component). Although, the number of clusters could be increased to a higher number to accommodate the possible diversity in the dataset, we leave that for future exploration as manually annotating large number of clusters is a time consuming task.

**Template Extraction.** Template extraction for both the components is done separately but follows the same method. For each sentence in the cluster, the entity mentions are extracted. If a sentence only contains one entity mention then the entity's performance stats is taken from the corresponding game. Based on the stats, an entity matching is performed to replace any occurrence of a entity's feature value to its feature name. For example, from the sentence: *"Henry Sims was able to notch a double - double, contributing 11 points (4 - 12 FG, 3 - 4 FT) and 12 rebounds."* where Henry Sims' performance stats are: $\{STARTER : no, PTS : 11, FGM : 4, FGA : 12, FG\_PCT : 33, FG3M : 0, FG3A : 0, FG3\_PCT : 0, FTM : 3, FTA : 4, FT\_PCT : 75, OREB : 5, DREB : 7, REB : 12, AST : 2, TO : 0, STL : 0, BLK : 0, PF : 2, MIN : 32, IS\_HOME : no, FIRST\_NAME : Nerlens, SECOND\_NAME : Noel\}$,

---

[2] https://github.com/UKPLab/sentence-transformers.

the template extracted is: *"FIRST_NAME SECOND_NAME was able to notch a double - double, contributing PTS points (FGM - FGA FG , FTM - FTA FT) and REB rebounds"*.

For teams' component templates we select sentences with more than two entity mentions (team names) for template extraction. This is done to take sentences that compare the performance of both teams, rather than just discussing one team's performance. Finally, we have two separate case-bases with their respective problem and solution representations. On the problem-side, the entities' performance (box-scores for player component and line-scores for team components) is used, while the extracted template is used for the solution-side representation.

### 4.2   Retrieval and Feature Weighting

After the case-base is created, the retrieval of similar cases for new problems is done by measuring euclidean distance. We also learn the feature weights for better similarity which is necessary because not all features have equal importance.

**Content Selection.** Central to a D2T task is selection of important contents from the input data. Most of the entities in the input data are not mentioned in the output summary. Even for the entities mentioned in the summary, not all of their features are mentioned. In the SportSett dataset, each game features around 25 players from both teams, but game summaries only discuss 5 to 6 players. Thus, similar to [4] we train a classifier to select important entities from the input data based on their feature values. We use this classifier to select important players from the game. In most cases, importance of an entity is not independent and is related to the feature values of other entities as well. Thus, to represent an entity, we concatenate it's feature values, with the feature values of other entities in the data. So, for an event with $e$ entities with $f$ features, an entity $E_1$ is represented as: $\{(E_{11}, E_{12}, \cdots, E_{1f}), (E_{21}, E_{22}, \cdots, E_{2f}), \cdots, (E_{e1}, E_{e2}, \cdots, E_{ef})\}$, where $E_{11}$ is $E_1$'s first feature value and $E_{ef}$ is $E_e$'s $f^{th}$ feature value.

An entity is given class 1 (important) if it was mentioned in the summary of that game, or class 0 (not important) if it wasn't mentioned. A classifier is then trained to learn if the entity should be selected for discussing in the summary or not. We train a logistic regression classifier which achieves 87% accuracy and 85% f1 score on the validation set of the SporSett dataset. Finally, the content selection is extended using templates, where after selecting important entities from an event by the classifier, important features are selected using the template of the most similar problem. For example, if a player has scored 'double-double' in the game, it is identified by a feature in the player's stats and a similar template such as *'FIRST_NAME SECOND_NAME lead the way for the PLAYER-TEAM-NAME, recording PTS points on FGM-of-FGA shooting, REB rebounds and AST assists in MIN minutes.'* is extracted which discusses some features in the sentence. This is often synonymous to how summaries are written (at-least in sports domains), as we first tend to select the important player from the game and then decide what features to discuss.

**Algorithm 1. Calculate the loss value for each candidate generated in PSO algorithm**

**Input:**  PSO candidate $\mathbf{W}$, Problem-side $\mathbf{P}_{CB}$ and Solution-side $\mathbf{S}_{CB}$ of case-base
**Output:**  The loss value for $\mathbf{W}$
1: $\mathbf{WP}_{CB} = \mathbf{P}_{CB} * \mathbf{W}$
2: $AlignScore_{CB} = 0$
3: **for each** $\mathbf{idx} \in range(| \mathbf{WP}_{CB} |)$ **do**
4:     $\mathbf{P}_T, \mathbf{S}_T = \mathbf{WP}_{CB}[idx], \mathbf{S}_{CB}[idx]$
5:     $\hat{\mathbf{P}}_{CB}, \hat{\mathbf{S}}_{CB} = \mathbf{WP}_{CB}[\sim idx], \mathbf{S}_{CB}[\sim idx]$
6:     Generate problem-side ranked list $\mathbf{PL}$ using $\mathbf{P}_T$ and $\hat{\mathbf{P}}_{CB}$
7:     Generate solution-side ranked list $\mathbf{SL}$ using $\mathbf{S}_T$ and $\hat{\mathbf{S}}_{CB}$
8:     $AlignScore_{idx} = nDCG(\mathbf{PL}, \mathbf{SL})$
9:     $AlignScore_{CB}+ = AlignScore_{idx}$
10: **end for**
11: $\mathcal{L}_W = 1 - (AlignScore_{CB}/ | \mathbf{WP}_{CB} |)$
12: **return** $\mathcal{L}_W$

**Feature Weighting.** For the players component data, a classifier is already trained to identify the important players from a game. This classification setting can also be used to learn the feature-importance of players' component (information gain feature weighting). It is noted that this method cannot be applied in a non-classification setting, such as in teams' component. Thus, a novel case-alignment based feature weighting method is proposed for non-classification settings. CBR systems are based on the principle of 'similar problems have similar solutions' and case-alignment can provide a measure of the extent to which this principle holds true for a specific design e.g. feature weighting scheme.

We use the method proposed in [30] for measuring the case-alignment of the case-base. The alignment score is then used as a loss function for a Particle-Swarm Optimiser whose parameters are the features' weights for a case-base. The loss function is formally defined in Algorithm 1. The ranked list on the problem side is generated using the euclidean distance between the problem-side of the target problem and the cases in the case-base, while the solution-side ranked list is generated using the cosine distance between the solution-side of target problem and cases in the case-base.

### 4.3   Generation

Generation again is done separately for different components. For a target problem, $k$-nearest neighbours are retrieved from the case-base using problem-side representation. $k$ solutions are generated by filling the tags with their corresponding values in the nearest neighbour solutions. A GPT-2 [23] language model is used to rank the five sentences based on perplexity score. The best among the five is chosen as the final solution for the given target problem. Since sentence ranking is a domain specific task, the GPT2 model is fine-tuned on the training set of the same dataset used in our experiments. Now several sentences are generated for different components and are fused into a paragraph using sentence fusing algorithm LaserTagger [16] trained on the same training data used in our experiments.

In the case of SportSett data, for the players' component: first, important players are selected using the classifier mentioned in an earlier section; then, for each important player selected, a sentence is generated using the process described above. Similarly a sentence is generated for the teams' component. A set of rules is used to generate the first component sentence describing which team won, and another set of rules to generate a sentence for both teams' next fixtures. Finally, all these sentences are fused into a paragraph using LaserTagger.

## 5    Experimental Setup

### 5.1    Dataset

The SportSett dataset [28] is used to evaluate our proposed and benchmark algorithm. It contains textual summaries combined with the box- and line-scores of NBA matches. The training set contains the matches from 2014, 2015 and 2016 seasons (4745 instances) while the dev and test sets contain matches from 2017 and 2018 seasons (1228 and 1229 instances) respectively.

We use the train set of SportSett for the creation of our case-bases. For training the important players classifier, we used the train set of SportSett with the dev set for testing. For fine-tuning the GPT2 and LaserTagger, we also used just the texts from the train set of SportSett. Similarly, only train set data is used for creating the case-bases for teams' and players' components. With all seasons used from train set, the teams' component case-base consists of 1200 cases, while players' component has 14985. With just 2014 season used for training, 360 and 4405 cases are available in teams' and players' case-bases

### 5.2    Baseline and Benchmark

We compare our system with a rule based baseline and a neural benchmark:

- **Rule-Based System** is the templatized generator used as baseline in [31]. The system has a standard template for winning team, another template for players stats which is filled with six highest scoring players' stats, and finally last template for teams next-game fixture. We extend the rule-based system to include day name and arena of the game, as well as next-opponent team names since this new information is now available in the SportSett data.
- **Neural System** is a sequence-to-sequence model proposed in [22]. It consists of an MLP encoder and LSTM decoder with copy mechanism. There's an added module to update the input record's representation during generation process. At each decoding step, a GRU is used to decide the record that needs to be updated and updates it's value.

Although, there are other neural systems with comparable performance to our selected benchmark, we use the selected one because of its reduced training time and ease in reusing the code. For example, authors in [24] proposed a hierarchical Transformer encoder model with standard LSTM decoder which achieves slightly better performance than our selected benchmark. But it takes 10 days to train with the hyper-parameters mentioned in the paper on a 16 GB Nvidia-P100 GPU compared to our selected benchmark which takes just 1 day.

### 5.3   Evaluation Methods

We use the family of Extractive Evaluation (EE) metrics for [31] for evaluating the models. These metrics are trained to extract entity names and numerical values from the text and predict the relation (feature name) between them. For example, from S05 in Fig. 1a, the IE models (ensemble of three LSTMs and three CNNs) can extract entity name *Ish Smith* and numerical value *15*. Then the model can predict its relation name as *PTS* (points) and will return a tuple as $t = (EntityName|FeatureValue|FeatureName)$ as $(IshSmith|15|PTS)$. The models extract several tuples from both human-written gold $(y)$ and system generated $(\hat{y})$ summaries. These tuples are then compared to calculate the following metric scores:

- **Relation Generation (RG)** is the precision of unique tuples $t$ extracted from generated summary $\hat{y}$ that also appeared in the input data. This metric can be used to measure the system's capability of generating factually correct texts supported by input data, i.e., accuracy of the system.
- **Content Selection (CS)** is the precision and recall between unique tuples $t$ extracted from gold summary $y$ and generated summary $\hat{y}$. Here, the systems ability of selecting content is measured in comparison with the human written summaries.
- **Content Ordering (CO)** is measured as the normalized Damerau Levenshtein Distance [5] between the sequences of tuples extracted from generated summary $\hat{y}$ and gold summary $y$. This demonstrates the systems ability of ordering the content in generated summary.

CS primarily targets the challenge of *what to say?*, while CO targets the *how to say it?* aspect. Apart from these metrics, we also use BLEU [19] score to compare the generations. BLEU score compares the n-gram overlap between the gold summary and generated summary and primarily rewards fluent texts rather than generations capturing more information from the input [31]. From all the metrics discussed here, RG can be used to measure the factual correctness of generations. While we acknowledge the fact that human judgement is the best evaluation practice for text generation systems, these autonomous metrics are a widely used proxy methods as crude surrogate for human judgement.

Initial versions of the Extractive Evaluation metrics [21, 31] only evaluated the numerical claims (such as points, rebounds, steals made by a player or team) mentioned in the text summaries. Authors in [29], proposed an extended version capable of evaluating day names, dates and game arenas as well. We further extend these metrics to evaluate the few more claims made in texts, for example, if a player was the leading scorer or if a player scored a double-double.

**Table 1.** Comparison of our CBR system with baseline and benchmark

| System | RG | CS-Precision | CS-Recall | CO | BLEU |
|---|---|---|---|---|---|
| Gold | 89.46 | – | – | – | – |
| Rule-Based | 95.35 | 55.20 | 23.65 | 10.61 | 6.50 |
| **Neural**$_{oneS}$ | 61.34 | 34.84 | 25.68 | 9.84 | 9.93 |
| **Neural**$_{allS}$ | 71.07 | 45.66 | 40.81 | 19.56 | 17.68 |
| **CBR**$_{oneS}$ | 73.22 | 50.18 | 24.78 | 10.91 | 10.40 |
| **CBR**$_{allS}$ | 77.22 | 46.46 | 33.53 | 11.92 | 11.93 |

## 6    Results and Discussion

### 6.1    Comparison with Benchmark and Baseline

We train our CBR system and neural system on two different sizes of training data. First, we use only 2014 season data for training the models, denoted by **(Neural/CBR)**$_{oneS}$. Then, we use all season data from train set for training which is denoted by **(Neural/CBR)**$_{allS}$. We compare these four systems and a rule-based baseline discussed in the previous section. The results are shown in Table 1. Results for the neural system are given as the average over 10 training runs with different random seeds.

First of all, we note that our **CBR**$_{oneS}$ outperforms **Neural**$_{oneS}$ on all metrics, except CS-recall. This demonstrates that our CBR system is much better at producing quality texts compared to neural system, even with fewer training samples. However, with the increase in training data, there is a huge gain in **Neural**$_{allS}$ across all metrics while **CBR**$_{allS}$ system improves very little. Still with any amount of data, our CBR system achieves better performance on RG (73% & 77% for CBR vs 61% & 71% for neural) and CS-precision (50% & 46% for CBR vs 34% & 45% for neural) metrics as compared to the neural system, indicating CBR system's benefit over neural on accuracy. We also note that rule-based system achieves best score on RG (95% ) and on CS-precision (55%). This is not surprising as the system is hard-coded with domain knowledge thus has very low accuracy errors[3]. But system fares poorly on mimicking the gold summaries, as it only recalls 23% of the contents from gold summaries. It also performs poor on the BLEU score which is conveys that rule-based system is not very fluent.

On the terms of fluency, we observe that BLEU score achieved by both CBR systems is better than the rule-based baseline. As compared to neural system, **CBR**$_{oneS}$ is slighlty better than **Neural**$_{oneS}$, while for **Neural**$_{allS}$, BLEU score is quite higher than **CBR**$_{allS}$. This reflects that our system is much fluent compared to rule-based system, as well as above-par to neural system when training data is scarce. To measure the diversity, we first calculate the vocabulary of texts

---

[3] Please note that the scores are not 100% because the metrics are based on trained models, which themselves achieve around 90% accuracy and f1 score while training.

**Table 2.** Ablation study results

|                        | RG    | CS-Precision | CS-Recall | CO    | BLEU  |
|------------------------|-------|--------------|-----------|-------|-------|
| Info Gain FtrW         | 75.22 | 49.53        | 28.73     | 11.78 | 10.84 |
| without GPT-2 scoring  | 76.09 | 44.64        | 34.88     | 12.33 | 9.26  |
| without LaserTagger    | 76.36 | 44.50        | 33.04     | 11.30 | 11.04 |
| **CBR**$_{allS}$       | 77.22 | 46.46        | 33.53     | 11.92 | 11.93 |

generated from different systems. We identify that the gold summaries from test set have a vocabulary of more than 5000 words, while both the CBR and neural systems have the vocabulary of 2000 words but rule-based system has vocabulary of only 900 unique words. In terms of content selected for output summary, the rule-based system is only able to discuss one-third of the unique record types discussed in gold summary, while our CBR system is able to discuss all of them. All this evidence suggests that the CBR system is able to decrease the trade-off between accuracy and diversity, especially in case of scarcity of training data.

## 6.2 Ablation Studies

We further perform three ablation studies on our **CBR**$_{allS}$ system. In the first study, we analyse the effect of our proposed case-alignment based feature weighting against the information gain based feature weighting (see Sect. 4.2). The results of this ablation are shown in the first row of Table 2. Here information gain from the important player classifier data is used to weight the features in players' component, however, no weighting is applied for teams' component. From the results, we can see that apart from CS-precision, there's at-least some drop in all metrics while a sizeable drop in CS-recall. The drop in CS-recall can mean that the system with case-alignment feature weighing is able to select templates that have contents closer to the human written summaries.

In the second study, we compare the effect of selecting the 'nearest neighbour' against 'best out of top-k nearest neighbour' for generating the new solution, of which results are shown in second row of the Table 2. Here, the **CBR**$_{allS}$ system is used without GPT2 solution ranking module. Again, we can see there's not much difference in EE metrics but there's a sizeable difference in BLEU score. This is expected as GPT2 scores the sentences based on perplexity that rewards fluency. So with the addition of an extra scoring component for solution reuse, we can improve the fluency of our generated text summaries.

Third study analyses the effects of applying LaserTagger for sentence fusion. Results are shown in the third row of the Table 2. We can see that there's slight drop in most metrics when LaserTagger is not used for sentence fusion. This is because that the texts generated from CBR systems have some incohereny such as: '*Bradley Beal led the way for* **the Wizards (32-48)** *with 25 points, complementing* **the Wizards (32-48)** *with five assists and two rebounds.*'. This incoherence is the result of using co-reference resolution while template extraction. With LaserTagger applied to the above sentence, it is modified into '*Bradley Beal led the way for* **the Wizards** *with 25 points, complementing* **them** *with five assists and two rebounds.*'. This is similar to the Referring Expression Generation phase of traditional D2T methods [25].

## 6.3 Qualitative Analysis

**Table 3.** A summary generated from our **CBR**$_{alls}$ system

The visiting Atlanta Hawks ( 13 - 29 ) defeated the host Philadelphia 76ers ( 27 - 16 ) 123 - 121 at Wells_Fargo_Center on Friday . The 76ers shot 52 percent from the field , including 33 percent from long range but were not able to hang on for the full 48 , as The Hawks surged back to get their revenge . Ben Simmons had a triple - double with 23 points ( 7 - 13 FG , 1 - 2 3Pt ) , 10 rebounds , 15 assists , three steals and one block in 43 minutes . Meanwhile , Jimmy Butler was the high - point man for Philadelphia , with 30 points on 9 - of - 19 shooting , in 40 minutes . JJ Redick was next in line with 20 points , three rebounds , an assist and a steal , as the only other 76ers player who managed double - digit points . Rookie DeAndre' Bembry shot 6 - for - 11 from the field to score 14 points , while also chipping in five rebounds . Kevin Huerter was the high - point man for the Hawks as he tied a season - high with 29 points on 11 - of - 17 shooting , including 5 - of - 8 from long range . John Collins was the 3 prong of the Hawks attack , as he finished with 25 points ( 10 - 17 FG , 1 - 1 3PT , 1 - 3 FT ) , along with five rebounds , two assists , two steals and one block , in 27 minutes . The 76ers now head to New_York for a Sunday night showdown versus the Knicks while the Hawks will return home to face the Bucks on Sunday .

A summary of the NBA match between 76ers and Hawks on 11th Jan, 2019 generated from the proposed system is shown in Table 3. The first and last sentences are generated from a set carefully crafted rules, while other sentences are generated through the CBR methodology. The accurate facts are shown in green while the inaccurate ones in red. The red inaccuracies are due to the imperfection in template extraction process, where sometimes an entity is replaced with wrong feature name because of more than one features having the same value. By addressing such cases in template extraction can improve the accuracy of the system in terms on numerical facts being conveyed. Another type of inaccuracy is shown in grey sentence, which wrongly mentions JJ Redrick being the only player with double-digit points after Jimmy Butler. These facts are much more complex to calculate and are grounded in the text, as this information is calculated across multiple entities (you need to know other players' scores as well to decide only two scored in double-digits). To address such errors, new features are needed to explicitly identify the information from across entities and/or events. Those new features will help the system in deciding the similarity of a template with across-entity information in such cases. One interesting observation is that no discourse is more than a sentence long, that's because the template extraction is done on a sentence level. Extraction on sub-paragraph level will make discourse longer and summaries much more human-like.

# 7 Conclusion and Future Work

In this work, we proposed a CBR system for Data-to-Text generation that aims to provide a good balance in the trade-off between accuracy and diversity of the text summaries produced by a D2T system. Our CBR system follows a modular approach to text generation in which content selection is performed first before surface realisation takes place via templates extracted from the training corpora. Experimentation results on a sports domain data-set show that our CBR system achieves better accuracy than a neural benchmark while better fluency and diversity than a rule-based baseline. We also introduce a novel case-alignment based feature weighting algorithm that is particularly effective in non-classification settings, such as text solutions. The benefit of our feature weighing algorithm over the information gain feature weighing baseline is also demonstrated.

In future, we would like to improve the template extraction process. In this paper, the extraction take place at a sentence level (micro-plan) which limits performance in relation to inter-sentence coherence in the generated texts. With an extraction process on sub-paragraph level (macro-plan) better coherence and discourse could be achieved for describing multiple entities in the summary. We would also like to investigate the possibility of having dynamic higher-level component organisation to increase the diversity in content structuring of the summaries. Finally, we plan to conduct more extensive evaluation of the system with alternative data-sets and using human judgement for evaluations.

# References

1. Adeyanju, I.: Generating weather forecast texts with case based reasoning. arXiv preprint arXiv:1509.01023 (2015)
2. Adeyanju, I., Wiratunga, N., Lothian, R.: Learning to author text with textual CBR. In: ECAI (2010)
3. Angeli, G., Liang, P., Klein, D.: A simple domain-independent probabilistic approach to generation. In: Proceedings of EMNLP, pp. 502–512 (2010)
4. Barzilay, R., Lapata, M.: Collective content selection for concept-to-text generation. In: Proceedings of HLT-EMNLP, pp. 331–338 (2005)
5. Brill, E., Moore, R.C.: An improved error model for noisy channel spelling correction. In: Proceedings of ACL, pp. 286–293 (2000)
6. Castro Ferreira, T., van der Lee, C., van Miltenburg, E., Krahmer, E.: Neural data-to-text generation: a comparison between pipeline and end-to-end architectures. In: Proceedings of EMNLP-IJCNLP, pp. 552–562 (2019)
7. Colin, E., Gardent, C., M'rabet, Y., Narayan, S., Perez-Beltrachini, L.: The WebNLG challenge: generating text from DBPedia data. In: Proceedings of INLG, pp. 163–167 (2016)
8. Dušek, O., Novikova, J., Rieser, V.: Evaluating the state-of-the-art of end-to-end natural language generation: the E2E NLG challenge, pp. 123–156 (2020)

9. Gatt, A., Krahmer, E.: Survey of the state of the art in natural language generation: core tasks, applications and evaluation. JAIR **61**, 65–170 (2018)
10. Goldberg, E., Driedger, N., Kittredge, R.: Using natural-language processing to produce weather forecasts. IEEE Expert **9**(2), 45–53 (1994)
11. Kar, D., Chakraborti, S., Ravindran, B.: Feature weighting and confidence based prediction for case based reasoning systems. In: Agudo, B.D., Watson, I. (eds.) ICCBR 2012. LNCS (LNAI), vol. 7466, pp. 211–225. Springer, Heidelberg (2012). https://doi.org/10.1007/978-3-642-32986-9_17
12. Kelly, C., Copestake, A., Karamanis, N.: Investigating content selection for language generation using machine learning. In: Proceedings of ENLG, pp. 130–137 (2009)
13. Kondadadi, R., Howald, B., Schilder, F.: A statistical NLG framework for aggregated planning and realization. In: Proceedings of ACL, pp. 1406–1415 (2013)
14. Kukich, K.: Design of a knowledge-based report generator. In: ACL, pp. 145–150 (1983)
15. Lamontagne, L.: Textual CBR authoring using case cohesion. In: Proceedings of TCBR Workshop at the 8th ECCBR, pp. 33–43 (2006)
16. Malmi, E., Krause, S., Rothe, S., Mirylenka, D., Severyn, A.: Encode, tag, realize: high-precision text editing. In: Proc of EMNLP-IJCNLP, pp. 5054–5065 (2019)
17. Massie, S., Wiratunga, N., Craw, S., Donati, A., Vicari, E.: From anomaly reports to cases. In: Weber, R.O., Richter, M.M. (eds.) ICCBR 2007. LNCS (LNAI), vol. 4626, pp. 359–373. Springer, Heidelberg (2007). https://doi.org/10.1007/978-3-540-74141-1_25
18. McInnes, L., Healy, J., Melville, J.: UMAP: uniform manifold approximation and projection for dimension reduction. arXiv preprint arXiv:1802.03426 (2018)
19. Papineni, K., Roukos, S., Ward, T., Zhu, W.J.: Bleu: a method for automatic evaluation of machine translation. In: Proceedings of ACL, pp. 311–318 (2002)
20. Portet, F.: Automatic generation of textual summaries from neonatal intensive care data. Artif. Intell. **173**(7–8), 789–816 (2009)
21. Puduppully, R., Dong, L., Lapata, M.: Data-to-text generation with content selection and planning. In: Proceedings of the AAAI Conference on Artificial Intelligence, vol. 33, pp. 6908–6915 (2019)
22. Puduppully, R., Dong, L., Lapata, M.: Data-to-text generation with entity modeling. In: Proceedings of ACL, pp. 2023–2035 (2019)
23. Radford, A., Wu, J., Child, R., Luan, D., Amodei, D., Sutskever, I.: Language models are unsupervised multitask learners. OpenAI Blog **1**(8), 9 (2019)
24. Rebuffel, C., Soulier, L., Scoutheeten, G., Gallinari, P.: A hierarchical model for data-to-text generation. In: Jose, J.M., et al. (eds.) ECIR 2020. LNCS, vol. 12035, pp. 65–80. Springer, Cham (2020). https://doi.org/10.1007/978-3-030-45439-5_5
25. Reiter, E., Dale, R.: Building Natural Language Generation Systems (2000)
26. Reiter, E., Robertson, R., Lennox, A.S., Osman, L.: Using a randomised controlled clinical trial to evaluate an NLG system. In: Proceedings of ACL, pp. 442–449 (2001)
27. Sripada, S., Reiter, E., Davy, I.: SumTime-Mousam: configurable marine weather forecast generator. Expert Update **6**(3), 4–10 (2003)
28. Thomson, C., Reiter, E., Sripada, S.: SportSett: basketball-a robust and maintainable dataset for natural language generation. In: IntelLanG (2020)
29. Thomson, C., Zhao, Z., Sripada, S.: Studying the impact of filling information gaps on the output quality of neural data-to-text. In: Proceedings of INLG, pp. 35–40 (2020)

30. Upadhyay, A., Massie, S., Clogher, S.: Case-based approach to automated natural language generation for obituaries. In: Watson, I., Weber, R. (eds.) ICCBR 2020. LNCS (LNAI), vol. 12311, pp. 279–294. Springer, Cham (2020). https://doi.org/10.1007/978-3-030-58342-2_18
31. Wiseman, S., Shieber, S.M., Rush, A.M.: Challenges in data-to-document generation. In: Proceedings of EMNLP, pp. 2253–2263 (2017)
32. Zhu, G.N., Hu, J., Qi, J., Ma, J., Peng, Y.H.: An integrated feature selection and cluster analysis techniques for case-based reasoning. Eng. Appl. Artif. Intell. **39**, 14–22 (2015)

# On Combining Knowledge-Engineered and Network-Extracted Features for Retrieval

Zachary Wilkerson$^{(\boxtimes)}$, David Leake, and David J. Crandall

Luddy School of Informatics, Computing, and Engineering, Indiana University, Bloomington, IN 47408, USA
{zachwilk,leake,djcran}@indiana.edu

**Abstract.** The quality of case retrieval in case-based reasoning (CBR) systems depends on assigning appropriate case indices. Defining feature vocabularies for indexing is an important knowledge acquisition problem for CBR, often addressed by hand. The manual process may result in high-quality vocabularies, but at considerable effort and expense, and it may be difficult for non-symbolic input such as images. Recently, the ability of deep learning (DL) to identify important features has made it appealing for learning to assign case features. However, such methods may miss features apparent to knowledge engineers. This paper presents a case study on methods for combining benefits of both engineered and DL-generated features. It considers case-based classification of cases described by both symbolic features and images. It evaluates the power of both types of features individually, examines how quality of engineered feature information affects their combined benefit, and tests network methods to generate weights for their combination. Experimental results show that in the test domain under suitable circumstances, the combined approach can outperform either method individually.

**Keywords:** Case-based reasoning · Deep learning · Indexing · Hybrid systems · Knowledge containers · Integrated systems

## 1 Introduction

The performance of CBR systems depends critically on retrieving the right cases. This depends on the indices used to organize and retrieve cases (e.g., [6,11,14,15,22]), which in turn depend on the vocabulary of features from which indices can be constructed. The feature vocabulary may be generated through a knowledge engineering process, sometimes reflecting deep analysis of a domain (e.g., [6,14,22]). However, relying on manual feature acquisition can be problematic. First, especially in instances where the domain is poorly understood, or in non-symbolic domains (e.g., classifying images), it may be hard to identify the right set of features for a feature vocabulary. Second, developing feature vocabularies may be highly expensive—and the expense may need to be repeated as

© Springer Nature Switzerland AG 2021
A. A. Sánchez-Ruiz and M. W. Floyd (Eds.): ICCBR 2021, LNAI 12877, pp. 248–262, 2021.
https://doi.org/10.1007/978-3-030-86957-1_17

vocabularies lose their appropriateness over time due to concept drift. Third, the situation assessment process required to characterize input problems in terms of the vocabulary may be difficult, resulting in partial, erroneous, or noisy case descriptions.

Some of the previous problems can be alleviated by applying machine learning (ML) to feature selection and similarity assessment. For example, learning techniques may be used to identify features to consider [7] or assign feature weightings [4]. Recently, substantial effort has focused on the potential of DL approaches to generate features and feature weightings. For example, convolutional neural networks (CNNs) have been used to extract feature information from images [24] and tri-axis sensors [21]. In that work, rather than relying on human-engineered features and situation assessment, the CBR system imports feature information from a network and uses it as the sole feature source during retrieval. Such methods facilitate feature generation and enable features to be tuned as data changes. However, they are not guaranteed to capture the deep relationships that may be contained in expert-generated features. Thus each approach has benefits and drawbacks.

In domains where a set of knowledge-engineered (KE) features exists, it is natural to consider combining human-engineered and network-learned (NL) features extracted using ML techniques. This paper presents a new method for extracting NL features and a case study on combining symbolic KE features with features extracted from CNNs for a classification task. It addresses how the benefit of combining such features varies with symbolic feature quality. As the effectiveness of retrieval depends strongly on feature weightings (e.g., [1]), it also studies how feature weight learning can be applied when merging the two sets of features, and its benefit. Results show that in the test domain, which combines symbolic and image information, the combined approach can outperform either method individually. This performance increase can be augmented with certain weight-learning strategies, though results also suggest that the benefit may be primarily for low-dimensional spaces, so new strategies may be necessary to accommodate feature-dense spaces created by NL techniques.

## 2   Convolutional Neural Networks for Classification

As a reference for the architecture described later in this paper, we begin with a brief description of convolutional neural networks for image classification. A CNN for image classification begins with alternating convolution and pooling layers that identify common shapes, contrasts, etc. present in similar regions of images with the same class during training; these extracted features then are "flattened" into a single layer and passed through a dense multilayer perceptron (MLP) section connected into the final output layer. A graphic representation of this process is shown in Fig. 1. CNNs may be applied broadly to multi-dimensional data (e.g., image data [24] or sensor data that tracks movement in three dimensions [21]), where their architecture enables processing and condensing of complex data into features based on data relationships. Such features then may be extracted from the CNN's internal structure and applied to the feature set in a case-based reasoner.

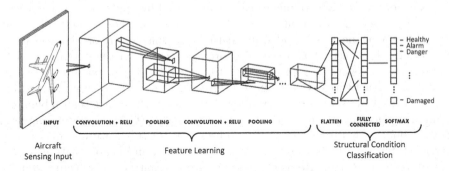

**Fig. 1.** Procedural diagram for a CNN in an aircraft sensor domain. Figure by Iuliana Tabian, Hailing Fu, and Zahra Sharif Khodaei is licensed under CC BY 4.0 [23].

## 3  Related Work

There has been much CBR research on feature learning using symbolic learning methods. Recently, there has been much interest in combining CBR and DL (e.g., [9,16,18,20]) Much of this work focuses on CBR-DL hybrids in which DL components provide capabilities such as feature extraction to a CBR system.

**Feature Learning.** A range of symbolic methods have been used to refine features/indices for CBR, often using knowledge-rich techniques. One example of feature learning strategies involves hybridizing with model-based learning to inform feature selection [3]. Bhatta and Goel apply a model-based system to select indices based on features simulated in the model. Barletta and Mark [2] propose explanation-based indexing. Cox and Ram [5] and Fox and Leake [7] apply introspective reasoning to refine features as expectation failures are encountered. Such methods rely on rich knowledge but can do powerful feature learning.

More recent research focuses on applying neural networks to directly infer similarity information from raw input data. Such methods do not require domain knowledge within the system (however, the dependence of network architecture on input structure makes many such methods domain-specific). Sani et al. present a system for human activity recognition that extracts features from a sensor and then uses a CNN to interpret the input data, which is represented in three dimensions [21]. The generated features are then compared against known wave form cases to infer the type, duration, etc. of activity that generated the sensory input data. Other approaches go a level of abstraction higher and look at the similarity functions themselves. Grace et al. [8] propose a hybrid system for creating plausible, yet unexpected, recipe designs. Their system applies DL techniques to infer relationships between cases in a case base; this provides additional knowledge that can be patterned to expectations when attempting to address the parameters of a presented goal. Mathisen et al. [17] use neural networks to learn similarity measures and also analyze different types of similarity metrics in depth. Outside of CBR, Kraska et al. [12] explore feature learning using linear models and neural networks to aggregate and discriminate between features, showing performance benefits over traditional structures like B-trees, hashmaps, and bloom filters; they also propose combinations of ML techniques or multi-dimensional indices as potential means to greater efficiency.

Inductive feature learning is especially applicable in domains such as image recognition, for which CNNs have been used to extract feature data from complex inputs to inform case-based reasoning systems. Turner et al. [24] apply this to novel object recognition. A CNN architecture classifies inputs that correlate with known classes with high confidence; when encountering "new" inputs with a correspondingly lower confidence, the image features are extracted from the CNN to be used in similarity calculations to group the new input with other similar images. As a result, the combined system can be sensitive to images that do not have known classification labels by loosely classifying them in terms of one another. Turner et al. extract features for their CBR system from between the convolution/pooling and dense layers of the CNN; we take a different approach by extracting features just before the output layer (details in Sect. 4).

**Learning Weights.** Many strategies exist to dynamically generate feature weights for case-based classification. Wettschereck et al. [25] present a survey of methods including hill-climbers, which modify feature weights according to a gradient to gradually maximize classification accuracy; genetic algorithms, which evaluate weights based on fitness as measured relative to the similarity calculation; and conditional probability models, which define weights based on the probability that a given class has the feature in question, among other wrapper and filter models. However, this is also a ripe domain for neural networks, which provide numerous opportunities to analyze relative importance of input features. In particular, Kenny and Keane [10] analyze multiple methods involving generating weights by taking advantage of neural network properties. One method generates weights by perturbing input elements individually and tracking the corresponding change in accuracy; this method builds on the assumption that feature importance correlates with the magnitude of accuracy change. We explore a version of this approach in Sect. 5, with slight modification for our test domain.

# 4 Bridging Engineered and Network-Extracted Features

This paper focuses on two aspects of bridging engineered features with features generated by DL:

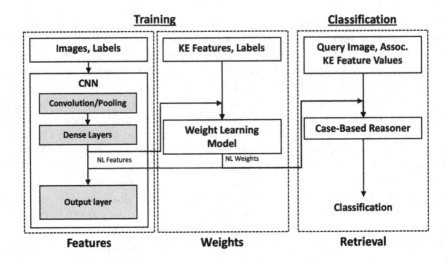

**Fig. 2.** Illustration of data flow through our model.

1. Extracting features from DL models to use in concert with KE features
2. Using neural networks to learn feature weights for both KE features and network-learned features.

It investigates these in the context of a case study of case-based classification. We propose a general model integrating three major components as shown in Fig. 2. In the model, features are learned by a CNN from training data, feature weightings for the new features and existing knowledge engineered features are learned by another network, and the features and weights are used in a case-based classifier. Specifically, components are:

1. A CNN that extracts features from input data (e.g., images) to be used for case-based classification
2. A neural network that generates weights for both learned and knowledge engineered features, for the classifier similarity calculation.
3. A case-based classifier that uses a combination of engineered features and features from (1), weighted according to (2), for case retrieval.

**CNN Architecture.** Our feature extraction CNN derives closely from the AlexNet architecture [13]. AlexNet is a foundational CNN architecture for image classification that employs a batch-normalized interleaving of five convolution and three pooling layers that are flattened into a network of two fully-connected dense layers that feed into the output layer. Our method deviates from other approaches on extracting CBR features from a CNN [21,24] by extracting features from the dense layer preceding the output layer in the CNN, rather than before the dense layers. The rationale for this approach is as follows. An output node's activation in a neural network is determined by a weighted sum of the outputs from the previous layer. Thus, extracting features immediately after the final convolution layer neglects intermediate layers' modifications to the feature set ultimately used to perform classification, motivating extracting features from later in the CNN structure. Also, we remove the bias node from the CNN output layer. This ensures that NL features are not skewed during training, because a bias node would factor into the weighted sum used for prediction but would not be extracted as a feature.

**Sequential Architecture and Weight Generation Approaches.** To generate network learned weights, we apply a sequential architecture (i.e., successive fully-connected layers) mapping inputs corresponding with each feature directly to the classifying output layer.

1. **Directly extracted weights:** After training, local feature weights are generated for each case in the case base. For each feature, the local feature weight is the normalized absolute value of the weight of the link leading into the output node corresponding to that case's class (for later similarity calculations, only magnitude is important). This produces a localized set of feature weights for the cases that are unique on a per-class basis. Both linear and RELU activation functions were considered for the output layer before applying softmax to select a class prediction, with comparative results reported in Sect. 5.
2. **Weights from Perturbation:** Calculating weight values based on the shift in prediction accuracy as feature inputs (KE features only, NL features only, or both combined into a single input set) are perturbed individually, according to the following equation derived from Kenny and Keane [10]:

$$w_i = \frac{\Delta acc(f_i, \sigma) + \Delta acc(f_i, -\sigma)}{2} \tag{1}$$

Here weight $w_i$ is the average change in prediction accuracy that results from perturbing feature $f_i$ by $\pm\sigma$. In contrast to extracting weights from the network directly, this generates a global set of feature weights applied to all cases, regardless of class.

## 5  Evaluation

Our evaluation addresses the following questions:

1. How is classification accuracy affected by degradation of reliability of input (KE features)?
2. How does using NL features in concert with KE features affect classification accuracy?
3. How do CBR retrieval weights based on NL weights influence classification accuracy for different combinations of NL and KE features?

## 5.1  Test Domain and Testbed System

*Test Domain:* As a test domain including both engineered features and non-symbolic information, we selected the Animals with Attributes 2 data set (AwA2) [26]. This data set, designed for one-shot learning, includes over 37000 images across 50 animal classes; each class also has an associated feature vector of 85 features corresponding to 85 symbolic descriptions (e.g., herbivorous, desert habitat, quadrupedal, etc.). Each feature is assigned a continuous value in $[-1.0, 100.0]$. Because all instances of a class are assigned the same feature vector, with no variance, these feature vectors yield "perfect" classification accuracy when used for retrieval. To simulate imperfect situation assessment assigning symbolic feature values and/or symbolic feature characterizations that are not 100% predictive, we use perturbation. This is defined by a multiplier $x$ that is applied to each feature value individually. The multiplier $x$ is generated by taking a random integer $k$ in the interval $[1, n]$ and randomly setting $x$ to $k$ or $1/k$ with equal probability. We consider values of $n$ on the interval $[1,10)$.

*Testbed System:* As case adaptation is beyond the scope of our work, the testbed case-based classifier has no adaptation component. The classifier retrieves the nearest neighbor (i.e., 1-NN) using a weighted Euclidean distance metric for similarity calculations, using either local feature weights (for directly extracted weights) or global weights (for weights extracted by perturbations).

Properties of the chosen data set were reflected in parameter choices for the networks. The CNN architecture was modified to use 1024 nodes in the dense layers (rather than the traditional 4096) to concentrate extracted information into fewer features in an effort to make comparisons between KE and NL features more one-to-one. However, this was only partially possible, because smaller layers increase training time and decrease accuracy, to the point where epoch training steps do not converge. Even though we found a one-to-one comparison impossible as a result, the number of nodes was still used as it did not appear to negatively impact classification accuracy. The output layers of both the CNN and sequential architecture contained 50 nodes based on the number of AWA2 classes, and the input layer of the sequential architecture contained one node for each feature. Specifically, this translated to 85 nodes when considering only KE features, 1024 when considering only NL features, and 1109 when considering both feature sets in tandem. Last, we found that $\sigma = 0.8$ led to the highest retrieval accuracy in preliminary tests when generating weights using perturbation, likely due to the lack of variance in the KE feature set.

## 5.2  Preliminary Experiments to Set Network Parameters

Both NL features and NL weights depend on training the networks from which they are generated. We first determined the number of epochs to use, to balance the trade-off between predictive accuracy and low training time. For the CNN, models are trained on ten randomly-selected images from each of the fifty classes in the AwA2 data set, for a total of 500 images. Sequential architecture models

**Table 1.** Comparing classification accuracy values ($\pm$ one standard deviation) for the sequential architecture for a given number of epochs, evaluated using the training set and an independent testing set.

| Epochs | Train accuracy | Test accuracy | Epochs | Train accuracy | Test accuracy |
|---|---|---|---|---|---|
| 10 | $0.176 \pm 0.028$ | $0.045 \pm 0.008$ | 60 | 1.0 | $0.085 \pm 0.015$ |
| 20 | $0.765 \pm 0.063$ | $0.064 \pm 0.009$ | 70 | 1.0 | $0.083 \pm 0.014$ |
| 30 | $0.974 \pm 0.010$ | $0.076 \pm 0.014$ | 80 | 1.0 | $0.088 \pm 0.014$ |
| 40 | $0.997 \pm 0.003$ | $0.076 \pm 0.012$ | 90 | 1.0 | $0.088 \pm 0.013$ |
| 50 | $1.000 \pm 0.001$ | $0.081 \pm 0.010$ | 100 | 1.0 | $0.092 \pm 0.013$ |

are trained on the 1024 NL features generated by the CNN and/or the 85 KE features. All epoch training evaluations are performed for a number of epochs on the interval $[10, 100]$ in increments of ten, with higher-resolution tests conducted for feature-dense spaces (i.e., requiring fewer then ten training epochs). Evaluations are performed thirty separate times and averaged to compute a sample mean and its standard deviation.

**Tuning Results.** From these procedures, we chose the following parameter settings. For learning NL features, the CNN model is trained for 50 epochs. The sequential architecture is trained for 80 epochs when learning weights for KE features only, 5 epochs when learning weights for NL features or both NL and KE features combined, and 50 epochs when learning weights using feature perturbation. These decisions reflect values that maximize classification accuracy on the training set while also minimizing the number of epochs. Further research could investigate finer tuning parameters, such as learning rate and early stopping.

We note that for our modified AlexNet architecture, training appears to hit a point of diminishing returns after fifty epochs. This pattern holds for prediction both on the training set and on an independent testing set of 500 new images (Table 1). Furthermore, the accuracy on the testing set is significantly lower, suggesting that a general set of NL features is difficult to learn from the training set, and/or that the model overfits to the training set. However, considering that the model is designed to learn features that discriminate between cases of different classes, overfitting relative to a given case base may be acceptable so long as network training can efficiently be redone as new cases are added.

### 5.3  How Retrieval Accuracy Changes with KE Feature Degradation

**Experiment Overview.** We explore relationships between retrieval accuracy and perturbation of the KE feature set. Specifically, retrieval accuracy is evaluated for leave-one-out experiments that are unweighted or weighted using linear or RELU activation functions for the sequential architecture to facilitate NL weight generation; each experiment is conducted for thirty iterations per value of $n$ to establish a sample mean and standard deviation.

**Table 2.** Comparing classification accuracy values (± one standard deviation) across the various perturbation levels. Results are shown for unweighted features and features weighted using linear and RELU output activation functions for the sequential network.

| $n$ | Unweighted | Linear | RELU |
|---|---|---|---|
| 9 | $0.440 \pm 0.028$ | $0.235 \pm 0.045$ | $0.238 \pm 0.041$ |
| 8 | $0.458 \pm 0.030$ | $0.251 \pm 0.044$ | $0.253 \pm 0.040$ |
| 7 | $0.506 \pm 0.025$ | $0.266 \pm 0.047$ | $0.287 \pm 0.035$ |
| 6 | $0.567 \pm 0.028$ | $0.291 \pm 0.056$ | $0.313 \pm 0.033$ |
| 5 | $0.651 \pm 0.028$ | $0.363 \pm 0.053$ | $0.358 \pm 0.057$ |
| 4 | $0.785 \pm 0.022$ | $0.481 \pm 0.059$ | $0.449 \pm 0.060$ |
| 3 | $0.931 \pm 0.014$ | $0.625 \pm 0.068$ | $0.642 \pm 0.046$ |
| 2 | $0.999 \pm 0.002$ | $0.941 \pm 0.035$ | $0.924 \pm 0.028$ |
| 1 | 1.0 | 1.0 | 1.0 |

**Sensitivity of Retrieval Accuracy to Feature Quality.** Results for these experiments are shown in Table 2. Predictably, retrieval accuracy decreases as the perturbation magnitude increases, because a higher degree of noise is present in the KE feature set. Given this relationship, it is interesting to consider the possibility of using retrieval accuracy with KE features alone as proxy for the comprehensiveness/completeness of the KE feature set (and by extension, under what conditions its combination with a NL feature set might provide the greatest benefit). More research is required to provide a finer-grained assessment. We observe that the linear and RELU activation strategies appear less performant than the unweighted strategy; we explore this result more deeply in Sect. 5.5.

### 5.4   How Using KE and NL Features in Concert Affects Accuracy

**Experiment Overview.** We evaluate classification accuracy using KE and NL features in tandem by considering various perturbations of KE features in concert with the NL feature set against retrieval accuracy using each set individually. Each experiment is performed using unweighted leave-one-out testing on the case base of 500 cases using uniform feature weights for thirty iterations to establish a sample mean and standard deviation.

**Benefits of Combining Features.** As shown in Fig. 3, there frequently exists an interesting–if not always statistically significant–increase in classification accuracy when considering a combination of KE and NL features over either feature set's individual classification accuracy. This accuracy increase is most evident and most significant when both feature sets considered individually lead to similar classification accuracy values (i.e., when the perturbation magnitude is such that their accuracy values are similar) and when classification accuracy using NL features alone is higher than when using KE features alone.

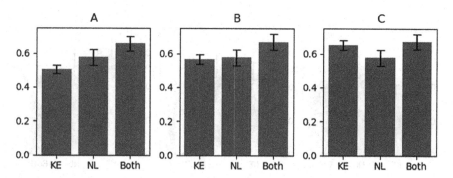

**Fig. 3.** Comparison of classification accuracy values for different feature perturbations ($n = 7$ A; $n = 6$ B; $n = 5$ C). Error bars represent one standard deviation relative to thirty iterations.

The existence of this "accuracy bump" has multiple potential causes. For one, general accuracy trends and standard deviation patterns appear to be dominated by the NL features; this is unsurprising given that many more NL features (1024) are considered than KE features (85). While it can be argued that this blunts the significance of the observed trend, it is important to provide further context. In particular, the modified AlexNet CNN produces novel features that capture aspects of the feature space not adequately represented in the KE feature set. That is, even though this trend may at least partially be attributed simply to the existence of more features, the NL features must also be significant/helpful in order to produce an increase in accuracy. The real question becomes whether the increase in accuracy when considering the union of the feature sets comes strictly from the existence of new features or from new interplay between the two feature spaces that creates a whole greater than the sum of its parts. This proved difficult to measure directly given the chosen domain. In preliminary tests, a CNN having only 85 nodes per dense layer never converged (i.e., it could never outperform a random baseline).

**Implications for Hybrid Systems.** These data suggest an interesting potential implication. Specifically, if this accuracy increase can be at least partially attributed to the nature of the two sets of features in a hybrid system (rather than simply an influx of new features alone), such a result could highlight direct hybridization of KE and NL features as a new avenue for accuracy improvement for CBR retrieval. That is, in the presence of additional environmental information (represented by the images in the AwA2 domain), a neural network may be able to generate features that are both novel when compared against the KE feature set and especially useful in concert with the KE feature set. This is naturally difficult to verify due to the well-documented inexplainability of neural network features, but future work focusing on detailed feature relationships and/or correlations, while likely computationally costly, might be able to identify useful correspondences between the feature sets for exactly this purpose.

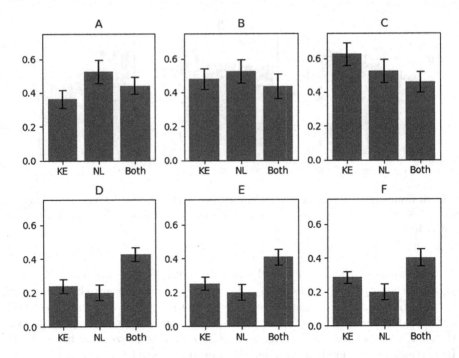

**Fig. 4.** Comparison of classification accuracy values for different feature perturbations ($n = 9$ D; $n = 8$ E; $n = 7$ F; $n = 5$ A; $n = 4$ B; $n = 3$ C). Error bars represent one standard deviation in experiments using a linear activation function (top) or RELU activation function (bottom) to generate NL weights using softmax.

### 5.5    How Learned Weights Further Influence Retrieval Accuracy

**Experiment Overview.** For these experiments, features are weighted based on the strategies described for NL weights in the model section. Classification accuracy values for combined NL and KE features are evaluated against using each set of features individually, based on leave-one-out experiments repeated thirty times to establish a sample mean and standard deviation.

**On Feature Weights and the "Curse of Dimensionality".** While previous research appears to achieve reasonable success generating weights by perturbing KE features [10], such methods may not be applicable to feature-dense spaces. Specifically, when generating NL weights using feature perturbation on NL features, we observe that perturbing a single feature seldom changes the overall classification accuracy of the model, even when considering large values of $\sigma$. Therefore, many of the generated weights are at or near zero, crippling similarity assessment. It is possible that perturbing features in batches or using more complex neural network models might address some of these shortcomings; however, we suspect that existing weighting algorithms are significantly less effective in high dimensional spaces created by generating NL features. Alternative weight generation algorithms may be applicable here (e.g., [1]), but research in additional domains is needed.

**Weighting Can Augment Retrieval Benefits.** In terms of overall retrieval accuracy, our initial results on using NL weights drawn directly from a network model in concert with combined NL and KE weights appear disappointing (Fig. 4). However, trend behavior in these experiments is interesting. First, we note that accuracy values for the linear activation function are consistently at least as high as those for the RELU activation function. This is reasonable given that RELU would likely favor large-magnitude negative correlation weights less strongly than positive correlation weights. Curiously, however, the linear activation accuracy values suggest that combining KE and NL features produces a harmful effect. Contrast this with the RELU activation function, where combining KE and NL features produces the most significant relative accuracy improvement across all tests (Fig. 4). So why did the weighting methods attempted not increase classification accuracy overall? This could be a result of the lack of variation in the raw KE features, so weighting provides little benefit for features that exist due to random perturbations; alternatively, this could simply be a symptom of the simplicity of the weighting algorithms investigated. However, the dramatic relative improvement in retrieval accuracy when generating weights using an RELU activation function suggests that deeper investigation into interplay between KE and NL features with NL weights is worthwhile.

## 6  Ramifications for Explainability

The previous experiments support the accuracy benefits of combining knowledge engineered and neural network features, especially for domains where additional features may be extracted from supplementary/environmental information. Unfortunately, while such features may be powerful and have the potential to capture aspects of the case base that humans cannot, this comes at a cost for explainability of retrieval. As the network-based features may be difficult to explain, it may be equally difficult to assess similarity judgments when they are based on network features.

Such a loss might not always be important. In a domain for which humans can assess similarity directly from the retrieved case, no explanation may be needed. In domains for which the combination of features results in substantial accuracy gains, the loss of explainability might be considered less important than gains in accuracy. However, the accuracy-explainability trade-off merits future research, and potential ways to mitigate it, such as integrating aides to interpreting feature assessments (e.g., CBR-LIME [19]) would be an interesting area for future research.

## 7  Conclusions

This paper presents results from a case study on methods for supplementing existing knowledge-engineered features with features learned from data with deep learning, with feature weightings for both learned by a neural network. The paper illustrates circumstances under which combining network-learned features with knowledge engineered features can produce classification accuracy values greater than either of the feature sets considered individually. It also points to challenges in weight generation for high-dimensional spaces, as may arise from learning large features sets from deep learning, and considers strategies to alleviate this difficulty.

These conclusions suggest numerous avenues for future work. First, testing across additional domains and network architectures and baselines is an essential next step. Also important exploring the tuning conditions under which combining KE and NL features produces maximum benefit, or under which the CNN generates features that are especially useful for retrieval. Investigating weighting strategies that perform better in feature-dense spaces is an another important step. Finally, an interesting question outside of the learning methods is how the inclusion of NL features and NL weights affects the explainability of the CBR model that applies them and how explanation issues might be addressed.

**Acknowledgments.** We acknowledge support from the Department of the Navy, Office of Naval Research (Award N00014-19-1-2655), and the US Department of Defense (Contract W52P1J2093009).

# References

1. Aha, D., Kibler, D., Albert, M.: Instance-based learning algorithms. Mach. Learn. **6**(1), 37–66 (1991)
2. Barletta, R., Mark, W.: Explanation-based indexing of cases. In: Kolodner, J. (ed.) Proceedings of a Workshop on Case-Based Reasoning, pp. 50–60. DARPA, Morgan Kaufmann, Palo Alto (1988)
3. Bhatta, S., Goel, A.: Model-based learning of structural indices to design cases. In: Proceedings of the IJCAI-93 Workshop on Reuse of Design, Chambery, France, pp. A1–A13. IJCAI (1993)
4. Bonzano, A., Cunningham, P., Smyth, B.: Using introspective learning to improve retrieval in CBR: a case study in air traffic control. In: Leake, D.B., Plaza, E. (eds.) ICCBR 1997. LNCS, vol. 1266, pp. 291–302. Springer, Heidelberg (1997). https://doi.org/10.1007/3-540-63233-6_500
5. Cox, M., Ram, A.: Introspective multistrategy learning: on the construction of learning strategies. Artif. Intell. **112**(1–2), 1–55 (1999)
6. Domeshek, E.: Indexing stories as social advice. In: Proceedings of the Ninth National Conference on Artificial Intelligence, pp. 16–21. AAAI Press, Menlo Park (1991)
7. Fox, S., Leake, D.: Introspective reasoning for index refinement in case-based reasoning. J. Exp. Theor. Artif. Intell. **13**(1), 63–88 (2001)
8. Grace, K., Maher, M.L., Wilson, D.C., Najjar, N.A.: Combining CBR and deep learning to generate surprising recipe designs. In: Goel, A., Díaz-Agudo, M.B., Roth-Berghofer, T. (eds.) ICCBR 2016. LNCS (LNAI), vol. 9969, pp. 154–169. Springer, Cham (2016). https://doi.org/10.1007/978-3-319-47096-2_11

9. Hegdal, S., Kofod-Petersen, A.: A CBR-ANN hybrid for dynamic environments. In: Proceedings of the ICCBR 2019 Workshop on Case-Based Reasoning and Deep Learning, September 2019

10. Kenny, E.M., Keane, M.T.: Twin-systems to explain artificial neural networks using case-based reasoning: comparative tests of feature-weighting methods in ANN-CBR twins for XAI. In: Proceedings of the Twenty-Eighth International Joint Conference on Artificial Intelligence (2019)

11. Kolodner, J.: Case-Based Reasoning. Morgan Kaufmann, San Mateo (1993)

12. Kraska, T., Beutel, A., Chi, E.H., Dean, J., Polyzotis, N.: The case for learned index structures. In: Sensors, pp. 489–504 (2019)

13. Krizhevsky, A., Sutskever, I., Hinton, G.E.: ImageNet classification with deep convolutional neural networks. In: Proceedings of the 25th International Conference on Neural Information Processing Systems, vol. 1, pp. 1097–1105 (2012)

14. Leake, D.: An indexing vocabulary for case-based explanation. In: Proceedings of the Ninth National Conference on Artificial Intelligence, pp. 10–15. AAAI Press, Menlo Park, July 1991

15. López de Mántaras, R., et al.: Retrieval, reuse, revision, and retention in CBR. Knowl. Eng. Rev. **20**(3) (2005)

16. Martin, K., Wiratunga, N., Sani, S., Massie, S., Clos, J.: A convolutional Siamese network for developing similarity knowledge in the selfBACK dataset. In: Sanchez-Ruiz, A.A., Kofod-Petersen, A. (eds.) Proceedings of the ICCBR 2017 Workshop on Case-Based Reasoning and Deep Learning, pp. 85–94. CEUR Workshop Proceedings (2017)

17. Mathisen, B.M., Aamodt, A., Bach, K., Langseth, H.: Learning similarity measures from data. Progress Artif. Intell. **9**, 129–143 (2019)

18. Nasiri, S., Helsper, J.F., Jung, M., Fathi, M.: Enriching a CBR recommender system by classification of skin lesions using deep neural networks. In: Proceedings of the ICCBR 2018 Workshop on Case-Based Reasoning and Deep Learning, July 2018

19. Recio-García, J.A., Díaz-Agudo, B., Pino-Castilla, V.: CBR-LIME: a case-based reasoning approach to provide specific local interpretable model-agnostic explanations. In: Watson, I., Weber, R. (eds.) ICCBR 2020. LNCS (LNAI), vol. 12311, pp. 179–194. Springer, Cham (2020). https://doi.org/10.1007/978-3-030-58342-2_12

20. Samakovitis, G., Petridis, M., Lansley, M., Polatidis, N., Kapetanakis, S., Amin, K.: Seen the villains: detecting social engineering attacks using case-based reasoning and deep learning. In: Proceedings of the ICCBR 2019 Workshop on Case-Based Reasoning and Deep Learning, pp. 39–48 (2019)

21. Sani, S., Wiratunga, N., Massie, S.: Learning deep features for kNN-based human activity recognition. In: Proceedings of ICCBR 2017 Workshops (CAW, CBRDL, PO-CBR), Doctoral Consortium, and Competitions co-located with the 25th International Conference on Case-Based Reasoning (ICCBR 2017), Trondheim, Norway, 26–28 June 2017. CEUR Workshop Proceedings, vol. 2028, pp. 95–103. CEUR-WS.org (2017)

22. Schank, R., et al.: Towards a general content theory of indices. In: Proceedings of the 1990 AAAI Spring Symposium on Case-Based Reasoning. AAAI Press, Menlo Park (1990)

23. Tabian, I., Fu, H., Khodaei, Z.S.: A convolutional neural network for impact detection and characterization of complex composite structures. In: IEEE Trans. Pattern Anal. Mach. Intell. (T-PAMI) **19** (2018)

24. Turner, J.T., Floyd, M.W., Gupta, K.M., Aha, D.W.: Novel object discovery using case-based reasoning and convolutional neural networks. In: Cox, M.T., Funk, P., Begum, S. (eds.) ICCBR 2018. LNCS (LNAI), vol. 11156, pp. 399–414. Springer, Cham (2018). https://doi.org/10.1007/978-3-030-01081-2_27
25. Wettschereck, D., Aha, D., Mohri, T.: A review and empirical evaluation of feature-weighting methods for a class of lazy learning algorithms. Artif. Intell. Rev. **11**(1–5), 273–314 (1997)
26. Xian, Y., Lampert, C.H., Schiele, B., Akata, Z.: Zero-shot learning - a comprehensive evaluation of the good, the bad and the ugly. IEEE Trans. Pattern Anal. Mach. Intell. (T-PAMI) **40**, 1–14 (2018)

# Task and Situation Structures
# for Case-Based Planning

Hao Yang[1]([⊠]), Tavan Eftekhar[1], Chad Esselink[1], Yan Ding[2], and Shiqi Zhang[2]

[1] Ford Motor Company, Dearborn, MI, USA
hyang1@ford.com
[2] State University of New York-Binghamton, Binghamton, NY, USA

**Abstract.** This paper introduces two new representation structures for tasks and situations, and a comprehensive approach for case-based planning (CBP). We focus on everyday tasks in open or semi-open domains, where exist a variety of situations that a planning (and execution) agent must deal with. This paper first introduces a new, generic structure for representing tasks and task plans. The paper, then, introduces a generic situation structure and a methodology of situation handling. The proposed structures support encoding all domain knowledge in *cases* while avoiding hard-coding domain rules.

**Keywords:** Case-based planning · Task structure · Task plan · Situation handling

## 1   Introduction

Case-based planning (CBP) [10] sees knowledge being embedded in cases, inside real-world life stories. Humans are able to directly use previous cases to solve new planning problems instead of employing domain rules to craft a new plan. This paper explores a comprehensive approach that encodes domain knowledge in cases, instead of relying on separate domain knowledge bases in a case-based planning system. This paper addresses a set of problems that resemble a service agent dealing with "everyday tasks", c.f., "problem solving tasks" [15]. A service agent faces a complicated world with a large variety of tasks and their variations, and it needs to deal with practically endless types of situations. Our research starts with a new, *generic* structure for representing task cases. On the other hand, a task plan is not to be static in the real world. It often needs to be revised in response to unexpected situations[1]. The paper then introduces a new, *generic* situation structure for representing situation cases. Accordingly, we develop a novel situation handling methodology that avoids hard-coding domain rules in applications while focusing on encapsulating knowledge in tasks and situation handling cases.

---

[1] The term "situation" in this paper has been frequently referred to as "anomaly" and "event" in the literature.

© Springer Nature Switzerland AG 2021
A. A. Sánchez-Ruiz and M. W. Floyd (Eds.): ICCBR 2021, LNAI 12877, pp. 263–278, 2021.
https://doi.org/10.1007/978-3-030-86957-1_18

Classical planning is to compute a sequence of *actions* for transforming the world from an initial *state* to a state that satisfies the *goals* [9]. To perform an action, the current state should satisfy the *preconditions* of the action. After performing the action, the *effects* of the action are expected to be realized so that the state will change accordingly. Classical planning assumes that a complete task plan is generated prior to the execution and it does not consider structures in a task plan [1,7]. In comparison, people plan at different levels of abstractions, e.g., a task can be divided into sub-tasks. Hierarchical Task Network (HTN) was introduced to reflect this intuitive planning technique [13,14,16,19]. In HTN planning, refinement rules, called *methods*, break down a task into sub-tasks, or High Level Actions (HLA) in HTN's term.

On the other hand, a plan may fail or stale during execution. It could be due to anomalies as the environment deviates from original assumptions. It could be due to an extraneous exogenous event that the agent has to handle. Or, there could be new demands from other agents that the agent needs to accommodate. All these are *Situations* that the agent needs to respond to by revising or repairing the task plan. A *Situation* is defined as "an unexpected event or demands that an agent needs to respond to" in this research.

Many researchers have addressed situation handling. For example, ASPEN has a plan repair mechanism developed for Mars rovers [4]. The plan repair unit keeps monitoring conflicts and applies repair methods when conflicts are detected. ASPEN has a total of ten repair methods. Goal Driven Autonomy (GDA) [6] was developed that includes a four-phase discrepancy detection and goal modification/reformulation process. It takes a control approach as it continuously monitors any deviations from expected states and has policies to address the discrepancy. More importantly and distinctively, it develops comprehensive methods to revise goals accordingly. In GDA implementations, the control logic and goal reasoning are largely rule-based and domain-specific [2]. Another school of plan repair methods is to use domain rules to remove or add actions to the existing plan [8,17].

Situations are unpredictable, especially in real-world applications. Rare *Situations* are often referred to as "corner cases" or "edge cases". Take robotaxi as an example. Assume a vehicle picks up a customer and sends the customer from location $A$ to location $B$. Many *Situations* can happen during the trip. At the pickup time, the vehicle may not find the customer showing up at the pickup location, or the vehicle could not access the prearranged pickup spot. During the trip, the customer may complain about the smell or spill in the car, or the customer needs to divert for an urgent errand. Those *Situations* that could be solved relatively easily by a human driver could be challenging to the Artificial Intelligence (AI) agent. Not only that *Situations* are numerous, but also the difference in the context of *Situations* compounds variations. The solution space is impossible to be exhaustively defined.

This paper illustrates a comprehensive design and practice that avoid hard-coding rules by introducing two new representation structures. It is unique that:

1. It has text-based, generic structures and syntax for tasks and situations.

2. It embeds domain knowledge in executed cases, not in "hard-coded" rules.
3. It uses context as additional attributes into guiding the search for solutions.

In the following, the paper first discusses *Task structure* and planning in Sect. 2, and then discusses *Situation structure* in Sect. 3. After that the paper discusses situation handling in Sect. 4. The paper presents a couple of examples in Sect. 5, where we use our prototype system Virtual Service Agent (VSA), an agent that serves passengers in a ride-hailing vehicle, as a research workbench for discussion and illustration. Finally, we summarize the significance of this work in Sect. 6.

## 2   Task Structure and Planning

In case-based planning (CBP), a task plan is viewed as a record of history, an episode of a story. Therefore, we use *Task* to refer to a task *case* in this paper. With this motivation, our *Task* structure encapsulates all parameter details in a task. Second, a task is an assignment given to an agent to perform. However, a subtask can also be viewed as an assignment derived from its parent task. They are analogous. "Abstraction" is an important concept studied in CBP [3,5]. Recognizing that tasks and subtasks are analogous helps us understand the levels of abstraction of *Tasks*. Third, the paper introduces *context* as an attribute of a *Task* that provides variations that will differentiate behaviors of a *Task*.

### 2.1   *Task* Structure

The *Task* structure is illustrated in Table 1, where it includes the conventional task plan information such as **Conditions** and **Effects**, with a few tweaks.

**Table 1.** Task structure

| Attribute | Explanation |
|---|---|
| Task_name: | string (could be considered as the task class name) |
| Parent_task: | (object or id of the parent task, null if no parent) |
| Sub-tasks: | (a list of sub-tasks of this task. Empty if it is a leaf) |
| Action: | (the action of the task) |
| Specs: | (detail specs of how the action is performed) |
| Conditions: | (conditions to satisfy before this task can be performed) |
| Effects: | (effects that will be assigned after the task is performed) |
| Context: | (a list of contexts of this task, each is in the form of "key: value") |
| Goals: | (goals to be verified if the task is performed successfully) |

**Task_name** is the name of a *Task* class (not a name for an instance of a *Task*). For example, it could be "drive_task". **Parent_task** is the parent of this *Task*, which is *null* if it is a root *Task*. **Sub tasks** is a list of sub-tasks, where each sub-task takes the same structure of a *Task*. **Action** is an abstract form of a *Task*. An *Action* has the action (verb), and a syntax of parameters of this *Action*, e.g., "`Robot-r drive from location-1 to location-2`". It should be noted that the idea a task being analogous with action is not new [11]. **Specs** contains the details of parameters that are used in the *Action*. For example, if `location-1` is a parameter in the above "*drive*" *Action*, then the location object is included in the **Specs**. Finally, **Context** contains any context information relevant to this *Task*. For example, if a drive *Task* is driving in rain, "raining" is among the context of this drive *Task*.

This schema is implemented in *json*[2]. We have *serialize* and *deserialize* functions in python that transform data to objects or objects to data when needed. We can encode a whole task object in "data" in a naturally understandable form.

## 2.2 Execution of *Tasks*

In our design, task planning is an integral part of the task execution process. It takes a variational approach which means that instead of applying rules to develop the plan tree, we use a *Task* template or a copy of a prior *Task* and replace the parameters (*spces*, *context*) of the *Task* with the parameters of the new Task. It is analogous whether it is from a *Task* template or a prior *Task* since a *Task* template is in structure the same as a real executed *Task*. The task planning agent keeps an agent-level global *state* stack. During task execution, checking conditions and checking goals will use the *state* information, while applying effects will change the state.

When initiated, a *Task* has a "status" (not shown in Table 1) of `unplanned`. After the planning, where a task develops its sub-tasks, the *Task* changes its "status" to `planned`. When a *Task* develops its sub-tasks, the "specs" in the sub-tasks will be mapped from the parent *Task*'s "specs." This is recorded in the "mapping" field of each sub-task. The "mapping" field was not shown in Table 1, but it is part of the *Task* structure. The following is an example of mapping:

```
{
  "spec.origin": "parent.specs.origin",
  "specs.destination": "parent.specs.destination"
}
```

It means that the `origin` in the *specs* of this *Task* is assigned the same as the `origin` of the parent *Task* specs. The `destination` in the specs of this *Task* is also assigned the same as the `destination` of the parent *Task* specs.

In the next execution stage, if there are sub-tasks, each sub-task is iterated and its `execution` function is recursively called the same way as the parent *Task*. If there are no sub-tasks, the *Action* is executed, which usually is sending the *Action* to another agent (the actor) for execution.

If there is an exception detected during the execution, the exception is handled based on the error message. Some of the exceptions will be considered as

---

[2] *json* is a lightweight data-interchange format. For details, please refer to this page: https://www.json.org/json-en.html.

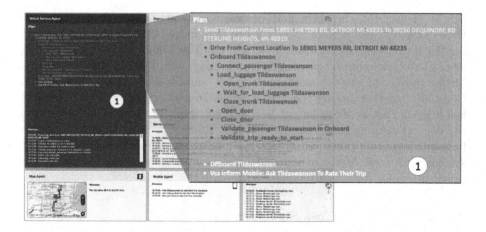

**Fig. 1.** VSA User Interface.

*Situations* and *Situations* will be handled by the agent. If the *Task* could not be executed, (e.g. when the conditions are not satisfied), and the *Situation* could not be handled successfully, the *Task* status will be changed to `failed`. When a *Task* is completed with no error, it is marked as `finished`. When a *Task* status is changed, the database record is updated. The database retains a rich *Task* plan repository, thus the *Task* plan case library.

### 2.3   Implementation of Our *Task* Structure

We implemented the *Task* structure and *Task* execution in our prototype system *Virtual Service Agent* (VSA). Figure 1 shows the graphic user interface of VSA. In this user interface, each window represents an agent. Within an agent window, there is an action panel at the upper and a message panel at the lower. The upper left window (Fig. 1①) is the VSA panel for monitoring the task plan at execution time. The lower left window is the *Map* agent that simulates the vehicle driving through a trip. Among other agents, a *Dialogue* agent communicates with the rider using natural language, a *Weather* agent retrieves live weather information, a *Mobile* agent emulates the communication to the rider through a mobile device, a *Vehicle* agent controls the vehicle mechanics and sensors, and a *Service Center* is the dispatch system that sends Trip *Tasks* to the vehicle.

Let us take a closer look at the *VSA* panel (Fig. 1①), we can find an example of the task hierarchy of a Trip *Task*. Each line prints an action of the *Task*. Light green represents "executing" tasks, dark green for "completed" tasks, and white for "unplanned" *Tasks*. We implement it to resemble a Trip *Task* handled by a vehicle agent sending a customer, `Tildaswanson`, a fictitious name, from location Meyers Rd to location Dequindre Rd. The Trip *Task* is received from a trip assignment agent (*Service Center*). A Trip *Task* has four top-level sub-tasks: A Drive *Task* that drives from where the vehicle is to the pickup location Meyers

Rd; at Meyers Rd, the agent performs the Onboard *Task*; it then performs a Drive *Task* that drives from Meyers Rd to Dequindre Rd; after arriving Dequindre Rd, it performs the Offboard *Task*. The sub-task Onboard *Task*, for example, has its sub-tasks: connect-passenger, load-luggage, etc. The load-luggage *Task* is further developed into sub-tasks: open-trunk, wait-for-load-luggage, close-trunk. Certainly, whether having the load-luggage *Task* depends on if the customer has luggage that needs to be put in the trunk. This information is captured in the context information of the parent task. Instead of using rules like "if has-luggage then ..." in the refinement method as you would expect in an HTN system, VSA uses *Task* attributes, including contexts, as indices to search for a previous similar *Task* as a template to develop the sub-tasks.

### 2.4 Discussions on Task Structure

A major motivation of developing *Task* structure is to avoid domain-specific data types and code and to avoid hard-coded rules. A task is decomposed into sub-tasks in an *instance* of a *Task*. The conventional approach usually includes a set of *domain rules* (refinement methods) that is separated from the plan data. In VSA, a *Task* carries domain rules in the *data* (the task plan). If a new variation of a *Task* refinement needs to be introduced, it is introduced by injecting a new *Task* instance into the system, leaving the old data (case) *untouched*. This *Task* structure design also serves the following purposes:

1. It collects task plan data naturally, with every detail of a task plan. It could potentially offer rich real-world data for machine learning. Machine learning is recognized as an important method to overcome the bottleneck of knowledge elicitation in planning systems and it has been used to learn actions and methods [18,20]. On the other hand, machine learning methods can also be used to sniff through the task plan data for discrepancies.
2. The proposed *Task* structure supports simulation well. CHEF [10] showed the importance of having a simulation system in a case-based planning system. Once an old plan is modified, it is not guaranteed to succeed. A robust simulator will be able to detect failures so that flaws in the modified plan can be repaired. In our implementation, a simulation function is invoked when a plan is modified to validate if the plan is feasible. The details of simulation and validation will be explained later in Sect. 4 and 5.

## 3    Situation Structure

As discussed in Sect. 1 (Introduction), *Situations* in an open or semi-open-world application are numerous and unpredictable. A *Situation* can happen as a result of a task failure. For example, it is a *Situation* when the vehicle cannot connect to the incoming customer. Or, it is a *Situation* when the passenger requests to divert the trip. For example, while en route to the airport, the passenger needs to go back home because he forgets to bring his passport. It is impractical to exhaustively enumerate all possible *Situations* plus all variations in the context of *Situations*.

Here, we present a new, generic *Situation* structure, similar to the *Task* structure, that is capable of describing all *Situations* and situation handling without domain-specific data types and code. *Situation* types and situation-handling knowledge are not hard-coded, but recorded in plain-text format (*json* strings) and are in data.

In a nutshell, a *Situation* will be handled using a **Remedy** to repair the plan. However, we do not expect to always apply the same **Remedy** to handle the same *Situation* (class, identified by the *Situation* name). A *Situation* has variations differentiated by **Context**. **Context** is an important attribute of a Situation. As Leake and Jalali (2014) [12] put it: there are three tenets of context and CBR: relevance, applicability, and preserving essential specifics of knowledge. Both our *Task* and *Situation* structures contain a **Context** attribute for this reason. There is also a **Logics** field in the *Situation* data structure. **Logics** is used to find additional **Context** information. It is intended to embed problem-solving knowledge in *Situation* data, not hard-coded rules. Table 2 is the *Situation* structure.

**Table 2.** *Situation* structure

| Attribute | Explanation |
| --- | --- |
| Name: | (name of this situation) |
| Time: | (time this situation occurred) |
| Task: | (the Task during which this situation is logged) |
| Context: | (a list of contexts under which this situation happened) |
| Remedy: | (a list of remedy actions to take) |
| Logics: | (knowledge of how to set the Context and the Remedy) |
| Goals: | (a list of new goals that the repaired plan should satisfy) |

When a *Situation* is detected or received (from another agent), it comes with **Name**, **Time**, **Task**, **Context**, and **Goals**. We call it the *Situation* header.

The agent is then to retrieve **Logics** of this *Situation* from the knowledge base and apply them. **Logics** is used to help determine the contexts that are most relevant to this *Situation*. The context information could be used for situation handling. For example, in a `car-window-broken` *Situation*, the **Logics** will inquire a sensor agent to find which window is broken, the severity of the damage, a weather agent to find out current weather condition. In the implementation, **Logics** is a list of functions that feed into the contexts. The following is an example of **Logics**. It is in the form of a (python) dictionary:

```
"logics": {
    "window_broken": "vda.checking_window",
    "weather": "weather.current_weather",
    "wetness": "chat.wetness"
}
```

In this example, the keys are attributes that will appear in the context. The values are the functions. The first is a function of the "vda" agent, referring to the vehicle agent, with sensors to tell if a window is malfunctioning or broken. The second corresponds to a "weather" agent function that returns the current weather condition. The third initiates a "chat" conversation that, through a Dialogue agent, provides how much of concern of the wetness in the cabin. These attributes are added to the Context information of this *Situation*. The functions could be more sophisticated, and examples of them are beyond the scope of this paper.

**Table 3.** *Remedy action* structure

| Attribute | Explanation |
|---|---|
| Operation: | (add/delete/modify) |
| Reference: | (a list defines references of attributes) |
| Mapping: | (a mapping function that fills the spec of the with_task) |
| With_task: | (the new task that will be added or modified) |

**Remedy** is a list of *remedy actions* used to alter the task plan so that the *Situation* is handled. A *remedy action* is simply adding/deleting/modifying a *Task* plan. Table 3 shows details of a *remedy action* structure. In the *remedy action* structure:

- **Operation**: the operation will be something like: "add after the drive_task"; or "modify this_task". It contains both an operation (add/modify/delete) and the target ("after the drive_task"/"after this_task", etc.). We adopt this natural syntax. It can be easily parsed with a set of vocabulary.
- **References**: A list of reference definitions. Through "references", the keys in the *mapping* are referenced to the actual object in the program. In the following example:

```
"references": {
    "drive_task": "executing task",
    "context": "situation context"
}
```

"drive_task" used in the *mapping* is referenced to the "executing task" (the **Task** in Table 2); "context" used in the *mapping* is referenced to the Context in the *Situation* (Table 2).

- **Mapping**: how the Specs of the new Task (the "With_task" in Table 3) is to be set. The following is an example of the *mapping*:

```
"mapping": {
    "specs.origin": "drive_task.specs.origin",
    "specs.dest": "context.current_location",
    "specs.actor": "drive_task.actor",
    "action.origin": "drive_task.specs.origin",
    "action.dest": "context.current_location",
    "estimated_time": "drive_task.actual_duration"
}
```

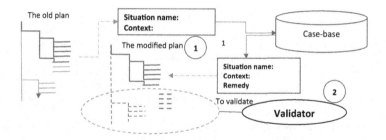

**Fig. 2.** An example of plan modification

In each mapping item, the *key* is the target of the parameter, the *value* is the source of the parameter. Please notice the source parameters "drive_task" and "context" are defined in the "references" described just above.

– **With_task**: the new *Task* that is to be added into the task plan.

## 4   Situation Handling

Sections 2 and 3 introduce our structures for representing *Tasks* and *Situations*. Leveraging the new structures we developed, we present our situation handling approach in this section.

When a *Situation* is detected, the agent will retrieve the **Logics** of this *Situation* (class) from the knowledge base. The **Logics** functions are invoked, and the returned values will populate additional Context information in the *Situation*. The *Situation* with its Context is then pushed to a Situation Queue. When the agent executes a Task, it also keeps monitoring if there is any *Situation* in the Situation Queue. If there is a *Situation* in the Queue, the agent will attempt to handle the *Situation*. The agent will first use the *Situation* name and Context to retrieve any prior *Situation* in the case library that matches best with the *Situation*. If a similar *Situation* is found in the case library, the **Remedy** of the old *Situation* will be used to repair the plan of the new *Situation*.

Once the **Remedy** is applied, the modified plan (Fig. 2①) will be validated using the Validator (Fig. 2②). In Fig. 2, we use solid lines to refer to the *Tasks* that have been executed in the modified plan. The dashed lines are those *Tasks* that have not been executed. The Validator is to validate the unexecuted *Tasks*. The validation is like a simulation. It starts with the current State. The agent simulates each *Task* by checking the conditions first, and then it applies the effects of the *Task* to the States, and finally it checks if the goals of the *Task* are met.

If the goals are met to the end, the modified plan is validated; otherwise, there are two options. One is to find another similar *Situation* case in the case library to repair the plan and validate the repaired plan again. If those attempts are failed, another option is to call in human assistance as described below.

What if there is a *Situation* that the system does not know before? What if there is no prior *Situation* that is similar enough (to pass a similarity threshold)

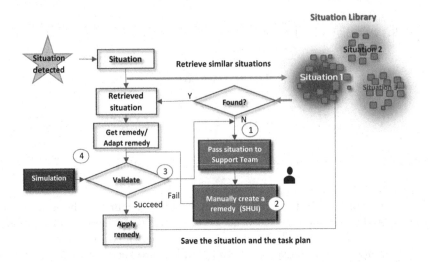

**Fig. 3.** Situation handling life cycle process

to the new *Situation*? In this case, human intervention is inevitable. However, what we want is that a new *Situation* class can be easily introduced, and a new Remedy can be easily constructed. We also want that the new situation handling case can be reused in the future. Figure 3 depicts this process. Figure 3① is when the remote customer assistant center is informed. A customer assistant will be able to quickly see the current status of the *Situation* ("what is the *Situation*?", "when did the *Situation* happen?", "the *contexts* of the *Situation*?", and "the Specs of the *Task*?"). The customer assistant can directly talk to the customer to find out additional context that helps him/her to resolve the *Situation*. The customer assistant will do all these through a system called Situation Handling UI (SHUI). SHUI is a comprehensive user interface representing what would be required for trained customer support experts (support specialists) to craft new Remedies in the integrated system.

Figure 4 is an example of the SHUI interface. The left panel (Fig. 4③) displays the real-time *Task* execution that is identical to what is in VSA (Fig. 1①). The lower-middle panel (Fig. 4①) displays the situation context. The support specialist can see exactly what is happening and what has happened at the vehicle remotely. The right panels are pallets that the support specialist pick, drag and drop "*Tasks*" and "remedy actions". The upper-middle panel (Fig. 4②) shows the revised **Remedy** and the "submit" button will send the revised **Remedy** to VSA to repair the plan.

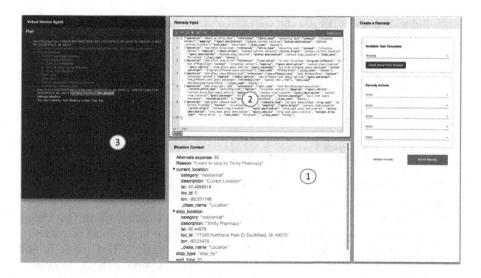

**Fig. 4.** Situation Handling UI

# 5   Illustrative Examples

The following are two examples to illustrate how situation handling works.

**Example 1: The story of "Window Leak"**

[*It starts raining. Passenger Annie saw water seep into the cabin. The window is not fully closed.*]

ANNIE: The water is getting in.

> [*The vehicle checks the window, one of them is open.*
> *The vehicle sends a command to the control unit to close the window. However, the window is not closed.*
> *Vehicle realizes that the window is in a malfunction.*
>
> *Vehicle recalls a case that the window that could not get closed because the window glass was blocked by a twig.*]

VEHICLE: Is there something that blocked the window glass?

ANNIE: Yes, looks like it is jammed

> [*The vehicle rolls down the window halfway and the rider cleaned a foreign object that jammed the window. The vehicle rolls up the window again and the window is closed this time.* ]

Here is what to happen: Passenger Kelly gets in the car and the vehicle starts the journey.

It starts to rain. However, water seeps through the window, and water drops onto Kelly.

"It is raining and it is wet here", Kelly claimed. A "wet-in-cabin" *Situation* is generated. The Logics of the *Situation* is:

```
"logics": {
    "window_broken": "vda.checking_window",
    "weather": "weather.current_weather",
    "wetness": "chat.wetness"
}
```

"window-broken" - It calls the vehicle agent to check if any window is broken; "weather" - from the weather agent, it returns current weather condition; "wetness" - it initiates a dialog with the passenger to obtain the following information: wherein the cabin is wet (seat? floor? on the person?).

The above information feeds into the Context of the *Situation*.

VSA finds a similar *Situation* from its *Situation* case library that has the following remedy:

```
"add close-window task"
"add confirm-problem-solved task"
```

The "close-window" *Task* is sent to the vehicle, and the vehicle sends a "close-window" command.

The "confirm-problem-solved" *Task* will trigger a dialog using the Dialog agent. It returns the confirmation and related response in the form of context.

Unfortunately, assuming, the confirmation is negative. The water is still pulling in. A new "window-fail-to-close" *Situation* is created with all the current context information.

The Logics under this *Situation* is:

```
"logics": {
    "close_window": "vda.close_wdw_status",
    "window_malfunc": "vda.wdw_malfunc_detect",
    "window_broken": "vda.broken_wdw_detect"
}
```

In the above Logics, the "close_window" context is already filled from the previous situation handling process. Therefore, the context is carried over.

Assume we have the following contexts (in addition to all other contexts we have had) after applying the Logics:

```
"context": {
    "close_window": true,
    "window_malfunc": false,
    "window_broken: false
}
```

A similar *Situation* was found that has Remedy:

```
"add confirm-passenger task: window-is-jammed"
```

The answer populates the Context. Assume that the Context is: "window-is-jammed": true.

A new similar *Situation* "window-is-jammed" is found and the remedy is:

```
"add open-window task"
"add request passenger task: remove foreign obj"
"add close-window task"
"add confirm-problem-solved task"
```

Assuming the final confirmation is positive, and the *Situation* is resolved. The newly logged *Situation* and the history will be saved to the *Situation* case library and Task case library. In case the final confirmation is negative, and the agent could not find a relevant *Situation*. In that case, VSA may send the *Situation* to SHUI, and human intervention will be called to resolve the *Situation*.

### Example 2: The story of "Pharmacy"

[*Passenger Joe went on a business trip. He rides in a vehicle towards the hotel. He passed by a pharmacy and realized that he can pick up a prescription there.*]

JOE:  Could you stop by that pharmacy?

    [*The vehicle requests the Map Agent to find a pharmacy that is on the way to the hotel. The vehicle shows the map location of a pharmacy on the screen in the vehicle.*]

VEHICLE:  Do you want to go to this pharmacy?.

JOE:  No, I'd like to go to the one we just passed. [*Joe only wants to go the pharmacy he just saw.*]

    [*The Map Agent presents more nearby pharmacies on the map on the screen.*]

VEHICLE:  How about these?

    [*Joe points to the one he wants to go on the touch screen.*]

VEHICLE:  Will you come back and continue your trip?

JOE:  Yes.

VEHICLE:  How long should I wait?

JOE:  Maybe 10 to 15 minutes.

VEHICLE:  I will wait for you at the front door of the store in 10 minutes.

    [*The vehicle turns around and drives to the pharmacy.
The vehicle offboards Joe at the pharmacy.*

    *10 minutes later, the vehicle will be back to resume the trip to the hotel.*]

Here is how this *Situation* proceeds in VSA:

The Dialogue Agent posts a "POI_dropoff" *Situation* (POI - point-of-interest) on the Situation Queue.

When VSA receives the "POI_dropoff" *Situation* on the Situation Queue, it attempts to handle the *Situation*.

The *Situation* Header looks like this:

```
Situation Name: POI_dropoff
Task: Drive_task
Context: {
    current_location: location ...,
    stop_location: location ...,
    stop_type: "stop_by",
    wait_time: 15
}
```

The situation handling finds a previous "POI_dropoff" *Situation* in the case library. The Context of the retrieved old *Situation* has "stop_type" of "final destination", which means the passenger would choose the "stop-location" as her final destination, she would not continue her original journey. The final destination of the trip was changed to the "stop_location" of the *Situation*, defined in the Context. The retrieved *Situation* has three *remedy actions* in the **Remedy**:

```
[
 {"operation": "abort at drive_task"...},
 {"operation": "add after current_drive_task"...},
 {"operation": "modify at next_offboard"...}
]
```

The **Remedy** is:

1. abort the current `drive_task`;
2. add a `drive` task to drive to the **stop-location**;
3. modify the `offboard_task` so that the offboard location is changed to the **stop-location**.

The final destination of the trip was changed to the new "**stop_location**", defined in the Context.

The new "`POI_dropoff`" *Situation*, however, is different such that the passenger will continue his journey to his original destination. This is defined in the goal of the *Situation*.

When **Remedy** of the retrieved *Situation* was adapted to the new "`POI_dropoff`" *Situation*, it encounters an exception in the validation (Fig. 2②, Fig. 3④), because the goals of the new *Situation* are different. One of the new goals is that the final destination should be the same as the original destination, instead of the "**stop_location**". This exception is captured and VSA will send the *Situation* to SHUI. A new **Remedy** is created manually in SHUI and is sent back (Fig. 4②) to VSA. The new **Remedy** has six remedy actions:

```
[
 {"operation": "abort at drive_task"...},
 {"operation": "add after current_drive_task"...},
 {"operation": "add after stop_drive"...},
 {"operation": "add after new_offboard_task"...},
 {"operation": "add after wait_task"...},
 {"operation": "add after onboard_task"...}
]
```

1. abort the current drive_task;
2. add a drive task (stop_drive) to drive to the "stop-location";
3. add an offboard task at the "stop-location";
4. add a wait task after the offboard task;
5. add an onboard task after the wait_task;
6. add a drive task after the onboard task that drives to the final destination.

Applying this **Remedy**, the new plan passes validation (Fig. 5). The new *Situation* with the revised **Remedy** is saved to the *Situation* case library so that next time, similar *Situation* will be handled without human intervention.

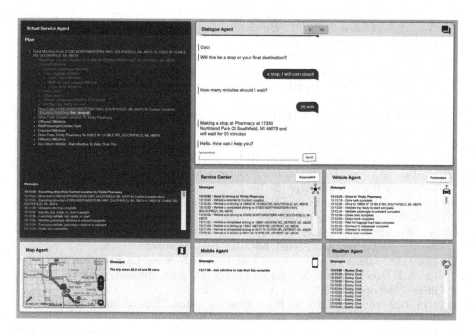

Fig. 5. The repaired plan

# 6  Conclusions

This paper focuses on solving planning problems for a service agent who faces possibly unlimited *Task* and *Situation* types, with additional context variations, in the real-world. Hard-coding domain-specific knowledge in such a system does not scale. This paper introduces a comprehensive solution that illustrates the possibility of adopting generic structures for tasks and situations, and completely embedding problem-solving knowledge in executed cases, both for task planning and situation handling. The cases can be reused to solve similar new problems. It enables the system easily expandable by continually injecting new *Task* plan cases and *Situation* handling cases.

# References

1. Aeronautiques, C., et al.: PDDL—The Planning Domain Definition Language. Technical report (1998)
2. Aha, D.W.: Goal reasoning: foundations, emerging applications, and prospects. AI Mag. **39**(2), 3–24 (2018)
3. Bergmann, R., Wilke, W.: On the role of abstraction in case-based reasoning. In: Smith, I., Faltings, B. (eds.) EWCBR 1996. LNCS, vol. 1168, pp. 28–43. Springer, Heidelberg (1996). https://doi.org/10.1007/BFb0020600
4. Chien, S.A., Knight, R., Stechert, A., Sherwood, R., Rabideau, G.: Using iterative repair to improve the responsiveness of planning and scheduling. In: AIPS, pp. 300–307 (2000)

5. Cox, M.T., Muñoz-Avila, H., Bergmann, R.: Case-based planning. Knowl. Eng. Rev. **20**(3), 283–288 (2005)
6. Dannenhauer, D., Muñoz-Avila, H.: Goal-driven autonomy with semantically-annotated hierarchical cases. In: Hüllermeier, E., Minor, M. (eds.) ICCBR 2015. LNCS (LNAI), vol. 9343, pp. 88–103. Springer, Cham (2015). https://doi.org/10.1007/978-3-319-24586-7_7
7. Fikes, R.E., Nilsson, N.J.: STRIPS: a new approach to the application of theorem proving to problem solving. Artif. Intell. **2**(3–4), 189–208 (1971)
8. Gerevini, A., Serina, I.: Fast plan adaptation through planning graphs: local and systematic search techniques. In: AIPS, pp. 112–121 (2000)
9. Ghallab, M., Nau, D., Traverso, P.: Automated Planning and Acting. Cambridge University Press, Cambridge (2016)
10. Hammond, K.J.: Case-Based Planning: Viewing Planning as a Memory Task. Academic Press, Cambridge (1989)
11. Kambhampati, S., Hendler, J.A.: A validation-structure-based theory of plan modification and reuse. Artif. Intell. **55**(2–3), 193–258 (1992)
12. Leake, D., Jalali, V.: Context and case-based reasoning. In: Brézillon, P., Gonzalez, A.J. (eds.) Context in Computing, pp. 473–490. Springer, New York (2014). https://doi.org/10.1007/978-1-4939-1887-4_29
13. Nau, D.S., et al.: SHOP2: an HTN planning system. J. Artif. Intell. Res. **20**, 379–404 (2003)
14. Sacerdoti, E.D.: Planning in a hierarchy of abstraction spaces. Artif. Intell. **5**(2), 115–135 (1974)
15. Scholnick, E.K., Friedman, S.L.: Planning in context: developmental and situational considerations. Int. J. Behav. Dev. **16**(2), 145–167 (1993)
16. Tate, A.: Generating project networks. In: Proceedings of the 5th International Joint Conference on Artificial Intelligence, vol. 2, pp. 888–893 (1977)
17. Van Der Krogt, R., De Weerdt, M.: Plan repair as an extension of planning. In: ICAPS, vol. 5, pp. 161–170 (2005)
18. Xiao, Z., Wan, H., Zhuo, H.H., Lin, J., Liu, Y.: Representation learning for classical planning from partially observed traces. arXiv preprint arXiv:1907.08352 (2019)
19. Yang, Q.: Formalizing planning knowledge for hierarchical planning. Comput. Intell. **6**(1), 12–24 (1990)
20. Zhuo, H.H., Muñoz-Avila, H., Yang, Q.: Learning hierarchical task network domains from partially observed plan traces. Artif. Intell. **212**, 134–157 (2014)

# Learning Adaptations for Case-Based Classification: A Neural Network Approach

Xiaomeng Ye⬤, David Leake(✉)⬤, Vahid Jalali, and David J. Crandall

Luddy School of Informatics, Computing, and Engineering, Indiana University, Bloomington, IN 47408, USA
xiaye@iu.edu, leake@indiana.edu, vjalalib@alumni.iu.edu, djcran@iu.edu

**Abstract.** Case-based Reasoning (CBR) solves a new problem by retrieving a stored case for a similar problem and adapting its solution to fit. Acquiring the required case adaptation knowledge is a classic problem. A popular method for addressing it is the case difference heuristic (CDH) approach, which learns adaptations from pairs of cases based on their problem differences and solution differences. The CDH approach was originally used to generate adaptation rules, but recent CBR research on case-based regression has investigated replacing learning rules with learning CDH-based network models for adaptation. This paper presents and evaluates a neural network-based CDH approach for learning adaptation models for classification, C-NN-CDH. It examines three variants, (1) training a single neural network on problem-solution differences, (2) segmenting adaptation knowledge by the classes of source cases, with a separate neural network to generate adaptations for each class, and (3) adapting from an ensemble of source cases and taking the majority vote. Experimental results demonstrate improved performance compared to previous research on statistical methods for computing CDH differences for classification. Additional results support that C-NN-CDH achieves classification performance comparable to that of multiple classic classification approaches.

**Keywords:** Case adaptation · Case difference heuristic · Classification · Ensemble learning · Neural network-based adaptation

## 1 Introduction

Case-based Reasoning (CBR) solves a new problem by retrieving a stored case with a similar problem and adapting the solution to accommodate the new problem (e.g., [1,16,20,26]). Case-based reasoning is appealing for properties such as enabling a natural knowledge capture process for cases in suitable domains (e.g., [21]), facilitating knowledge acquisition, and interpretability of cases to justify solutions [6,16].

However, obtaining the adaptation knowledge needed to adapt prior solutions is a classic challenge. In response, extensive research has explored the use

© Springer Nature Switzerland AG 2021
A. A. Sánchez-Ruiz and M. W. Floyd (Eds.): ICCBR 2021, LNAI 12877, pp. 279–293, 2021.
https://doi.org/10.1007/978-3-030-86957-1_19

of machine learning methods to acquire case adaptation knowledge for both classification (e.g., [8,12]) and regression (e.g., [9,11,19,24,25]). An interesting recent direction is the use of neural network methods for case difference heuristic (CDH) learning of adaptations for case-based regression. This paper presents and evaluates a neural network method for learning the adaptation knowledge needed for case-based classification.

The case difference heuristic approach, first proposed by Hanney and Keane [10], is one of the most-used methods to learn adaptation knowledge. It takes pairs of cases from the case base and from each pair learns a rule to adapt one case to another. As a simplified example, consider applying CBR to apartment rental price prediction. Suppose two apartments A and B are very similar, except that A has a carpeted floor while B has a wooden floor, and that the rent for B is $200 more. By comparing apartments A and B, a CDH approach might learn the rule that changing from carpet to wooden floor increases an apartment's rent by $200. An issue for CDH approaches is how to generalize the difference between case pairs. For example, rather than learning an absolute price difference from the two apartments, a CDH approach could learn the percent change or another characterization of the observed difference.

Recent work by Liao, Liu, and Chao [19] learns adaptations for regression tasks with a case difference heuristic approach using a neural network to learn the difference characterization. Their approach trains a network to map problem differences to differences in output values, avoiding the need to pre-define generalization strategies. Jalali, Leake, and Forouzandehmehr [12] apply the CDH approach to classification, using a statistical method to generate case adaptation rules for classification.

To our knowledge, this study presents the first neural network-based case difference heuristic approach to classification. Our approach, which we refer to as case-based **C**lassification with **N**eural **N**etwork-based **CDH** (C-NN-CDH), uses neural networks to learn adaptation knowledge from pairs of cases. Experimental results on multiple data sets show that the C-NN-CDH approach outperforms the statistical approach of [12], the previous state of the art for statistical adaptation for classification.

This study investigates five variants of the C-NN-CDH approach. Some variants segment the pairs of cases based on their classes and train a separate model per segment. The segmented variants offer faster training but slightly lower accuracy. We also tested variants using an ensemble of adaptations; these provide roughly comparable performance to their non-ensemble counterparts. A variant taking a majority vote of one adaptation from each class provides additional accuracy for certain data sets. Comparisons with a sampling of standard classification approaches supports that the accuracy of the C-NN-CDH approaches is competitive with those methods. In particular, this hybrid method provides accuracy comparable to that of a network-only method dedicated to the classification task, while its use of CBR provides at least two benefits: inertia-free lazy learning (enabling online learning and avoiding the need for costly retraining with new data), and the ability to provide cases that can be considered when assessing a classification. Similar cases can be useful for explanation [6], and less similar cases—for example, whose classifications are changed by adaptation—may be useful as nearest unlike neighbors [23] or counterfactual explanations [13].

## 2   Background

### 2.1   Learning Case Adaptation Knowledge

Adaptation is arguably the most difficult process in CBR. Much CBR research has applied machine learning to acquire adaptation knowledge of different forms. Some approaches apply case-based reasoning to adaptation. For example, Leake et al. [17] present a method in which a case base of adaptation cases is populated from past successful adaptations, and Craw et al. [5] present a method to assemble pairs of stored cases, retrieve the pair most similar to the pair of a retrieved case and the query, and adapt the retrieved case to the current query.

The case difference heuristic (CDH) approach [10] is a widely used knowledge-light method for learning case adaptation rules from knowledge contained in the case base. For each pair of cases, the CDH approach generates an adaptation rule capturing the transformation needed to adapt the solution of one case into the solution of the other. Specifically, it attributes the difference in the solution descriptions of the cases to the difference in their problem descriptions. When deciding applicability of the generated rules in the future, the rule is considered to apply if the difference between the retrieved case's problem and the new problem is similar to the difference from which the rule is generated. Thus the similarity to the original difference becomes the antecedent for the rule. When triggered, the rule adapts the solution of the retrieved case according to the previous solution difference.

Applying the CDH approach requires addressing several design questions. One concerns how to calculate problem differences; another concerns how to select the case pairs for training (e.g., from pairs of neighboring cases or from random pairs); another concerns how to translate a raw solution difference into the change to be effected by the rule, for example, as an additive, multiplicative or other change. Such design questions lead to many variations of CDH-based systems (e.g. [5,11,19,25]). This paper considers both standard pair selection methods and a new training pair selection approach (based on class-to-class classification, described in Sect. 2.2), and proposes a general approach for neural network CDH-based classification.

**Augmenting CDH with Network Methods.** Adaptation rule generation using the CDH approach is often shaped by pre-defined criteria for generating rules from differences. In contrast, machine learning-based approaches [4,5,19,25,32] provide increased flexibility. Liao, Liu, and Chao [19], Policastro, Carvalho, and Delbem [25], and Zhang et al. [32] propose a network approach in which the network generates adaptations from a problem and retrieved solution passed to the network. Part of the appeal of network-based approaches is that network learning facilitates the generation of more complex transformation functions. Outside of CBR but in similar spirit, Wetzel et al. [27] use a Siamese network to predict target value differences given two data points, and predict a target value using an ensemble of training data points. Leake, Ye, and Crandall [18] follow the CDH approach of Craw, Wiratunga, and Rowe [5], by considering both problem difference and adaptation context, and follow Liao, Liu, and Chao [19] in proposing a neural network-based case difference heuristic approach, NN-CDH, as a general technique for regression tasks.

**Applying the CDH Method to Classification.** The traditional CDH approach has been successfully applied in solving regression tasks. However, less research has addressed classification tasks or dealing with nominal attributes in generating adaptation rules. Early CDH approaches for classification or dealing with nominal attributes relied on exact matching and binarization (e.g., [3,8]) which can only express the relationship between nominal values using one bit of information (i.e. 0 and 1). More recently, CDH was enhanced with the Value Difference Metric (VDM) and ensemble methods [12]. VDM is a probabilistic method to measure similarity that enables comparing nominal values in a one dimensional numeric space. That work used an ensemble of adaptations for classification (EAC), retrieving multiple source cases, generating needed adaptation rules, adapting from all retrieved source cases, and producing a final solution by the majority vote of all adapted solutions.

To illustrate the importance of expressive power in comparing nominal values, consider a sample classification task in which the goal is to decide whether a given fruit is an apple, based on its color. In this case, similarity based on exact match of color results in identical treatment of the difference between the colors red and yellow and the colors red and blue. However, using a more expressive method such as VDM enables recognizing the relative proximity of red and yellow compared to that of red and blue. Recent advances in deep neural networks have made it possible to take the expressive power in comparing nominal values a step further by expressing them in multi-dimensional space as embedding vectors. To the best of our knowledge this has not been exploited previously for CDH learning.

## 2.2    Class-to-Class Classification

The Class-to-class (C2C) approach to classification is a difference-based method that classifies a query based on instances from multiple classes [28–30]. A C2C model first learns the similarity and difference patterns between pairs of classes. Given two cases, the trained C2C model can determine whether their similarity and difference conform to learned patterns. If they do, the C2C model can provide evidence for their belonging to the corresponding classes. The C2C approach inspires a method for choosing CDH training pairs bridging each pair of classes, as described in Sect. 3.

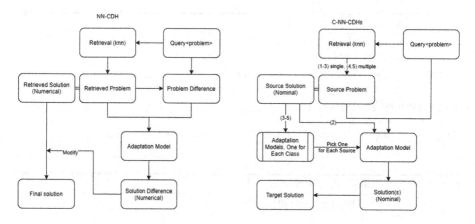

**Fig. 1.** Workflows of NN-CDH (left) and the C-NN-CDHs (right). Variants (1–5) of C-NN-CDHs involve different procedures and arrows are marked with the corresponding numbers to reflect that.

Traditional case-based classification methods explain their conclusion by the most similar case retrieved and its adaptation, while C2C methods have the benefit of explaining with both supportive and contrastive evidence. The contrastive evidence from the C2C approach is one type of counterfactual explanation (cf. Keane and Smyth [13] and Kenny and Keane [14]). For example, for an applicant rejected for a loan, the supportive explanation is that another applicant with similar attributes is also rejected, and the contrastive explanation is that this applicant shares many attributes of an accepted applicant but has a worse credit history. The C2C-inspired adaptation variant presented in this paper captures the information needed for contrastive explanation of classifications.

## 3 An NN-CDH Approach for Classification

NN-CDH and other CDH methods learn from pairs of cases. One of the pairs is treated as the source case (with its source problem and solution) and the other as the target (with its target problem and solution), where the source is to be adapted toward the target. For simplicity, we will refer to these as a case pair. A CDH method learns an adaptation rule to adapt the solution of the source case to provide a solution for the target case. For an NN-CDH approach, the CBR system first retrieves a source case similar to the query (target case), and calculates the problem difference between the source problem and the target problem. The problem difference is then passed to a neural network, which is previously trained on problem and solution differences of training case pairs. The neural network predicts the solution difference between the source solution and the target solution. Finally, the CBR system applies the predicted solution difference to the source solution, and uses the adapted result as the final prediction. This process is shown in Fig. 1.

---

**Algorithm 1.** C-CDH, variant (0)

---

1: **for** each $i$ in all classes **do**
2:     case_pairs[$i$] $\leftarrow$ {}
3: **for** each *source* and *target* in *pairs* **do**
4:     case_pairs[$sol(source)$].append([$prob(source), prob(target){:}sol(target)$])
5: **procedure** Testing(*query*)
6:     $retrieved \leftarrow$ 1-nn($CB, query$)
7:     $r \leftarrow$1-nn(case_pairs[$sol(retrieved)$], [$prob(retrieved), prob(query)$])
8:     **return** $r$

---

**Algorithm 2.** C-NN-CDH, variant (1-2)

---

1: $adapt\_NN \leftarrow$ new classification neural network
2: case_pairs $\leftarrow$ {}
3: **for** each *source* and *target* in *pairs* **do**
4:     **if** using variant (1) **then**
5:         case_pairs.append([$prob(source), prob(target){:}sol(target)$])
6:     **else if** using variant (2) **then**
7:         case_pairs.append([$prob(source), prob(target), sol(source){:}sol(target)$])
8: $adapt\_NN$.fit(case_pairs)
9: **procedure** Testing(*query*)
10:     $retrieved \leftarrow$ 1-nn($CB, query$)
11:     $r \leftarrow adapt\_NN$.predict($prob(retrieved), prob(query), sol(retrieved)$)
12:     **return** $r$

---

Calculating the difference between problem or solution values requires a difference function, which is especially difficult to define for nominal values. Our method replaces the traditional CDH difference calculation by the implicit calculation of a machine learning technique (e.g. neural network), potentially taking into account not only the difference, but the context of the source case itself. We name this general approach of handling pairs of cases as the case difference heuristic approach for classification ("C-CDH").

As a baseline testbed system, we implemented a C-CDH system that stores case pairs treated as adaptation rules. The case pairs are grouped based on source solution. We refer to this as variant (0) and describe it in Algorithm 1. The system performs classification by retrieving the most similar source case, and retrieving the case pair, which is selected to share the same source solution and has the most similar source problem and target problem (cf. [22]). The target solution of the retrieved case pair is used as the final classification.

Our implementation of C-CDH uses a classification neural network to learn and predict the target solution based on information from the source problem and the target problem. This is the basic version of classification with a neural network-based case difference heuristic (C-NN-CDH) approach and will be referred to as variant (1). As a direct extension, we built variant (2) in which the adaptation neural network also takes in the source solution as input. Variants (1) and (2) are described in Algorithm 2.

Variants (3–5) add grouping of case pairs based on their source solutions. The target solutions in a group are not restricted. With grouping, an adaptation neural network can be trained on a specific group of case pairs to learn the adaptation knowledge where the source solution is determined. In other words, each **specialized adaptation neural network** learns how to adapt cases of a specific solution toward all solutions (including the source solution). By segmenting the pairs of cases based on source solutions, we naturally incorporate the source solution as an important input, as it determines which specialized adaptation neural network to use. The training is also easier as one group of case pairs is more homogeneous and the knowledge to learn is more specific. Variants (3–5) share the same training procedure for their specialized adaptation neural networks but differ in their testing procedures.

Variant (3) predicts the target solution by retrieving the most similar source case and using one specialized adaptation neural network. Inspired by the ensemble of adaptations for classification approach (EAC) [12], variant (4)—named EAC-NN-CDH—retrieves $k$ multiple similar source cases (we used $k = 3$), adapts all source cases using corresponding specialized adaptation neural networks, and selects a classification by majority vote. Ties are broken arbitrarily.

Variant (5) is inspired by the class-to-class (C2C) approach and is named C2C-NN-CDH. C2C-NN-CDH retrieves one most similar source case from each class, adapts all source cases using their corresponding networks, and uses the majority vote to decide the final classification. The voting process is similar to the all-versus-all approach in multiclass classification [2]. In principle, the classification from this approach could be explained contrastively by reference to cases for other classes that support the majority vote. Variants (3–5) are described in Algorithm 3.

Variants (1–5) are illustrated in Fig. 1. All variants are summarized below:

(0) Non-network C-CDH.
(1) C-NN-CDH with one adaptation neural network that considers source problem and target problem.
(2) Based on (1), but also considers source solution.
(3) Uses multiple specialized adaptation neural networks.
(4) Based on (3), but uses an ensemble of adapted solutions from multiple cases.
(5) Based on (4), but uses an ensemble of adapted solutions from multiple cases of all classes.

## 4    Evaluation

We evaluated all variants (0–5) on two groups of data sets. The first group of data sets follows those used in Jalali, Leake, and Forouzandehmehr [12], to enable comparison with the previous state-of-the-art on statistical CDH classification. Experiments on this group allow comparison with the ensemble approach EAC and EAC-retrieval, an ablated EAC removing the adaptation component. The second group is a subset of data sets in the comparative evaluation of classification algorithms by Zhang et al. [31]. Experiments on this group allow comparison

**Algorithm 3.** C-NN-CDH, variant (3-5)

1: **for** each *case* in $CB$ **do**
2:     $CB\_by\_class[sol(case)]$.append($case$)
3: **for** each $i$ in all classes **do**
4:     case_pairs[$i$] ← {}
5: **for** each *source* and *target* in *pairs* **do**
6:     case_pairs[$sol(source)$].append([$prob(source), prob(target):sol(target)$])
7: **for** each $i$ in all classes **do**
8:     $adapt\_NN[i]$ ← new classification neural network
9:     $adapt\_NN[i]$.fit(case_pairs[$i$])
10: **procedure** Testing($query$)
11:     **if** using variant (3) **then**
12:         $retrieved$ ← 1-nn($CB, query$)
13:         $r$ ← $adapt\_NN[sol(retrieved)]$.predict($prob(retrieved), prob(query)$)
14:     **else if** using variant (4) **then**
15:         **for** each *retrieved* in k-nn($CB, query$) **do**
16:             $rs$.append( $adapt\_NN[sol(retrieved)]$.predict($prob(retrieved), prob(query)$))
17:         $r$ ← majority_vote($rs$)
18:     **else if** using variant (5) **then**
19:         **for** each $i$ in all classes **do**
20:             $retrieved$ ← 1-nn($CB\_by\_class[i], query$)
21:             $rs$.append( $adapt\_NN[i]$.predict($prob(retrieved), prob(query)$))
22:         $r$ ← majority_vote($rs$)
23:     **return** $r$

with algorithms evaluated in that paper, including: Extreme Learning Machine (ELM), Sparse Representation-based Classification (SRC), Deep Learning (DL), Support Vector Machine (SVM), Random Forests (RF), AdaBoost (AB), C4.5, Naive Bayes classifier (NB), K Nearest Neighbours classifier (KNN), and Logistic Regression (LR). We compare the results of runs of our systems with the reported results from Jalali, Leake, and Forouzandehmehr [12] and Zhang et al. [31]. We note that the data preprocessing steps are not described in detail in the two papers. This may result in minor variations.

All data sets are for classification tasks, taken from the UCI repository [7]. All nominal values are converted to one-hot encoding and all numeric values are standardized by removing the mean and scaling to unit variance. For most data sets, five 10-fold cross validations are carried out, where 10% of the total cases are used for testing and 90% are used for training (only two 10-fold cross validations are run for two larger data sets with excessive training time). The average accuracies and balanced accuracies (with their standard deviations) of all runs for each data set are recorded. Standard deviations are omitted in reports below as almost all are less than 0.05. Balanced accuracy in general is comparable to accuracy but not shown due to space limitations.

## 4.1   Assembling Case Pairs

As discussed in Sect. 2.1, the collection of case pairs is a design problem for CDH. Given a training data set of $n$ cases and $m$ classes, our test implementations learn adaptation knowledge from three kinds of pairs from the case base:

- Neighboring pairs: Each case is paired with its nearest neighbor using 1-NN (k-nearest neighbor with $k = 1$). There are $n$ neighboring pairs.
- Random Pairs: Each case is paired with 10 random cases. There are $10n$ random pairs.
- Class-to-class Pairs (C2C Pairs): Each case is paired with its nearest neighbor in every other class. There are $n(m - 1)$ pairs.

Each type of pair provides one specialized form of adaptation knowledge to the adaptation model: neighboring pairs provide minor adaptations to cover small problem differences, random pairs provide random and bigger adaptations, and C2C pairs provide adaptations needed to change one case into other classes. The number of each kind of pair is a design parameter that could be fine-tuned. Pairs might also be selected according to other criteria such as generality or applicability, but this is beyond the scope of this study.

## 4.2   Implementation Details

For the C-CDHs, the retrieval component retrieves a single case. As baselines for performance without adaptation, we also implemented nearest neighbor algorithms 1-NN and 3-NN, with 3-NN averaging the classifications of the three most similar cases.

The adaptation neural network is a feedforward network with 2 hidden layers (128 and 64 nodes with ReLU activation functions) and an output layer with softmax activation function. The loss function is categorical cross entropy and the model is optimized using Adam [15]. For comparison, a neural network classifier is implemented with the same configuration. Note that the adaptation neural network is a component in the CBR system (C-NN-CDH) and produces the final classification based on a retrieved case and the query, while the neural network classifier directly produces the final classification based solely on the query. All networks are trained until their parameters converge.

For 3-NN and 1-NN, all training cases are used as the case base. For the neural network classifier, 10% of the training cases are separated out for training validation. For all the C-CDHs, pairs are assembled using methods described in Sect. 4.1. For non-network variant (0), all case pairs are stored for future search. For the adaptation neural networks in variants (1–5), 95% of the case pairs are used for training and 5% for validation.

## 4.3   C-NN-CDH vs. EAC

Table 1 compares the accuracy of C-NN-CDH with EAC. The accuracy of the best performing system and the best performing C-CDH for every data set is highlighted. We observe:

**Table 1.** Accuracies of systems compared with EAC [12]

| | EAC | | Baseline Systems | | | C-CDHs | | | | | |
|---|---|---|---|---|---|---|---|---|---|---|---|
| | EAC-retrieval | EAC | NNet | 3-NN | 1-NN | (0) | (1) | (2) | (3) | (4) | (5) |
| Credit | 81.89 | 84.36 | **86.33** | 83.33 | 79.78 | 70.04 | 82.05 | 81.53 | 82.29 | 82.53 | *84.53* |
| Balance | 74.74 | 84.02 | 97.37 | 79.64 | 77.69 | 65.61 | **97.92** | 97.85 | 97.63 | 97.37 | 97.60 |
| Car | 93.50 | 96.05 | 99.74 | 83.13 | 78.10 | 68.22 | **99.93** | 99.91 | 98.9 | 99.36 | 99.46 |

- The non-network C-CDH often makes the retrieval result of 1-NN worse. Because all variants use the same retrieval, this can be ascribed to variant (0) actually impairing performance. We hypothesize that this is due to using untuned retrieval for case pairs.
- The C-NN-CDHs have slightly different performance, the best of which is on par with that of the neural network classifier.
- The C-NN-CDHs consistently improve the retrieval result of 1-NN. The C-NN-CDHs also outperform EAC in many experiments. Note that the EAC-retrieval, by using a probability-guided metric, is often better than 1-NN. This means that C-NN-CDHs build on worse retrieval than EAC but end with better results, demonstrating the value of their learned adaptation capability.

### 4.4    C-NN-CDH vs. Other Classification Algorithms

We compare variants of C-CDH with 11 state-of-the-art classification algorithms (referred to as "other algorithms" in the next paragraph) that are not necessarily related to CBR [31]. Data sets are chosen to be compatible with the our baseline and proposed systems. In other words, they require no additional preprocessing and are not complicated data such as images or structured sequences.

In Table 2, for each data set, the accuracies of the baseline systems and C-CDHs are listed. In Table 3, the best and the worst of other algorithms and their corresponding accuracies are listed for comparison. Last, the best performing C-NN-CDH is chosen and its projected rank is shown in Table 3—i.e., the rank if it were ranked among other algorithms. We observe:

- The average rank of our best C-NN-CDHs is 3.4. In Zhang et al. [31], SVM has an average rank of 3.5 and is the third best among the 11 classifiers in terms of average rank (however we do not test C-NN-CDHs on all the data sets as in Zhang et al. [31]).
- C-NN-CDHs do not always improve the final result compared to the simple retrieval of 1-NN. For example, when testing on white wine quality, all C-NN-CDHs perform worse than 1-NN. We hypothesize that this is due to the high number of classes in this data set and the subjective nature of wine quality. When the relation between problem and solution is highly volatile, nearest neighbor is already a good guess while any adaptation might alter the prediction for worse.

**Table 2.** Variant characteristics and variant accuracies compared with classifiers in [31]

| | Baseline systems | | | C-CDHs | | | | | |
|---|---|---|---|---|---|---|---|---|---|
| | NNet | 3-NN | 1-NN | (0) | (1) | (2) | (3) | (4) | (5) |
| Neural Network | Yes | | | | Yes | Yes | Yes | Yes | Yes |
| Segmented Training | | | | | | | Yes | Yes | Yes |
| Ensemble | | Yes | | | | | | Yes | Yes |
| Class-to-class | | | | | | | | | Yes |
| Yeast | **59.83** | 50.10 | 53.17 | 42.30 | 52.18 | 51.76 | 46.48 | 49.23 | *58.33* |
| Seeds | 94.0 | 91.52 | 92.85 | 85.23 | **95.33** | **95.33** | 93.61 | 95.14 | 94.85 |
| Pima | **75.36** | 73.20 | 70.57 | 64.08 | 69.40 | 69.55 | 68.02 | 69.08 | *72.16* |
| Page-blocks | **96.85** | 96.65 | 96.46 | 94.73 | 95.95 | 95.66 | 94.89 | 95.15 | *96.75* |
| Contraceptive | **54.39** | 41.03 | 43.46 | 38.91 | 47.44 | 47.78 | 48.70 | 48.59 | *51.39* |
| White Wine | 57.63 | 54.25 | **65.69** | 46.93 | 63.53 | 63.44 | 55.04 | 60.35 | *63.54* |
| Balance | 97.37 | 79.64 | 77.69 | 65.61 | **97.92** | 97.85 | 97.63 | 97.37 | 97.60 |
| Car | 99.74 | 83.13 | 78.10 | 68.22 | **99.93** | 99.91 | 98.90 | 99.36 | 99.46 |

**Table 3.** Rank of best C-CDHs among classifiers in Zhang et al. [31]

| | Best | | Best C-CDH | | | Worst | |
|---|---|---|---|---|---|---|---|
| | Name | Accuracy | Name | Accuracy | Rank | Name | Accuracy |
| Yeast | ELM | 64.87 | (5) | 58.33 | 7 | DL | 33.11 |
| Seeds | KNN | 95.24 | (1,2) | 95.33 | 1 | DL | 23.81 |
| Pima | AB | 83.12 | (5) | 72.16 | 6 | DL/SRC | 59.74 |
| Page-blocks | SVM | 94.44 | (5) | 96.75 | 1 | ELM/DL | 87.04 |
| Contraceptive | GBDT | 55.41 | (5) | 51.39 | 7 | AB/DL | 41.22 |
| White Wine | AB/DL | 56.94 | (5) | 63.54 | 1 | NB | 39.59 |
| Balance | SRC | 1.0 | (1) | 97.92 | 2 | DL | 46.03 |
| Car | GBDT | 1.0 | (1) | 99.93 | 2 | AB/DL | 67.05 |

## 4.5  C-NN-CDH vs. Baseline Neural Network

Because standard deviation is not reported in the work being compared to C-NN-CDH [12, 31], we are not able to calculate a P-value stating the significance of the difference between C-NN-CDHs and their methods. However, we are able to do so for the difference between the best performing C-NN-CDH and the baseline neural network in Table 4. The P-value is the probability of obtaining the observed difference between the samples if the null hypothesis were true. The null hypothesis states that the two distributions of results are the same. The calculation is based on the assumption that the distributions are normal. As Table 4 shows, the neural network wins in three data sets, C-NN-CDH wins

**Table 4.** The significance of the difference between best performing C-NN-CDH and baseline neural network

|  | NNet Better <————————> C-NN-CDH Better | | | | | | | | |
|---|---|---|---|---|---|---|---|---|---|
|  | Contra. | Pima | Credit | Yeast | Page-b. | Seeds | Balance | Car | White W. |
| NNet | .5439 | .7536 | .8633 | .5983 | .9685 | .9400 | .9737 | .9974 | .5763 |
| C-NN-CDH | .5139 | .7216 | .8453 | .5833 | .9675 | .9533 | .9792 | .9993 | .6354 |
| P-value | .0004 | .0006 | .0525 | .0687 | .6761 | .1407 | .1373 | .0174 | <.0001 |

in two, and there are no significant differences between the two in the remaining half of the data sets.

It is expected that Table 4 does not show a clear advantage of C-NN-CDH over neural network in terms of accuracy, because the two use the same architecture and are naturally of similar power. However, C-NN-CDH is a component generally applicable to CBR classification systems, which can offer benefits such as lazy learning and explainability, in contrast to a neural network.

## 4.6   Evaluation Summary

From experiments on both groups of data sets, we answer the following questions:

- **Can the neural network effectively learn adaptation knowledge?** Yes. One or more C-NN-CDHs can always provide performance comparable to or even better than that of the neural network classifier. In principle the adaptation neural networks might learn to discard the source problem and solely use the target problem to predict the target solution, effectively performing as a neural network classifier. However, our experiments reveal that this is not the case, because (1) the weights associated with the source problem are non-zero, and (2) as shown in Table 4, C-NN-CDHs perform significantly differently from the neural network in multiple experiments. C-NN-CDHs are indeed learning adaptation knowledge. This demonstrates that if a neural network is powerful enough to tackle the classification task directly, it may also be powerful enough to learn the adaptation knowledge or the relation between pairs of cases in the task domain.
- **Is the source solution an important attribute to consider in adaptation?** Not necessarily. A surprising result is that variant (1) actually performs almost identically to variant (2), which also considers the source solution in adaptation. We speculate that this is because the source solution is heavily coupled with the source problem, and therefore does not provide additional information useful in adaptation.
- **Does segmenting pairs of cases by source case solution lead to better performance?** This depends. In terms of accuracy, variants (1,2) actually perform better than variants (3,4) on most data sets. We speculate that this is because a single adaptation neural network in (1,2) is well trained with all pairs of cases, while a specialized adaptation neural network in (3,4) is trained

with a segmented group of examples. In terms of efficiency, the training time needed for variants (1,2) is several times higher than for variants (3,4). This is expected as variants (3,4) train on segmented training examples, and therefore converge faster.

- **Does an ensemble of adaptations improve accuracy?** Not really, for these data sets. EAC-NN-CDH (variant (4)) performs about the same as its counterpart variant (3) without ensemble. Jalali et al. [12] showed that EAC is a better adaptation method than applying a single adaptation, while we do not observe a significant benefit of EAC-NN-CDH over C-NN-CDH. We attribute this to the generalization power of C-NN-CDH, which produces predictions stable enough that an ensemble version does not appreciably alter its prediction.

- **Is a class-to-class approach useful for adaptation?** Yes. C2C-NN-CDH (variant (5)) performs differently from and, in many scenarios, better than the other C-NN-CDHs. C2C-NN-CDH reaches its prediction by collecting evidence from diverse source cases from all classes, which can provide more global support, especially when there are multiple classes. Moreover, the C2C approach offers the possibility of explanation with contrastive evidence.

## 5 Conclusion

The flexibility of a case-based reasoning system to solve novel problems depends on its ability to adapt prior solutions to new circumstances. The generation of knowledge for adapting cases is a classic challenge for case-based reasoning. The case difference heuristic approach is a knowledge-light method for learning adaptation knowledge. Neural network-based CDH has been successfully applied to case-based regression but not previously to classification.

This paper presents a method with multiple variants for extending network CDH for classification tasks with three contributions beyond the prior methods. First, variants (3–5) group the pairs of cases used for learning by the solutions of the source problems they adapt, generating per-category adaptation knowledge. Second, they apply one or multiple neural networks to learn the adaptation knowledge for classification. Third, variant (5) utilizes cross-class adaptation to reach a conclusion from cases of diverse classes.

In our experiments, the C-NN-CDH approach achieves better performance than EAC, the state-of-the-art statistical CDH adaptation method, and is on par with standard classification methods from outside of case-based reasoning. As a form of CBR, case-based classification using C-NN-CDH also preserves other benefits of CBR including lazy learning, suitability for online learning, and explainability.

**Acknowledgments.** We acknowledge support from the Department of the Navy, Office of Naval Research (Award N00014-19-1-2655), and the US Department of Defense (Contract W52P1J2093009).

# References

1. Aamodt, A., Plaza, E.: Case-based reasoning: foundational issues, methodological variations, and system approaches. AI Commun. **7**(1), 39–52 (1994)
2. Aly, M.: Survey on multiclass classification methods. Technical report, Caltech (2005)
3. Badra, F., Cordier, A., Lieber, J.: Opportunistic adaptation knowledge discovery. In: McGinty, L., Wilson, D.C. (eds.) ICCBR 2009. LNCS (LNAI), vol. 5650, pp. 60–74. Springer, Heidelberg (2009). https://doi.org/10.1007/978-3-642-02998-1_6
4. Corchado, J.M., Lees, B.: Adaptation of cases for case based forecasting with neural network support. In: Pal, S.K., Dillon, T.S., Yeung, D.S. (eds.) Soft Computing in Case Based Reasoning, pp. 293–319. Springer, London (2001). https://doi.org/10.1007/978-1-4471-0687-6_13
5. Craw, S., Wiratunga, N., Rowe, R.: Learning adaptation knowledge to improve case-based reasoning. Artif. Intell. **170**, 1175–1192 (2006)
6. Cunningham, P., Doyle, D., Loughrey, J.: An evaluation of the usefulness of case-based explanation. In: Ashley, K.D., Bridge, D.G. (eds.) ICCBR 2003. LNCS (LNAI), vol. 2689, pp. 122–130. Springer, Heidelberg (2003). https://doi.org/10.1007/3-540-45006-8_12
7. Dua, D., Graff, C.: UCI machine learning repository (2017)
8. D'Aquin, M., Badra, F., Lafrogne, S., Lieber, J., Napoli, A., Szathmary, L.: Case base mining for adaptation knowledge acquisition. In: Proceedings of the Twentieth International Joint Conference on Artificial Intelligence (IJCAI-07), pp. 750–755. Morgan Kaufmann, San Mateo (2007)
9. Fuchs, B., Lieber, J., Mille, A., Napoli, A.: Differential adaptation: an operational approach to adaptation for solving numerical problems with CBR. Knowl.-Based Syst. (2014, in press)
10. Hanney, K., Keane, M.T.: Learning adaptation rules from a case-base. In: Smith, I., Faltings, B. (eds.) EWCBR 1996. LNCS, vol. 1168, pp. 179–192. Springer, Heidelberg (1996). https://doi.org/10.1007/BFb0020610
11. Jalali, V., Leake, D.: Enhancing case-based regression with automatically-generated ensembles of adaptations. J. Intell. Inf. Syst. **46**(2), 237–258 (2015). https://doi.org/10.1007/s10844-015-0377-0
12. Jalali, V., Leake, D., Forouzandehmehr, N.: Learning and applying case adaptation rules for classification: an ensemble approach. In: Proceedings of the Twenty-Sixth International Joint Conference on Artificial Intelligence, IJCAI 2017, pp. 4874–4878 (2017)
13. Keane, M.T., Smyth, B.: Good counterfactuals and where to find them: a case-based technique for generating counterfactuals for Explainable AI (XAI). arXiv:2005.13997 (2020)
14. Kenny, E.M., Keane, M.T.: On generating plausible counterfactual and semi-factual explanations for deep learning. arXiv:2009.06399 (2020)
15. Kingma, D., Ba, J.: Adam: a method for stochastic optimization. arXiv:1412.6980 (2017)
16. Leake, D.: CBR in context: the present and future. In: Leake, D. (ed.) Case-Based Reasoning: Experiences, Lessons, and Future Directions, pp. 3–30. AAAI Press, Menlo Park (1996)
17. Leake, D., Kinley, A., Wilson, D.: Acquiring case adaptation knowledge: a hybrid approach. In: Proceedings of the Thirteenth National Conference on Artificial Intelligence, pp. 684–689. AAAI Press, Menlo Park (1996)

18. Leake, D., Ye, X., Crandall, D.: Supporting case-based reasoning with neural networks: an illustration for case adaptation. In: AAAI Spring Symposium on Combining Machine Learning and Knowledge Engineering (AAAI-MAKE) (2021)
19. Liao, C., Liu, A., Chao, Y.: A machine learning approach to case adaptation. In: 2018 IEEE First International Conference on Artificial Intelligence and Knowledge Engineering (AIKE), pp. 106–109 (2018)
20. López de Mántaras, R., et al.: Retrieval, reuse, revision, and retention in CBR. Knowl. Eng. Rev. **20**(3) (2005)
21. Mark, W., Simoudis, E., Hinkle, D.: Case-based reasoning: expectations and results. In: Leake, D. (ed.) Case-Based Reasoning: Experiences, Lessons, and Future Directions, pp. 269–294. AAAI Press, Menlo Park (1996)
22. McSherry, D.: An adaptation heuristic for case-based estimation. In: Smyth, B., Cunningham, P. (eds.) EWCBR 1998. LNCS, vol. 1488, pp. 184–195. Springer, Heidelberg (1998). https://doi.org/10.1007/BFb0056332
23. Nugent, C., Doyle, D., Cunningham, P.: Gaining insight through case-based explanation. J. Intell. Inf. Syst. **32**, 267–295 (2009)
24. Patterson, D., Rooney, N., Galushka, M.: A regression based adaptation strategy for case-based reasoning. In: Proceedings of the Eighteenth National Conference on Artificial Intelligence, pp. 87–92. AAAI, Menlo Park (2002)
25. Policastro, C., Carvalho, A., Delbem, A.: Automatic knowledge learning and case adaptation with a hybrid committee approach. J. Appl. Log. **4**(1), 26–38 (2006)
26. Riesbeck, C., Schank, R.: Inside Case-Based Reasoning. Lawrence Erlbaum, Hillsdale (1989)
27. Wetzel, S.J., Ryczko, K., Melko, R.G., Tamblyn, I.: Twin neural network regression. arXiv:2012.14873 (2020)
28. Ye, X.: The enemy of my enemy is my friend: class-to-class weighting in k-nearest neighbors algorithm. In: Proceedings of the Thirty-First International Florida Artificial Intelligence Research Society Conference, FLAIRS 2018, pp. 389–394 (2018)
29. Ye, X.: C2C trace retrieval: fast classification using class-to-class weighting. In: Proceedings of the Thirty-Second International Florida Artificial Intelligence Research Society Conference, FLAIRS 2019, pp. 353–358 (2019)
30. Ye, X., Leake, D., Huibregtse, W., Dalkilic, M.: Applying class-to-class Siamese networks to explain classifications with supportive and contrastive cases. In: Watson, I., Weber, R. (eds.) ICCBR 2020. LNCS (LNAI), vol. 12311, pp. 245–260. Springer, Cham (2020). https://doi.org/10.1007/978-3-030-58342-2_16
31. Zhang, C., Liu, C., Zhang, X., Almpanidis, G.: An up-to-date comparison of state-of-the-art classification algorithms. Expert Syst. Appl. **82** (2017)
32. Zhang, F., Ha, M., Wang, X., Li, X.: Case adaptation using estimators of neural network. In: Proceedings of 2004 International Conference on Machine Learning and Cybernetics (IEEE Cat. No.04EX826), vol. 4, pp. 2162–2166, August 2004

# Similar Questions Correspond to Similar SQL Queries: A Case-Based Reasoning Approach for Text-to-SQL Translation

Wei Yu, Xiaoting Guo, Fei Chen, Tao Chang, Mengzhu Wang,
and Xiaodong Wang[✉]

College of Computer, National University of Defense Technology,
Kaifu District, Changsha 410073, China
{yuwei19,guoxiaoting18,chenfei14,changtao15,
wangmengzhu19,xdwang}@nudt.edu.cn

**Abstract.** Based on the natural truth that similar questions correspond to similar SQL queries, a CBR-based approach is proposed to deal with the Text-to-SQL task in this paper. We follow the traditional CBR processes: similarity assessment, case retrieval, and case reuse. First, we introduce a neural classifier in the similarity assessment stage and comprehensively uses classification probability and literal cosine similarity to measure similarity. Then, based on the results of the similarity assessment, our model retrieves a case template. Finally, our model fills the columns and values generated by the Ranker module and Question Answering (QA) module into the solution template. At this point, a SQL query suitable for the new case is generated. We evaluate our models on a large-scale Text-to-SQL dataset—WikiSQL. Experimentally, our model has a competitive performance compared with the baseline and significantly improves the accuracy of the aggregation function prediction.

**Keywords:** Case-based reasoning · Text-to-SQL · Semantic parsing

## 1 Introduction

As a significant branch of Artificial Intelligence (AI), Case-Based Reasoning (CBR) has received more and more research attention. It is a kind of analogical reasoning that focuses on reasoning based on previous experience. The essence of CBR can be summarized as two principles: the real-world regularities and the tendency to encounter similar problem [1]. Over the years, research on CBR has led to a large number of applications in various fields from recommendation systems [4,5] to design [10,13], education [11] and health [3,12]. In this paper, CBR is applied to a novel scenario known as Text-to-SQL translation.

As we all know, humans use speech and text to communicate, but machines take logical expressions or structured data to process information. Hence, there is a huge gap between natural language and forms that machines can understand. In this case, the task of semantic parsing is proposed. The semantic parser

© Springer Nature Switzerland AG 2021
A. A. Sánchez-Ruiz and M. W. Floyd (Eds.): ICCBR 2021, LNAI 12877, pp. 294–308, 2021.
https://doi.org/10.1007/978-3-030-86957-1_20

Question: What's the minimum total attendance of the Premier League association football?      **Original data**

| column_name | League | Sport | Country | Season | Games | Average attendance | Total attendance |
|---|---|---|---|---|---|---|---|
| column_type | text | text | text | text | real | real | real |
| | ... | ... | ... | ... | ... | ... | ... |
| values | Premier League | association football | England/Wales | 2011 | 380 | 34601 | 13148465 |
| | ... | ..... | ... | ... | ... | ... | ... |

SQL:   *SELECT MIN*(Total attendance) *WHERE* League='Premier League' *AND* Sport='association football'      **Target**

**Fig. 1.** A sample of Text-to-SQL translation in WikiSQL. *Question* is the natural language description of the query about *Table*, and *SQL* is the structured query language corresponding to *Question*.

can associate the non-uniform and non-standard human natural language with abstract and rigorous logical expressions. Text-to-SQL is an important application of semantic parsing. As shown in Fig. 1, it is a sample of Text-to-SQL task in WikiSQL [27]. The task aims to translate the given natural language into an executable SQL query statement according to the table schema. It tries to build a bridge between unstructured natural language and structured SQL. In this way, users can manipulate the database using natural language without a professional background. Due to the high practicability of this task, it has a large number of application scenarios and the related research has been extremely hot in recent years.

In previous works, the mainstream approach is to treat SQL query like the following structure: SELECT *AGG(C)* FROM *T* WHERE *[ C OP V ]*.[1] By this means, the task is transformed into a slot filling problem and many sophisticated neural network structures are used to predict the slots (blue tokens). Despite the extensive works [8,14,16,19,23] that have been conducted, the *AGG* prediction is still a bottleneck for the Text-to-SQL model in WikiSQL [27] and the error of the current best model is as high as 10%.

Inspired by CBR technology, we firmly believe that similar questions should correspond to similar SQL queries. Hence, we reorganize the task from the perspective of CBR and follow the basic process of traditional CBR to deal with Text-to-SQL task. As shown in Fig. 2, our model first retrieves the solution template based on the similarity assessment, then generates the slot value required by the solution template, and finally fills the slot value into the solution template to synthesize a new solution (SQL).

As with other CBR-based tasks, our approach also has to deal with four main challenges: case representation, similarity assessment, case retrieval, and case reuse. For the case representation, the pair of question and corresponding SQL is a natural representation of the case. On this basis, our approach further divides SQL into 30 categories according to the aggregation function of the SQL

---

[1] *AGG* is an aggregator (NULL, SUM, COUNT, AVG, MIN and MAX), *T* is table name, *C* is column name, *OP* is an operation ($>, <, =$), *V* is value and $[]^*$ represents one or more conditions.

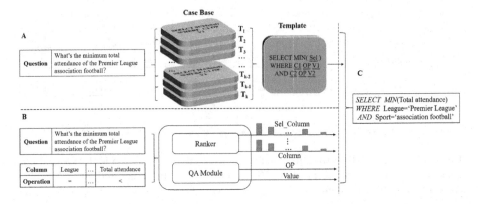

**Fig. 2.** Illustration of the proposed model architecture. Part A, the model retrieves the case template based on the similarity assessment results. The value of the slot is generated through the Ranker module and the QA module in part B. The new solution is assembled in part C.

statement and the number of Where clause conditions, see Sect. 3.1 for details. For the similarity assessment, it is difficult to measure the similarity between the two questions. The existing textual similarity assessment methods only pay attention to the literal similarity and ignore the similarity of the query intent. The retrieved cases can easily have a high degree of literal similarity, but the corresponding SQL is completely different. To this end, we introduce a neural classifier in the similarity assessment stage and comprehensively use classification probability and traditional text cosine similarity to measure similarity. In case retrieval, our model adopts the solution template corresponding to the higher classification probability or text cosine similarity. According to the retrieved template, our approach introduces a Ranker and QA module to generate and fill the value of slot.

The main contributions of the work are as follows:

i. we design a novel CBR-based translator for Text-to-SQL task. To our best knowledge, this is the first time the CBR-based method has been used for Text-to-SQL task;

ii. a composite similarity assessment method is proposed to comprehensively measure the literal cosine similarity and the classification probability of questions;

iii. we demonstrate the effectiveness of the approach with experiments on WikiSQL dataset and significant improvement of aggregation function prediction has been achieved.

The rest of the paper is organized as follows. The related works are discussed in the Sect. 2. The proposed case-based methodology for Text-to-SQL is highlighted in Sect. 3. The experimental design and results are discussed in Sect. 4, before concluding the paper and looking at future works in Sect. 5.

# 2   Related Work

## 2.1   Text-to-SQL

Text-to-SQL is a sub-area of semantic parsing that has received an intensive study recently. Depending on the constrained setting of the task, it can be classified into three categories: single-table, cross-domain, and context-dependent. Single-table Text-to-SQL constrained the problem by two factors: each question is only addressed by a single table, and the table is known [27]. Cross-domain means that every query is conditioned on a multi-table database schema, and the databases do not overlap between the train and test sets [25]. Context-dependent setting requires the model not only to focus on the precise generation of SQL queries, but also to consider the comprehensive utilization of contextual information [26].

Although some complex Text-to-SQL settings have been proposed, generating SQL for individual queries in the single table setup (WikiSQL [27]) is the most fundamental. The existing single-table Text-to-SQL models can be divided into two categories: sequence-to-sequence or sequence-to-set. They mainly share a similar encoder-decoder architecture, and the difference lies in the decoder part. The sequence-to-sequence methods [7,27] decode SQL sequentially, mainly using the attention and copying mechanism. Such models suffer from the *order-matters*, since they do not sufficiently enforce SQL syntax. The sequence-to-set models decompose SQL generation procedure into sub-modules, e.g., SELECT column, AGG function, WHERE value, etc. In this way, they can avoid the *order-matters* and achieve better performances on WikiSQL. Firstly, SQLNet [23] performs classification on those sub-modules. Based on that, TypeSQL [24] introduces the type information to better understand rare entities in the input. A progressive decoding is performed on Coarse-to-Fine model [8]. Further more, SQLova [16], X-SQL [14] and Hybrid [19] utilize pre-trained models in encoder and significantly improve the performance.

Compared to them, our method belongs to the sequence-to-set type, and the pre-training model is also used to empower our model. Differently, our model uses CBR technology to translate text into the corresponding SQL, and comprehensively uses classification probability and vector similarity to improve model performance.

## 2.2   Textual Case-Based Reasoning

Textual case-based reasoning provides a method in which text from previously solved examples with similar inputs is reused as a template solution to generate text for the current problem. In the work [2], the authors propose a CBR system that uses examples with similar weather states in previous cases to generate weather forecast text. In [9], a Reviewer's Assistant is designed to help people to write reviews on sites. In [22], they propose a case-based method for reusing text to automatically generate obituaries from a set of input attribute-value pairs.

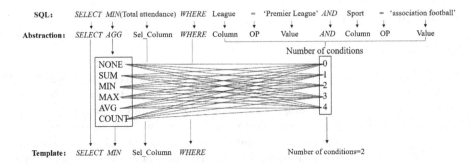

**Fig. 3.** Illustration of categorizing the solution template.

Although these works are applied in various fields, the research mainly focuses on the difficulty of mapping unstructured text from previous experience to structured representation and measuring semantic similarity to retrieve and reuse previous cases. Similarly, our work is also doing similar things in the Text-to-SQL scenario. The difference is that problems and solutions are all unstructured text in the above-mentioned related work, while the solution in our work is structured SQL.

## 3   Case-Based Methodology

The core of developing a CBR system is the availability of experienced knowledge that can provide successful examples of previous solutions for reusing to solve new problems. The first task is the case representation which is used to capture case knowledge as associated problem and solution components. Then, the similarity assessment is developed to measure the similarity of the problem representation to support retrieval. Finally, fine-tuning the retrieved case template according to the current problem, that is, case reuse. Following this process, the rest of this section is organized as:

Case Representation $\rightarrow$ Similarity Assessment $\rightarrow$ Case Retrieval $\rightarrow$ Case Reuse.

### 3.1   Case Representation

Cases are generally represented in two parts: problem and solution. For Text-to-SQL task, the problem is a given natural language question $Q = \{q_1, q_2, \cdots, q_n\}$ about the table $T = \{c_1, c_2, \cdots, c_m\}$, where $q_i$ represents the $i$-th token in the question $Q$, $c_l$ represents the $l$-th column in the table $T$, and $n, m$ is the total number of tokens and columns. The solution is the SQL query $S = \{s_1, s_2, \cdots, s_k\}$ corresponding to the question $Q$, where $s_j$ represents the $j$-th token in the solution $S$.

Generally, the CBR system retrieves the solution template based on the similarity between the new case and the cases in the case base. However, the problem $Q$ is one of 64,776 questions about 26,521 tables in the WikiSQL. It is difficult to calculate the similarity between the two questions. For this reason, we divide highly similar solutions (SQL) into 30 categories to facilitate better calculation of similarity. In this way, the solution can be retrieved as long as the text of the question meets the characteristics of a certain type of problem, and it does not necessarily have to be highly literally similar. Hence, the classification probability of the problem can be used as a measurement of similarity. Figure 3 is an illustration of categorizing the solution template, it is categorized based on the aggregation functions and the number of conditions. For all SQL queries in the WikiSQL dataset, there are six types of aggregation functions and five types of conditions in the where clause. Let $Temp = \{temp_1, temp_2, \cdots, temp_{30}\}$ denote 30 solution templates.

For the neural networks in our approach, we follow the work [19] to build a column-wise model via an explicit head on Roberta [18]. The inputs are organized as:

$$[CLS], x_1, x_2, \cdots, x_m, [SEP], q_1, q_2, \cdots, q_n, [EOS],$$

where $x_1, x_2, \cdots, x_m$ is the token sequence of column $c$, and $q_1, q_2, \cdots, q_n$ is the token sequence of question $Q$. $CLS$, $SEP$, and $EOS$ are special tokens used to organize text, which are the abbreviation of classification, separation and end of sentences respectively. Those token sequences are encoded by Roberta to form the final inputs of our neural networks (Classifier, Ranker, and QA module). The final inputs contains two hidden states: sequence_output $-h$ and pool_output $-h_{CLS}$. $h$ is the hidden states of each token and $h_{CLS}$ is the hidden states of the whole sequence.

### 3.2 Similarity Assessment

**Cosine Similarity.** When it comes to text similarity, the cosine similarity of text vectors is usually preferred. Let $V_q = \{v_{q1}, v_{q2}, \cdots, v_{qn}\}$ denote the vector of the current question, $V_{base_k} = \{v_{base_1}, v_{base_2}, \cdots, v_{base_n}\}$ denote the vector of $k$-th question in the case base. The cosine similarity(CS) of two text vectors is calculated as follows:

$$CS = \frac{V_q \cdot V_{base_k}}{\|V_q\| \times \|V_{base_k}\|}$$
$$= \frac{\sum_{i=1}^{n} (v_{qi} \times v_{base_i})}{\sqrt{\sum_{i=1}^{n} (v_{qi})^2} \times \sqrt{\sum_{i=1}^{n} (v_{base_i})^2}} \tag{1}$$

However, cosine similarity does not work well in the Text-to-SQL task. There are two reasons: (1) the word vector is sparse, as shown in Fig. 4.a. (2) the word vector pays more attention to the similarity of the token level. It easily leads to a high degree of literal similarity, but the corresponding SQL is not.

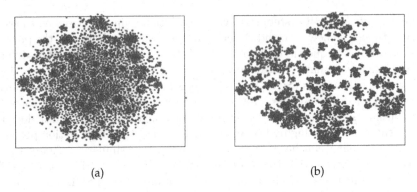

(a)                                     (b)

**Fig. 4.** Two-dimensional presentation of the text vector after dimensionality reduction by t-SNE [20]. Part a is the original text data and part b is the data with normalized column names and values.

**Cosine Similarity Without Specific Values.** To address the above issue, we propose an improved similarity assessment without specific values. Specifically, we replace the words in the question that implicate the table columns and values with fixed tokens. In this way, the interference related to the table content is reduced. As shown in Fig. 4.b, this change makes the vector of problem text more aggregated.

**Classification Probability as Similarity.** The above similarity assessment calculates the cosine similarity between the new problem and the other problem in the case base. Although this one-to-one model is straightforward and concise, finding the most similar problem is error-prone.

For a more comprehensive measure of similarity, a neural classifier is introduced. According to the categorization in Sect. 3.1, the classifier determines what type of problem the new problem belongs to. In this way, the classification probability $P(Temp_s|Q)$ of the neural classifier can be used as a valid similarity assessment. It is a one-to-group mode, which is a measure of similarity from the distribution level. The classification probabilities are calculated as follows:

$$P(Temp_s|Q) = \sum_{c_i} P(Temp_s|c_i, Q)P(c_i|Q) \tag{2}$$

$P(c_i|Q)$ is the similarity between column $c_i$ and question $Q$ and can be computed:

$$P(c_i|Q) = sigmoid(w \cdot h_{[CLS]}), \tag{3}$$

where $w$ is weight matrix.

### 3.3   Case Retrieval

In each retrieval, we calculate the classification probability $(CP)$ and cosine similarity $(CS)$ separately, and take the case template of the larger corresponding to. It should be noted that the value range of cosine similarity and classification probability is different, i.e., $CS \in [-1, 1]$ and $CP \in [0, 1]$. To unify the range, the cosine similarity is normalized.

$$Temp = \arg\max(CP_{Temp_s}, CS_{Temp_s})$$

$$CP_{Temp_s} = \max_{Temp_s} P(Temp_s \mid Q) = \max_{Temp_s} \sum_{c_i} P(Temp_s \mid c_i, Q) P(c_i \mid Q)$$

$$CS_{Temp_s} = \max_{m \in case\ base} (\frac{1 - CS(Q, Q_m)}{2})$$

(4)

### 3.4 Case Reuse

For the retrieved case template, the aggregation function $AGG$ and the number of conditions $WHERE_{NUM}$ have been fixed. It only needs to fill in the $Sel\_Column$, $Column$, $OP$, and $Value$ according to the current question $Q_c$. To address this issue, a Ranker is introduced to rank and choose the candidate columns and a QA module is introduced to predict objects associated with the specific column. For the current question $Q_c$, we denote $S_{Q_c}$ to be the set of columns that are in select clause, and $W_{Q_c}$ as the set of columns that are in the Where clause.

i. For $Sel\_Column$, let $P(c_i \in S_{Q_c} | Q_c) = sigmoid(w_{sc} \cdot h_{[CLS]})$. The top candidate column is chosen to form select clause.
ii. For $Column$, let $P(c_i \in W_{Q_c} | Q_c) = sigmoid(w_{wc} \cdot h_{[CLS]})$. The top $WHERE_{NUM}$ columns are chosen to form where clause.
iii. For a condition operator $o_i$, let $P(o_i | c_i, Q_c) = softmax(w_{op}[i, :] \cdot h_{[CLS]})$.
iv. For $Value$ start and end indices, let $P(q_i = start | c_i, Q_c) = softmax(w_{start} \cdot h_i^{Q_c})$ and $P(q_i = end | c_i, Q_c) = softmax(w_{end} \cdot h_i^{Q_c})$.

where $w_{sc}$, $w_{wc}, w_{op}$, $w_{start}$, $w_{end}$ are a linear transformation matrices.

## 4    Experiments

In this section, we demonstrate the results of our approach on the WikiSQL [27] dataset and compare it to the other state-of-the-art approaches.

### 4.1 Dataset and Evaluation Metrics

**Dataset.** WikiSQL [27] is the largest Text-to-SQL dataset with single-table setting. It contains 56,355, 8421, and 15878 question-SQL pairs about 26,521 tables for training, development, and testing. All the SQL queries have one select column and aggregation operator, and 0 to 4 conditions.

**Evaluation Metrics.** Generally, the single-table Text-to-SQL task takes the Logical Form Accuracy and Execution Accuracy as evaluation metrics. Execution accuracy is calculated as follows: $ExecutionAccuracy = \frac{N_{ex}}{N}$, where $N_{ex}$ is the number of SQL statements whose execution result is correct, $N$ denotes the total number of examples in the dataset. $Logical\ Form\ Accuracy = \frac{N_{lf}}{N}$ is the calculation method of logical form accuracy, where $N_{lf}$ is the number of queries that has exact string match with the ground truth. The two evaluation metrics have different concerns. Execution Accuracy is to evaluate the model in terms of execution results, while Logical Form Accuracy is more concerned about the output whether has an exact string matching.

### 4.2    Baselines

We compare the proposed method to the following state-of-the-art models:

- Seq2SQL [27] takes attentional sequence to sequence neural semantic parser [7] as baseline model.
- SQLnet [23] fundamentally solves the *order-matters* problem by employing a sequence-to-set model to generate SQL queries. It is the first sequence-to-set model.
- TypeSQL [24] utilizes type information to better capture rare entities and numbers in the question.
- SQLova [16] takes a full advantage of BERT [6] through an effective table contextualization method.
- X-SQL [14] proposes to enhance the structural schema representation with the contextual output from BERT-style pre-training model [6], and together with type information to learn a new schema representation for down-stream tasks.
- HydraNet [19] breaks down the problem into column-wise ranking and decoding and finally assembles the column-wise outputs into a SQL query.
- SDSQL [15] presents the Schema Dependency guided multi-task Text-to-SQL model to guide the network to effectively capture the interactions between questions and schemas.

### 4.3    Implementation Details

We utilize PyTorch [21] and Python 3.6 to implement our proposed model. For the input representation, we use Roberta-large [18] version and fine-turn it with a 3e−5 learning rate during training. The batch size is set to 16 and we use Adam [17] optimizer to minimize loss. The system is Ubuntu 18.04 with two 2080ti graphic cards.

### 4.4    Results

**Overall Performance.** We first compare the performance with other state-of-the-art models. As shown in Table 1, we can see that our model outperforms all

**Table 1.** Performance of various methods on WikiSQL dataset. LF is an abbreviation for Logical Form Accuracy and EX is an abbreviation for Execution Accuracy.

| Model | Dev | | Test | |
|---|---|---|---|---|
| | LF | EX | LF | EX |
| Seq2SQL [27] | 49.5 | 60.8 | 48.3 | 59.4 |
| SQLNet [23] | 63.2 | 69.8 | 61.3 | 68.0 |
| TypeSQL [24] | 68.0 | 74.5 | 66.7 | 73.5 |
| SQLova [16] | 81.6 | 87.2 | 80.7 | 86.2 |
| X-SQL [14] | 83.8 | 89.5 | 83.3 | 88.7 |
| HydraNet [19] | 83.6 | 89.1 | 83.8 | 89.2 |
| SDSQL [15] | 86.0 | 91.8 | 85.6 | 91.4 |
| Ours | 84.5 | 90.3 | 84.7 | 90.2 |

**Table 2.** Fine-grained analysis for various methods on dev set of WikiSQL dataset. *Sel_Column* is the column name in the select clause, *AGG* is the aggregation function, and *WHERE_NUM* is the number of conditions in the where clause. *Column*, *OP*, and *Value* refer to the column name, operator, and value in a condition, respectively.

| Model | Sel_Column | AGG | WHERE_NUM | Column | OP | Value |
|---|---|---|---|---|---|---|
| SQLova [16] | 96.8 | 90.6 | 98.5 | 94.3 | 97.3 | 95.4 |
| X-SQL [14] | 97.2 | 91.1 | 98.6 | 95.4 | 97.6 | 96.6 |
| HydraNet [19] | **97.6** | 91.4 | 98.4 | 95.3 | 97.4 | 96.1 |
| SDSQL [15] | 97.3 | 90.9 | 98.5 | **98.1** | **97.7** | **98.3** |
| Ours | 97.4 | **92.1** | **98.6** | 95.6 | 97.6 | 96.8 |

existing models on all evaluation metrics except SDSQL. It should be noted that SDSQL is a multi-task Text-to-SQL model that integrates additional auxiliary task (schema dependency) to capture the complex interaction between schemas and questions. As you can see from the last two rows of the Table 1, despite the absence of additional auxiliary tasks, our model is competitive with SDSQL.

**Fine-Grained Analysis.** In order to further investigate the performance of each sub-module in the various methods, as shown in Table 2, we perform a fine-grained analysis. Firstly, it can be seen that HydraNet has the highest accuracy of *Sel_Column* and SDSQL has the best performance in conditions prediction (*Column*, *OP*, and *Value*). Secondly, the accuracy of our model is outstanding in all sub-modules, especially *AGG* and *WHERE_NUM*. The reason is mainly due to our novel composite similarity assessment could comprehensively use classification probability and text cosine similarity to improve the model performance.

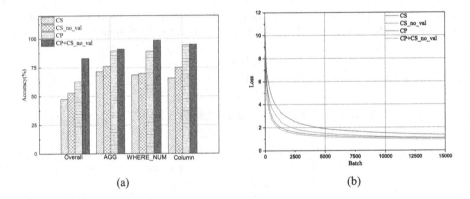

(a)                                    (b)

**Fig. 5.** Accuracy (part a) and Loss (part b) of our model (CP+CS_no_val) and ablated models.

### 4.5    Ablation Study

To understand the importance of different similarity assessments in our approach, we conduct a series of ablation studies, *i.e.*, Cosine Similarity (CS), Cosine Similarity without specific value (CS_no_val), Classification Probability as similarity (CP), and our model (CP+CS_no_val).

Accuracy of our model and ablated models are presented in Fig. 5.a, we focus on indicators related to similarity assessment, *i.e.*, *Overall*, *AGG*, and *WHERE_NUM*. Firstly, we notice that CS_no_val performs better than CS, demonstrates that Cosine Similarity without specific value is an effective method to improve performance. Secondly, we observe that CP outperforms both CS_no_val and CS. This observation shows that classification probability as a similarity assessment is superior to cosine similarity in Text-to-SQL task. Finally, it can be seen that our model(CP+CS_no_val) is the best performer. It indicates that there is no conflict and interference between CP and CS_no_val, and they work together to further improve the performance of the model.

In order to explore the convergence of our model and ablated models, the loss data is recorded during training. Closer inspection of the loss curves in Fig. 5.b shows that: (1) the loss value of CS is the largest and converges the slowest; (2) CS_no_val and OP are similar and the convergence speed is in the middle; (3) our loss value is the smallest and converges the fastest. Overall, these results indicate that using classification probability and cosine similarity comprehensively can accelerate the convergence of the model.

### 4.6    Case Study

We present typical error cases of three state-of-the-art approaches (HydraNet [19], SDSQL [15], and Ours) and analyze the strengths and weaknesses of them in this section. As can be seen from the Table 3, there are three types of typical error cases: Few-shot Errors, Schema Dependency Errors, and Lack of Common Sense.

**Table 3.** Error cases of three state-of-the-art approaches. Incorrect tokens are marked red.

| Case One: Few-shot Errors | |
|---|---|
| Question | Name the number of total votes for # of seats won being 30? |
| HydraNet | SELECT COUNT(# of total votes) WHERE # of seats won=30 |
| SDSQL | SELECT COUNT(# of total votes) WHERE # of seats won=30 |
| Ours | SELECT # of total votes WHERE # of seats won=30 |
| Groundtruth | SELECT # of total votes WHERE # of seats won=30 |
| **Case Two: Schema Dependency Errors** | |
| Question | How many times is denmark ranked in technology? |
| HydraNet | SELECT COUNT(Rank) WHERE Technology=denmark |
| SDSQL | SELECT COUNT (Technology) WHERE Rank=denmark |
| Ours | SELECT COUNT(Rank) WHERE Technology=denmark |
| Groundtruth | SELECT COUNT (Technology) WHERE Rank=denmark |
| **Case Three: Lack of Common Sense** | |
| Question | What is the sum of week number(s) had an attendance of 61,985? |
| HydraNet | SELECT SUM(Week) WHERE Attendance = 61,985 |
| SDSQL | SELECT SUM(Week) WHERE Attendance = 61,985 |
| Ours | SELECT SUM(Week) WHERE Attendance = 61,985 |
| Groundtruth | SELECT COUNT(Week) WHERE Attendance=61,985 |

**Few-Shot Errors.** What stands out in Case One is HydraNet and SDSQL use the COUNT aggregation function incorrectly. The essential reason is that there are a large number of '*the number of* → *COUNT*' pairs in the training corpus. Encountering the phrase of '*the number of*', the model mistakenly believes that the *COUNT* aggregation function should be used. Why can our model translate the question into the correct SQL query? Because there is a highly similar case in the case base—'*Name the number of candidates for # of seats won being 43?*'. Our model chooses the template corresponding to this highly similar case, instead of blindly following the choice of the neural model.

**Schema Dependency Errors.** Admittedly, SDSQL is indeed better than other models in terms of Schema dependency. From Case Two, it can be seen that our model and HydraNet cannot distinguish whether *denmark* is the value of the *Rank* column or the *Technology* column. However, this is a piece of cake for SDSQL which has an auxiliary task—schema dependency.

**Lack of Common Sense.** This result of Case Three is somewhat counterintuitive. In general, for the question in Case Three, most models may choose the *SUM* aggregation function. However, the value of the *Week* column is a string that cannot be added. Hence, the $COUNT$ aggregation function is the best choice.

Through the analysis of the above cases, it can be found that the three models have their own strengths, and our model has more advantages in terms of fewer samples.

## 5    Conclusion and Future Work

This paper proposes a novel CBR-based model to solve the Text-to-SQL task. For the similarity assessment, classification probability and text cosine similarity are comprehensively used to improve the model performance. The proposed model has competitive performance and significantly improves the aggregation function prediction on the WikiSQL dataset. It is a successful attempt of the CBR-based method on the task of Text-to-SQL translation.

In future work, we will further apply advanced technologies in the field of CBR to more complex Text-to-SQL task, such as cross-domain and context-dependent.

**Acknowledgments.** This work was supported by the Natural Science Foundation of China under Grant 61572513.

## References

1. Aamodt, A.: Foundational issues, methodological variations, and system approaches. AI Commun. **7**(1), 39–59 (1994)
2. Adeyanju, I.: Generating weather forecast texts with case based reasoning. CoRR abs/1509.01023 (2015). http://arxiv.org/abs/1509.01023
3. Bartlett, C.L., Liu, G., Bichindaritz, I.: Classifying breast cancer tissue through DNA methylation and clinical covariate based retrieval. In: Watson, I., Weber, R. (eds.) ICCBR 2020. LNCS (LNAI), vol. 12311, pp. 82–96. Springer, Cham (2020). https://doi.org/10.1007/978-3-030-58342-2_6
4. Bridge, D.G., Göker, M.H., McGinty, L., Smyth, B.: Case-based recommender systems. Knowl. Eng. Rev. **20**(3), 315–320 (2005). https://doi.org/10.1017/S0269888906000567
5. Caro-Martinez, M., Recio-Garcia, J.A., Jimenez-Diaz, G.: An algorithm independent case-based explanation approach for recommender systems using interaction graphs. In: Bach, K., Marling, C. (eds.) ICCBR 2019. LNCS (LNAI), vol. 11680, pp. 17–32. Springer, Cham (2019). https://doi.org/10.1007/978-3-030-29249-2_2
6. Devlin, J., Chang, M., Lee, K., Toutanova, K.: BERT: pre-training of deep bidirectional transformers for language understanding. In: Proceedings of the 2019 Conference of the North American Chapter of the Association for Computational Linguistics: Human Language Technologies, NAACL-HLT 2019, Minneapolis, MN, USA, 2–7 June 2019, Volume 1 (Long and Short Papers), pp. 4171–4186. Association for Computational Linguistics (2019)
7. Dong, L., Lapata, M.: Language to logical form with neural attention. In: Proceedings of the 54th Annual Meeting of the Association for Computational Linguistics, ACL 2016, Berlin, Germany, 7–12 August 2016, Volume 1: Long Papers. The Association for Computer Linguistics (2016)

8. Dong, L., Lapata, M.: Coarse-to-fine decoding for neural semantic parsing. In: Gurevych, I., Miyao, Y. (eds.) Proceedings of the 56th Annual Meeting of the Association for Computational Linguistics, ACL 2018, Melbourne, Australia, 15–20 July 2018, Volume 1: Long Papers, pp. 731–742. Association for Computational Linguistics (2018)
9. Dong, R., Schaal, M., O'Mahony, M.P., McCarthy, K., Smyth, B.: Harnessing the experience web to support user-generated product reviews. In: Agudo, B.D., Watson, I. (eds.) ICCBR 2012. LNCS (LNAI), vol. 7466, pp. 62–76. Springer, Heidelberg (2012). https://doi.org/10.1007/978-3-642-32986-9_7
10. Dufour-Lussier, V., Ber, F.L., Lieber, J., Nauer, E.: Automatic case acquisition from texts for process-oriented case-based reasoning. Inf. Syst. 40, 153–167 (2014)
11. Eisenstadt, V., Langenhan, C., Althoff, K.-D., Dengel, A.: Improved and visually enhanced case-based retrieval of room configurations for assistance in architectural design education. In: Watson, I., Weber, R. (eds.) ICCBR 2020. LNCS (LNAI), vol. 12311, pp. 213–228. Springer, Cham (2020). https://doi.org/10.1007/978-3-030-58342-2_14
12. Feely, C., Caulfield, B., Lawlor, A., Smyth, B.: Using case-based reasoning to predict marathon performance and recommend tailored training plans. In: Watson, I., Weber, R. (eds.) ICCBR 2020. LNCS (LNAI), vol. 12311, pp. 67–81. Springer, Cham (2020). https://doi.org/10.1007/978-3-030-58342-2_5
13. Goel, A.K., Craw, S.: Design, innovation and case-based reasoning. Knowl. Eng. Rev. 20(3), 271–276 (2005). https://doi.org/10.1017/S0269888906000609
14. He, P., Mao, Y., Chakrabarti, K., Chen, W.: X-SQL: reinforce schema representation with context. CoRR abs/1908.08113 (2019). http://arxiv.org/abs/1908.08113
15. Hui, B., et al.: Improving text-to-SQL with schema dependency learning. CoRR abs/2103.04399 (2021). https://arxiv.org/abs/2103.04399
16. Hwang, W., Yim, J., Park, S., Seo, M.: A comprehensive exploration on WikiSQL with table-aware word contextualization. CoRR abs/1902.01069 (2019). http://arxiv.org/abs/1902.01069
17. Kingma, D.P., Ba, J.: Adam: a method for stochastic optimization. In: 3rd International Conference on Learning Representations, ICLR 2015, San Diego, CA, USA, 7–9 May 2015, Conference Track Proceedings (2015)
18. Liu, Y., et al.: Roberta: a robustly optimized BERT pretraining approach. CoRR abs/1907.11692 (2019). http://arxiv.org/abs/1907.11692
19. Lyu, Q., Chakrabarti, K., Hathi, S., Kundu, S., Zhang, J., Chen, Z.: Hybrid ranking network for text-to-SQL. Technical report. MSR-TR-2020-7, Microsoft Dynamics 365 AI, March 2020. https://www.microsoft.com/en-us/research/publication/hybrid-ranking-network-for-text-to-sql/
20. van der Maaten, L., Hinton, G.: Visualizing data using t-SNE. J. Mach. Learn. Res. 9(86), 2579–2605 (2008). http://jmlr.org/papers/v9/vandermaaten08a.html
21. Paszke, A., et al.: PyTorch: an imperative style, high-performance deep learning library. In: Advances in Neural Information Processing Systems 32: Annual Conference on Neural Information Processing Systems 2019, NeurIPS 2019, Vancouver, BC, Canada, 8–14 December 2019, pp. 8024–8035 (2019)
22. Upadhyay, A., Massie, S., Clogher, S.: Case-based approach to automated natural language generation for obituaries. In: Watson, I., Weber, R. (eds.) ICCBR 2020. LNCS (LNAI), vol. 12311, pp. 279–294. Springer, Cham (2020). https://doi.org/10.1007/978-3-030-58342-2_18
23. Xu, X., Liu, C., Song, D.: SQLNet: generating structured queries from natural language without reinforcement learning. CoRR abs/1711.04436 (2017). http://arxiv.org/abs/1711.04436

24. Yu, T., Li, Z., Zhang, Z., Zhang, R., Radev, D.R.: TypeSQL: knowledge-based type-aware neural text-to-SQL generation. In: Proceedings of the 2018 Conference of the North American Chapter of the Association for Computational Linguistics: Human Language Technologies, NAACL-HLT, New Orleans, Louisiana, USA, 1–6 June 2018, Volume 2 (Short Papers), pp. 588–594. Association for Computational Linguistics (2018)
25. Yu, T., et al.: Spider: a large-scale human-labeled dataset for complex and cross-domain semantic parsing and text-to-SQL task. In: Riloff, E., Chiang, D., Hockenmaier, J., Tsujii, J. (eds.) Proceedings of the 2018 Conference on Empirical Methods in Natural Language Processing, Brussels, Belgium, 31 October–4 November 2018, pp. 3911–3921. Association for Computational Linguistics (2018)
26. Yu, T., et al.: SParC: cross-domain semantic parsing in context. In: Korhonen, A., Traum, D.R., Màrquez, L. (eds.) Proceedings of the 57th Conference of the Association for Computational Linguistics, ACL 2019, Florence, Italy, 28July–2 August 2019, Volume 1: Long Papers, pp. 4511–4523. Association for Computational Linguistics (2019)
27. Zhong, V., Xiong, C., Socher, R.: Seq2SQL: generating structured queries from natural language using reinforcement learning. CoRR abs/1709.00103 (2017)

# Deciphering Ancient Chinese Oracle Bone Inscriptions Using Case-Based Reasoning

Gechuan Zhang[1] , Dairui Liu[1,2(✉)] , Barry Smyth[1,2] ,
and Ruihai Dong[1,2]

[1] School of Computer Science, University College Dublin, Dublin, Ireland
{gechuan.zhang,dairui.liu}@ucdconnect.ie
[2] Insight Centre for Data Analytics, University College Dublin, Dublin, Ireland
{barry.smyth,ruihai.dong}@ucd.ie

**Abstract.** Ancient oracle bone inscriptions (OBIs) are important Chinese cultural artefacts, which are difficult and time-consuming to decipher even by the most expert paleographers and, as a result, a large proportion of excavated OBIs remain unidentified. In practice, OBIs are deciphered by translating between different writing systems; Chinese writing systems have evolved over time and ancient OBIs can be deciphered by translating their inscriptions to a known inscription in an *adjacent* writing system, but this is a complex and time-consuming process. In this paper we propose a novel case-based system, to support this task, allowing a paleographer to present an unknown inscription (image) as a query, to receive a set of similar images from an adjacent writing system with associated scholarly information, and so help guide the deciphering of the query. One important contribution of this work involves the use of an auto-encoder to learn suitable image representations to capture the relationship between two adjacent writing systems. We demonstrate the effectiveness of this approach using a novel, purpose-built case base, and discuss its use in a paleographic setting.

**Keywords:** Machine learning · Oracle bone inscriptions · Case-based reasoning

## 1 Introduction

Oracle bone inscription (OBI) is the earliest known form of Chinese writing and an ancestor of modern Chinese characters. It was engraved on oracle bones—animal bones (see Fig. 1) or turtle plastrons used in pyromantic divination and story-telling [2]—in the late 2nd millennium BC and the vast majority of more than 50,000 inscribed items, have been found at the Yinxu site located in Xiaotun Village, Anyang City, Henan Province.

These items are vital Chinese cultural artefacts and as one of the earliest systematic writing systems in the world, OBIs are of great international scholastic import. Beginning with the OBI, the modern Chinese writing system has

© Springer Nature Switzerland AG 2021
A. A. Sánchez-Ruiz and M. W. Floyd (Eds.): ICCBR 2021, LNAI 12877, pp. 309–324, 2021.
https://doi.org/10.1007/978-3-030-86957-1_21

evolved, uninterrupted, through several different stages as shown in the insert in Fig. 1. Indeed, this evolutionary continuity is an important feature of the Chinese writing system when it comes to deciphering unidentified OBIs, as we shall see, because it means that paleographers can match unidentified inscriptions with identified characters from an adjacent writing system based on shared visual features.

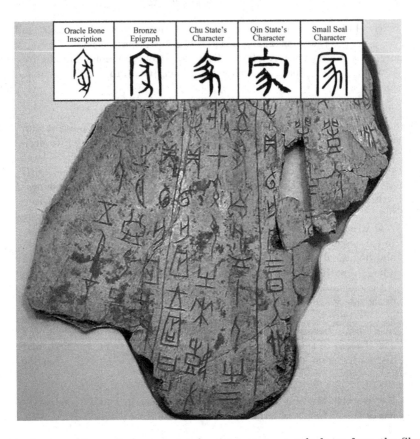

**Fig. 1.** An example oracle bone script fragment on a scapula bone from the Shang Dynasty, and five evolution stage examples of the Chinese writing system. Image courtesy of BabelStone, CC BY-SA 3.0 (https://commons.wikimedia.org/w/index.php?curid=16189953)

Since OBIs were first discovered, more than 4,500 characters have been excavated, but only one-third of these characters have been deciphered or translated [14]. In part, this is because of the complexity of the task: it requires considerable paleographic expertise and is time-consuming in the extreme as paleographers seek to identify subtle matches between complex inscriptions and more recent writings. For some time, paleographers have been attempting to accelerate this translation process, but with only limited success to date. In this paper, we argue that modern machine learning techniques are up to the task [12,15] and we describe a novel case-based reasoning system designed to support paleographers during this translation process.

In the following section, we describe relevant related work, paying close attention to some recent efforts that have been conducted to support paleographers during this important task. Then, in Sect. 3 we present our case-based reasoning approach and describe the basic operations of the system. In Sect. 4 we present a new large-scale public data-set purpose-built for this task. Section 5 describes one important and novel aspect of this approach, namely the use of an auto-encoder architecture to automatically learn suitable image representations to facilitate cross-domain retrievals between adjacent writing systems. Finally, before concluding, In Sect. 6 we describe the results of an initial evaluation.

## 2   Related Work

The work presented in this paper is, at its core, a version of a familiar machine translation task, which has been the subject of natural language processing and machine learning research for many decades [6]. Indeed, case-based reasoning ideas have a storied history in this field [18] since statistical machine translation approaches were first presented by IBM's Peter Brown when he infamously declared that *"Every time I fire a linguist, my system's performance improves!"* as he described an empirical corpus-based approach to machine translation that, flew in the face of the conventional linguistic-based approaches that prevailed at the time [3]. This led to something of a deluge of novel approaches that came to be viewed as *example-based machine translation* (EBMT) which relied primarily on case-like *translation pairs* as the basis for translation, rather than linguistically inspired translation rules and other forms of linguistic knowledge [4,19]. Today these EBMT approaches are every bit as mainstream as their linguistic complements and more recent neural-network based approaches [10].

The present is distinguished by the visual nature of the translation task: the translation targets are complex and varied visual images rather than well-defined character-based representations. Early work on the automated analysis of OBIs is characterised by image processing based approaches to inscription recognition and classification. For example, Zhou et al. represented OBIs as non-directional graphs and extracted their topological properties, in addition to relevant stroke-based features, for the recognition tasks [20]. In 2010, Li et al. proposed a DNA-based encoding and retrieve method for OBIs [16], while other researchers [13] proposed a grid point feature extraction algorithm when dealing with oracle bone rubbings. In 2015, Feng et al. attempted to recognise OBIs based on a statistical analysis of context and methods combining with the Hopfield network [7]. The results of these early efforts helped to demonstrate the potential for computational techniques and machine learning to support the deciphering of OBIs, even though recognition accuracy was not yet high enough to greatly accelerate the translation process.

To be clear, the OBI translation task is an extremely challenging one because it is based on unusual, domain-specific images without a robust feature representation, and because conventional feature extraction techniques have proved to be only partially effective. In this regard, recent advances in deep neural networks in the area of machine vision are especially important [12] because they provide a means to learn powerful representations from raw data such as images. For example, in this work, we rely on ideas from Rumelhart's auto-encoders (AE) [17] to learn a representation for the pair of images that make up cases. In fact, we use a variational auto-encoder (VAE) [11] to learn a suitable representation because VAEs have been shown to avoid many over-fitting problems associated with standard AEs by regularising the training process. We will show that these learned representations provide a much more powerful representational basis for determining similarity and relevance.

## 3   A Case-Based Reasoning Perspective

Before we describe our approach further, it is worthwhile to clarify some terminology that will be used in what follows. First, when we refer to two *adjacent* writing systems, we refer to two styles of writing that appeared consecutively or even overlapped in time. In this work, we will focus on three writing systems— OBIs are the earliest, followed by Bronze Epigraphs (BE), followed by Chu State Characters (CSC)—as shown in Fig. 2(a). OBIs and BEs are adjacent, as are BEs and CSCs. This concept of adjacency is important because adjacent writing systems share some common forms and patterns that will be exploited during representation learning and case retrieval.

Next, an *inscription* is an *image* and different inscriptions may be variations on a specific *category* (e.g. 'Home') but not every category is known (deciphered); we will often refer to a *category* group as a set of inscriptions that related to a specific known category.

### 3.1   From Inscriptions to Cases

There are two types of cases in our system, *known* cases and *unknown* cases. In Fig. 2(a) the known cases are shown as filled markers and the unknown case as unfilled markers. A known case corresponds to an inscription that has been deciphered, which means it is associated with a category and, typically other inscriptions from other writing systems, as well as the relevant scholarly information that is important to the paleographer such as, the origins and provenance of the inscription, information about how the inscription was deciphered, links to other inscriptions associated with the same category, possible from other writing systems; thus known cases represent deciphered categories. Notice how this figure suggests that there are proportionally fewer known cases associated with earlier writing systems. This is in fact the case and it suggests that its retrievals from later writing systems may be more likely to lead to deciphered categories.

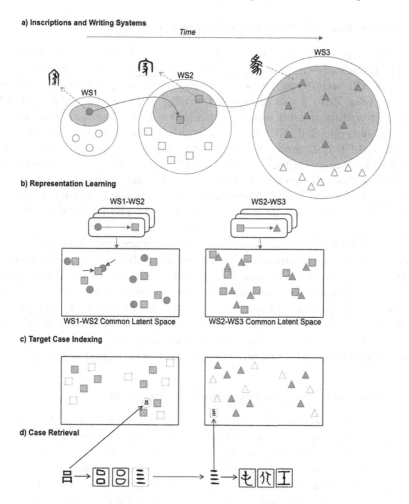

**Fig. 2.** The overall design of the deciphering process.

Conversely, an unknown case represents an inscription (or a set of inscriptions thought to refer to the same unknown concept) in a single writing system that has not yet been deciphered. Unknown cases are incomplete: they may include some relevant scholarly information (the origins of the inscription, previous attempts to decipher it, etc.), but there will be no linkages to related inscriptions in adjacent writing systems. Nonetheless, these unknown inscriptions may still be useful because they may serve as useful follow-up queries to identify cases from later writing systems, which are more likely to be associated with already deciphered categories, as we shall see.

## 3.2 From Adjacent Writing Systems to Case Bases

Thus, we have access to a large number of cases containing inscriptions from different (adjacent) writing systems. Some of these inscriptions will be known, but many will be unknown, especially for older writing systems. In this work, each case base *spans* a pair of adjacent writing systems, thus for a sequence of $n$ adjacent writing systems we can generate $n-1$ case bases. Here we focus on 3 writing systems, OBI, BE, and CSC, which means there are two case bases, OBI-BE and BE-CSC; in what follows, for a case base such as OBI-BE, we refer to OBI as the *source* writing systems and BE as the *target* writing system. For a given case base, the query will come from the source writing system, and the cases in the case base will correspond to cases from the target writing system. For example, for the OBI-BE, queries will be OBIs, and retrieved cases will be BEs.

This highlights one of the unique features of cases and case bases: retrieval depends on the form of cross-domain reminding in the sense that the query and retrieved cases are from different (but adjacent, and therefore related) writing systems. But how can we compare an inscription (image) from a writing system $i$ to an inscription from another writing system $i+1$? It is more usual in CBR for the query and (problem descriptions of) cases to be based on a shared set of features, which facilitates comparison and similarity assessment. Here though, the query and the case are images from different writing systems.

## 3.3 Representation Learning and Case Indexing

One approach to solving this problem might be to use conventional image analysis techniques to extract features from the inscription images from the source and target domains. However, in this work, we adopt an alternative strategy by using an extended encoder-decoder architecture to learn suitable representations from pairs of known inscriptions drawn from writing system $i$ and $i+1$ as outlined in Fig. 2(b) and detailed in Sect. 5. Thus, we learn a representation for the OBI-BE case base and a separate representation for BE-CSC.

Once a suitable representation is learned, its target cases are indexed using this representation so that they are available for retrieval; effectively, this means that target cases (including unknown cases) are encoded through the same encoder used during the representation learning, as shown in Fig. 2(c). Notably, this facilitates the indexing of known and unknown cases, as the encoding process operates on a raw inscription image regardless of whether it has been deciphered. This is important because it means that unknown cases can also be made available for retrieval, which means they can be used as subsequent queries into the $i+2$ writing system.

## 3.4 An Example Use-Case

An example use-case is shown in Fig. 2(d). The paleographer submits an unknown (OBI) inscription as a query to the OBI-BE case base, returning the

$k = 3$ most similar BE cases. In this example, the first two cases are known BE inscriptions, and the third case is an unknown BE inscription. The hope is that these known cases may help the paleographer to gain some insights into the meaning of the query inscription.

Importantly, the retrieval of an unknown case is not a retrieval failure. Indeed unknown cases can be used follow-up queries, this time into the BE-CSC case base. In this example, the unknown BE case results in the retrieval of three further known (CSC) cases. In this way, the unknown case serves as a *bridge* into the CSC writing system, which is better populated with known/deciphered cases.

## 4    A Novel Oracle Bone Inscription Data-Set

As mentioned already, progress in supporting the OBI translation task using machine learning ideas has been impeded by the lack of suitable training data. Hence, for this work, we have developed a suitable data-set, which goes far beyond that has been available to date; this data-set was needed for the work presented in this paper but has now also been made available as an important resource to the ancient Chinese research community, which will be released in Mendeley Data soon.[1]

### 4.1    Data Collection

The original data for our data-set came from two specialised websites—the Chinese Etymology[2] website and the ZDIC[3] website—both of which provide access to character images for categories in multiple writing systems using a common hierarchical labelling system. The data were collected between January and November 2020. From these data, we can generate the source-target pairs between writing systems that we need for representation learning.

A total of 5,138 character categories, associated with five writing systems were collected as shown in Table 1. Writing systems were selected based on the timeline of culture development and cover the typical evolutionary stages from OBI. Three of the earliest writing systems (OBI, BE, and CSC) were selected for training and testing purposes (as mentioned in Sect. 3) because they provided access to the largest collection of images. We refer to these data as the *Ancient-3* data-set. The remaining two writing systems (Qin State's Character and Small Seal Style Character) were not used further in this work, but they are included in the larger *Ancient-5* data-set because they include data from important stages in OBI evolution, which we believe will be useful for the research community.

---

[1] http://dx.doi.org/10.17632/ksk47h2hsh.2.

[2] https://hanziyuan.net/.

[3] https://www.zdic.net/.

**Table 1.** The overall statistics of our ancient Chinese data-set.

| Writing system | Category number | Inscription number |
|---|---|---|
| Oracle Bone Inscription (OBI) | 1,186 | 39,009 |
| Bronze Epigraph (BE) | 1,394 | 35,140 |
| Chu State's Character (CSC) | 1,640 | 32,203 |
| Qin State's Character (QSC) | 1,537 | 3,652 |
| Small Seal Style Character (SSSC) | 5,138 | 5,138 |

### 4.2 Data Pre-processing

The Ancient-3 data-set comprises a total of 106,352 image inscriptions for 1,640 character categories albeit with significant variation in category coverage, as shown in Fig. 3; for example, just over 86% of categories contained less than 50 inscriptions. We deal with this imbalance in two ways:

1. First, we used the Fast Gradient Sign Method (FGSM) [8] to augment categories with fewer than 50 inscriptions. FGSM is widely used to generate adversarial training examples to improve the robustness of neural networks; it provides a practical means to generate additional examples for sparse categories by perturbing the pixels of existing examples in a manner that is not obvious to the human eye. In this way, we can effectively over-sample sparse categories without compromising image fidelity.

2. Conversely, we under-sample categories with an abundance of inscriptions by using the Mean Shift clustering algorithm [5] to divide a category's inscriptions into several clusters, only retaining those inscriptions associated with the largest cluster.

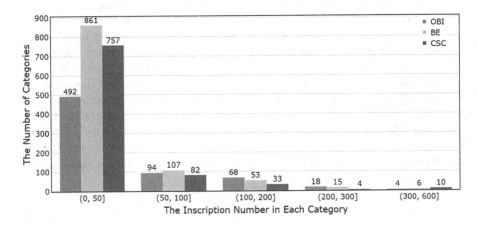

**Fig. 3.** The distribution of inscriptions by character category in the *original* Ancient-3 data-set.

In what follows, we will distinguish between these two variations of the Ancient-3 data-set, where appropriate, by using the term *original* to indicate the original, unbalanced version and *balanced* to refer to the balanced, over/under-sampled version.

## 5   Cross-Domain Representation Learning

As mentioned in Sect. 3.3, the major challenge of our solution for the OBI deciphering task is: how to compare the inscriptions from two adjacent writing systems (aka the cross-domain query and case in CBR) in a unified format which reflects both the common features and the evolution pattern of the inscriptions so that we can directly compare the similarities/relevance of cross-domain inscriptions. Our idea is to project adjacent writing systems' inscription images into a common latent space where known inscriptions in each pair appear similar/close to each other (see Fig. 2(b)). In this section, we first introduce the Vanilla VAE architecture [11], which is the basic architecture used for representation learning; then, we propose two novel approaches for capturing image features from cross-domain inscriptions. Both approaches can be treated as extensions of the VAE architecture.

### 5.1   Vanilla Variational Auto-encoder

This basic architecture, vanilla VAE, is an efficient implementation of the basic VAE idea, which aims to capture the image features from character inscriptions through its encoder module. Both the vanilla VAE's symmetric neural networks (encoder and decoder) use the same number of convolutional blocks. Each block contains three stacked layers: a convolution layer, a batch normalisation layer, and a rectified linear unit (ReLU) layer. The number of blocks in each network is a modifiable parameter for the architecture. By training the vanilla VAE through inscriptions from different writing systems, the encoder itself can be used to generate the latent variable $z$ as the abstract representations for the inscription images, to map all the abstract representations into the common high-dimensional common space for further analysis (see Fig. 4(a)). Its similarity loss function, $l(input, output)$, consists of two parts: the reconstruction loss between the input and the output inscriptions and the latent loss defined as KL-divergence through the latent variable $z$ [11]. However, the drawback of the vanilla VAE's encoder is that, except for its function of capturing the common features of different inscriptions, it still lacks the capability to express the evolution patterns among pairs of known inscriptions drawn from the adjacent writing system, i.e., the abstract representations of inscriptions in a pair may be far away from each other in the latent space.

### 5.2   MMD-Enhanced Architecture (MA)

To address the drawback discussed above, we propose the first technical approach, MMD-enhanced Architecture (MA), which uses the maximum mean discrepancy (MMD) distance [9] between the abstract representations of known inscriptions to reflect the linkage among adjacent writing systems. Precisely, during the encoder training process, pairs of known inscriptions drawn from adjacent writing systems are input into the MA. In Fig. 4(b), $S$ represents an image from the source writing system, $T$ represents an image from the target writing system. The encoder network extracts two latent variables $Z_S$ and $Z_T$. The MMD between $Z_S$ and $Z_T$ in the final loss function (Eq. 1) is to make sure the two latent variables will be mapped close to each other in the high-dimensional common space.

$$L_{MA} = l(input_S, output_S) + l(input_T, output_T) + w * mmd(Z_S, Z_T) \qquad (1)$$

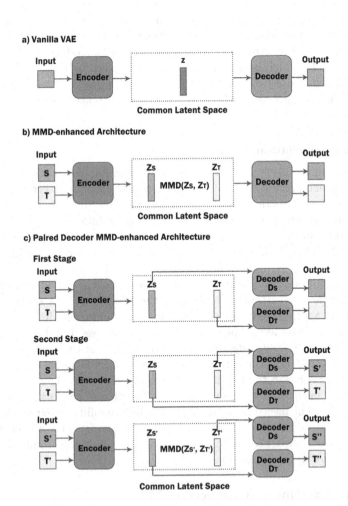

**Fig. 4.** Different architectures of representation learning.

### 5.3  Paired Decoder MMD-Enhanced Architecture (PDMA)

Inspired by the work from Artetxe et al., which proposed a framework to generate parallel corpora through mapping monolingual corpora embedding [1], we further designed the second technical approach, the paired decoder MMD-enhanced architecture (PDMA). Instead of only constraining the representations in adjacent writing systems to approach each other, the PDMA learns the dual nature from two writing systems and handles bidirectional knowledge transferring between them simultaneously. As displayed in Fig. 4(c), together with a shared encoder, a pair of decoders, $D_S$ and $D_T$, are used as the decoder module in the PDMA for decoding abstract representations of different writing systems (target and source) respectively.

We designed two stages to train the PDMA encoder module. In the first stage, each decoder ($D_S$ and $D_T$) respectively restores its input latent variable to an inscription that should be similar to the original inscription where the latent variable is generated; and the two decoders' VAE reconstruction losses are calculated individually. In the second stage, each decoder takes the latent variable extracted from a different writing system ($D_S$ takes $Z_T$, $D_T$ takes $Z_S$) as the input and generates a corresponding pseudo inscription ($S'$ generated by $D_S$, $T'$ generated by $D_T$). Then these pseudo inscriptions ($S'$ and $T'$) are fed into the encoder again to get their latent representations ($Z'_S$, $Z'_T$). The two decoders restore them to pseudo inscriptions ($S''$, $T''$) once more. The final loss (Eq. 2) consists of the similarity losses between original inscriptions and pseudo inscriptions, together with the MMD distance between the latent variables $Z'_S$ and $Z'_T$, which have the major influence for the feature representation, received from the encoder module.

$$L_{PDMA} = l(S, S'') + l(T, T'') + w * mmd(Z'_S, Z'_T) \qquad (2)$$

By using a pair of decoders in the decoder module, we believe the PDMA encoder module can be enhanced in two aspects: first, for each inscription, an abstract representation that is independent from its original writing system can be generated by the encoder module; second, the encoder module pays more attention to extract the evolution features of adjacent writing systems (such as OBI and BE), for example, the leading features, which exist in an OBI and are the main factors leading to its evolution to its corresponding BE; or the inherited features, which exist in the BE but inherited from its origin OBI.

## 6  Evaluations

Due to the novelty of the paleography translation task, we have to design our own evaluation scheme. We conducted two experiments, the inscription-level and the category-level, to evaluate our architecture. In our first experiment (inscription-level), as mentioned in Sect. 3.4, for an unknown query inscription, we retrieved top-n target cases (individual inscription images from the target writing system) for paleographers as references; In our second experiment (category-level), we focused on directly linking the unknown query inscription to its corresponding category in the target writing system. In this section, we first introduce how to split the train/test data sets, then we present our accuracy metrics for two experiments followed by the final results. The details of the experiments are presented in an GitHub repository[4].

## 6.1 Train-Test Division

Compared with traditional machine learning tasks, the train-test division for this OBI deciphering task is more complicated because the data contains multiple writing systems, and the pseudo unknown inscriptions are specially required. We first separated the data-set on its category-level, and then we used the category-level division to collect inscriptions for training and testing. Precisely, for each two adjacent writing systems, we randomly chose 10% (categories) for the test-set (pseudo-unknown inscriptions, which will not be used by the architecture for training). Based on the categories in the train-set, we collected their associated inscriptions from the *balanced* Ancient-3 data-set for training; and then based on the categories in the test-set, we collected their associated inscriptions from the *original* Ancient-3 data-set for testing (described in Sect. 4). We only used practical inscription images for testing to avoid bias.

In the training process, we generated the source-target pairs of known inscriptions from the train-set based on their category information to train the representation learning architecture (see Sect. 5). After this, we used the representation learning architecture to index target cases as described in Sect. 3.3 (i.e., projecting all the target cases into the common latent space). The scale of the target candidate cases will affect the deciphering accuracy. In our experiment, we tested on two scales: 1) common scale: only used the inscriptions from common categories in both adjacent writing systems as candidates (e.g., OBI has 1,186 categories, and BE has 1,394 categories, but only 676 of them are common categories). 2) full scale: all the inscriptions in the target writing system were used as potential candidates (e.g., all the inscriptions from 1,394 categories in BE). The second scale helped us estimate the ability of our architecture against interference in real-world applications.

## 6.2 Accuracy Calculation Scheme

We measured the possibilities of retrieved cases through the distances between the unknown inscription and the potential target known inscriptions. We also sorted the ranking according to the distances. If a candidate inscription is closer to the unknown inscription, it is more likely to be the deciphered result. In general, we considered a retrieved case as true-positive if the correct translation inscription is contained in the top-n candidates. We defined the accuracy of the translation task as Eq. 3, where $hit_n$ is the number of successful cases, and $T$ is the number of inscriptions used as unknown ancient characters.

---

[4] https://github.com/ICCBR/AncientDiscovery.

$$Top\ N\ accuracy\ =\ \frac{hit_n}{T} \tag{3}$$

The accuracy calculation scheme on the category-level is similar to the previous one. The only different operation is, as a category may contain multiple inscriptions, to represent it, we calculated the centre of all the inscriptions' encoded latent variables for each category from the target writing system.

## 6.3    Evaluation Results

Here, we report both inscription-level and category-level results over the two case bases, OBI-BE and BE-CSC. The top-n accuracy in Eq. 3 is denoted as $acc@n$ in each table. We selected eight top-n values for each experiment. Each table displays the performances of our architecture under different settings of approaches (MA and PDMA), and candidate scales (common scale and full scale).

**Table 2.** The **inscription-level** accuracy(%) of translation task from OBI to BE on 26,958 (676 categories) and 35,140 (1,394 categories) BE candidate inscriptions.

|  |  | acc@1 | acc@10 | acc@20 | acc@50 | acc@100 | acc@200 | acc@400 | acc@600 |
|---|---|---|---|---|---|---|---|---|---|
| 26,958 | PDMA | 15.57 | 29.56 | 33.55 | 40.07 | 46.72 | 53.67 | 62.72 | 68.47 |
|  | MA | 10.42 | 23.42 | 28.10 | 36.38 | 43.93 | 51.95 | 60.02 | 65.17 |
| 35,140 | PDMA | 14.80 | 28.44 | 31.53 | 38.22 | 43.97 | 50.62 | 58.39 | 63.58 |
|  | MA | 8.79 | 21.79 | 26.34 | 33.38 | 40.41 | 48.22 | 56.11 | 61.43 |

**Table 3.** The **category-level** accuracy(%) of translation task from OBI to BE on 676 and 1,394 BE candidate categories.

|  |  | acc@1 | acc@10 | acc@20 | acc@50 | acc@100 | acc@200 | acc@400 | acc@600 |
|---|---|---|---|---|---|---|---|---|---|
| 676 | PDMA | 5.71 | 20.42 | 27.50 | 43.46 | 56.89 | 70.83 | 85.46 | 97.73 |
|  | MA | 5.02 | 19.48 | 28.70 | 42.60 | 53.84 | 67.70 | 85.50 | 96.57 |
| 1,394 | PDMA | 4.38 | 16.60 | 22.65 | 34.66 | 48.99 | 61.99 | 74.56 | 82.45 |
|  | MA | 3.82 | 12.78 | 19.09 | 31.15 | 42.13 | 53.97 | 66.62 | 76.88 |

The results of the two case bases tasks OBI-BE and BE-CSC meet our expectations. Both MA and PDMA approaches performed well. PDMA performed better in most situations, but we observed it spent more time in both training and translation processes. For the inscription-level, in the translation task from OBI to BE (see Table 2), the probability of finding the correct answer attained almost 50%, when we retrieved top-100 cases; in the translation task from BE to CSC (see Table 4), this probability also reached over 40%, when top-200 cases are recalled. For the category-level, it is noteworthy that the performance of our architecture was not influenced greatly, even when we used a much larger target data-set. In both Table 3 and Table 5, when the target changed from the common scale to the full scale, the number of candidate categories almost doubled, but our architecture remained over 80% of its common-scale accuracy.

**Table 4.** The **inscription-level** accuracy(%) of translation task from BE to CSC on 25,055 (886 categories) and 32,203 (1,640 categories) CSC candidate inscriptions.

|        |      | acc@1 | acc@10 | acc@20 | acc@50 | acc@100 | acc@200 | acc@400 | acc@600 |
|--------|------|-------|--------|--------|--------|---------|---------|---------|---------|
| 25,055 | PDMA | 4.35  | 13.34  | 18.41  | 27.27  | 34.33   | 43.79   | 53.78   | 59.17   |
|        | MA   | 2.52  | 9.52   | 12.80  | 21.19  | 28.72   | 37.23   | 46.75   | 53.47   |
| 32,203 | PDMA | 3.91  | 11.57  | 15.64  | 24.02  | 31.75   | 39.79   | 49.72   | 54.48   |
|        | MA   | 2.21  | 8.13   | 11.44  | 17.65  | 24.53   | 33.01   | 43.32   | 48.90   |

**Table 5.** The **category-level** accuracy(%) of translation task from BE to CSC on 886 and 1,640 CSC candidate categories.

|       |      | acc@1 | acc@10 | acc@20 | acc@50 | acc@100 | acc@200 | acc@400 | acc@600 |
|-------|------|-------|--------|--------|--------|---------|---------|---------|---------|
| 886   | PDMA | 4.82  | 15.83  | 21.34  | 31.27  | 40.83   | 53.37   | 68.47   | 80.80   |
|       | MA   | 3.97  | 13.40  | 18.88  | 28.37  | 37.52   | 50.03   | 67.15   | 79.76   |
| 1,640 | PDMA | 4.45  | 14.50  | 18.47  | 28.03  | 36.89   | 47.29   | 60.25   | 69.29   |
|       | MA   | 3.12  | 11.79  | 16.30  | 25.35  | 33.98   | 44.70   | 58.13   | 68.16   |

## 7   Conclusions and Future Work

We presented and evaluated a case-based architecture developed for automatically deciphering unknown OBIs. We assessed the performance of the proposed architecture over a specially collected ancient character data-set, which builds solid foundations and should assist other researchers for the interdisciplinary study of machine learning and archeology. The architecture uses unsupervised machine learning methods and case-based reasoning to simulate different stages of the OBI evolution process.

In this work, we mainly focused on the inscriptions' shape features and the linkages between adjacent writing systems. Although OBIs are hieroglyphs, many of them have advanced functions and have abstracted away from the shape. Considering not only the shape but also the semantic and contextual information over different writing systems would be a natural next towards developing a more powerful decipher architecture. Combining both computer vision and natural language process methods to support the identification and translation of ancient pictographs, may give rise to a new generation of digital assistants for the archeology community.

**Acknowledgments.** This publication has emanated from research conducted with the financial support of Science Foundation Ireland under Grant number 12/RC/2289_P2. For the purpose of Open Access, the author has applied a CC BY public copyright licence to any Author Accepted Manuscript version arising from this submission.

# References

1. Artetxe, M., Labaka, G., Agirre, E., Cho, K.: Unsupervised neural machine translation. arXiv preprint arXiv:1710.11041 (2017)
2. Boltz, W.G.: Early Chinese writing. World Archaeol. **17**(3), 420–436 (1986)
3. Brown, P.F., et al.: A statistical approach to French/English translation. In: RIAO, pp. 810–829 (1988)
4. Collins, B., Cunningham, P., Veale, T.: An example-based approach to machine translation. In: Conference of the Association for Machine Translation in the Americas (1996)
5. Derpanis, K.G.: Mean shift clustering. In: Lecture Notes, p. 32 (2005)
6. Dorr, B.J., Jordan, P.W., Benoit, J.W.: A survey of current paradigms in machine translation. Adv. Comput. **49**, 1–68 (1999)
7. Feng, G., Jing, X., Yong-ge, L.: Recognition of fuzzy characters on Oracle-bone inscriptions. In: 2015 IEEE International Conference on Computer and Information Technology; Ubiquitous Computing and Communications; Dependable, Autonomic and Secure Computing; Pervasive Intelligence and Computing, pp. 698–702. IEEE (2015)
8. Goodfellow, I.J., Shlens, J., Szegedy, C.: Explaining and harnessing adversarial examples. arXiv preprint arXiv:1412.6572 (2014)
9. Gretton, A., Borgwardt, K., Rasch, M., Schölkopf, B., Smola, A.J.: A kernel method for the two-sample-problem. In: Advances in Neural Information Processing Systems, pp. 513–520 (2007)
10. Johnson, M., et al.: Google's multilingual neural machine translation system: enabling zero-shot translation. Trans. Assoc. Comput. Linguist. **5**, 339–351 (2017)
11. Kingma, D.P., Welling, M.: Auto-encoding variational Bayes. arXiv preprint arXiv:1312.6114 (2013)
12. LeCun, Y., Bottou, L., Bengio, Y., Haffner, P.: Gradient-based learning applied to document recognition. Proc. IEEE **86**(11), 2278–2324 (1998)
13. Lei, G.: Research on feature extraction algorithm for characters on oracle bone rubbings in meticulous and neat written style. Comput. Appl. Softw. 06 (2014)
14. Li, X.: Study on the pictographic characters in the inscriptions on bones and tortoise shells. Master's thesis, Fujian Normal University (2008)
15. Pan, S.J., Yang, Q.: A survey on transfer learning. IEEE Trans. Knowl. Data Eng. **22**(10), 1345–1359 (2009)
16. Li, Q., Yang, Y., Wang, A.: A DNA-based encoding and retrieving method for Jiaguwen. In: 2010 3rd International Conference on Computer Science and Information Technology, vol. 4, pp. 51–55. IEEE (2010)
17. Rumelhart, D.E., Hinton, G.E., Williams, R.J.: Learning representations by back-propagating errors. Nature **323**(6088), 533–536 (1986)
18. Somers, H.: Example-based machine translation. Mach. Transl. **14**(2), 113–157 (1999)

19. Sumita, E., Iida, H., Kohyama, H.: Translating with examples: a new approach to machine translation. In: The Third International Conference on Theoretical and Methodological Issues in Machine Translation of Natural Language, no. 3, pp. 203–212. Citeseer (1990)
20. Zhou, X.L., Hua, X.C., Li, F.: A method of Jia Gu Wen recognition based on a two-level classification. In: Proceedings of 3rd International Conference on Document Analysis and Recognition, vol. 2, pp. 833–836. IEEE (1995)

# Author Index

Printed in the United States
by Baker & Taylor Publisher Services